Routledge Revivals

Regulation of the Natural Gas Producing Industry

Originally published in 1972, *Regulation of the Natural Gas Producing Industry* combines several papers prepared for a seminar in Washington D.C in 1970. The purpose of this seminar was to exchange views between specialists on the issues related to regulating natural gas production. As such these papers explore issues such as producer price regulations, arguments for and against the regulation of natural gas, the changing environment of natural gas supply and what the future holds in terms of the natural gas producing industry. This title will be of interest to students of Environmental Studies and policy makers.

Routledge Revivals

Regulation of the Natural Gas Producing Industry

[illegible]

Regulation of the Natural Gas Producing Industry

Edited By
Keith C. Brown

Routledge Revivals
Routledge
Taylor & Francis Group

First published in 1972
by Resources for the Future, Inc.

This edition first published in 2016 by Routledge
2 Park Square, Milton Park, Abingdon, Oxon, OX14 4RN
and by Routledge
711 Third Avenue, New York, NY 10017

Routledge is an imprint of the Taylor & Francis Group, an informa business

Publisher's Note
The publisher has gone to great lengths to ensure the quality of this reprint but points out that some imperfections in the original copies may be apparent.

Disclaimer
The publisher has made every effort to trace copyright holders and welcomes correspondence from those they have been unable to contact.

A Library of Congress record exists under LC control number: 71186502

ISBN 13: 978-1-138-94879-2 (hbk)
ISBN 13: 978-1-315-66944-1 (ebk)

REGULATION OF THE NATURAL GAS PRODUCING INDUSTRY

REGULATION OF THE NATURAL GAS PRODUCING INDUSTRY

Papers presented at a Seminar conducted by Resources for the Future, Inc. in Washington, D. C., 15-17 October 1970

Edited by **KEITH C. BROWN**

Published by
RESOURCES FOR THE FUTURE, INC.

Distributed by
THE JOHNS HOPKINS UNIVERSITY PRESS
Baltimore and London

RESOURCES FOR THE FUTURE, INC.
1755 Massachusetts Avenue, N.W., Washington, D.C. 20036

Resources for the Future is a nonprofit corporation for research and education in the development, conservation, and use of natural resources and the improvement of the quality of the environment. It was established in 1952 with the cooperation of the Ford Foundation. Part of the work of Resources for the Future is carried out by its resident staff; part is supported by grants to universities and other nonprofit organizations. Unless otherwise stated, interpretations and conclusions in RFF publications are those of the authors; the organization takes responsibility for the selection of significant subjects for study, the competence of the researchers, and their freedom of inquiry.

The charts for this book were drawn by Frank and Clare Ford.

RFF editors: Henry Jarrett, Vera W. Dodds, Nora E. Roots, Tadd Fisher.

Manufactured in the United States of America
Distributed by The Johns Hopkins University Press, Baltimore, Maryland 21218
Library of Congress Catalog Card Number 71-186502

ISBN 0-8018-1383-2

Price $8.50

FOREWORD

The papers in this volume were prepared for a seminar held in Washington, D.C. on October 15–17, 1970. The seminar, attended by about 50 specialists drawn mainly from universities and government agencies, was intended to encourage an exchange of views among the writers of the papers and others with experience and interest in natural gas regulatory problems.

The idea for the seminar originated with the Office of Science and Technology (OST) of the Executive Office of the President. S. David Freeman, at the time director of the Energy Policy Staff of the OST, requested that RFF take responsibility for conducting this project, and we were pleased to accept. Although the effort was financed by the OST, this book is an RFF publication and is in no sense to be regarded as a government document. OST did, however, agree that the importance of the subject and the value of the papers were such as to make publication desirable in order that a wider audience be reached.

The project was led by Keith Brown, associate professor of economics at the Krannert Graduate School of Industrial Administration, Purdue University. Professor Brown selected the subjects to be dealt with, arranged for the preparation of the papers, presided at the seminar, and wrote the editor's introduction to this book. RFF is grateful to him for his performance of all these demanding tasks.

The controversial history of producer price regulation of natural gas, the nature of the Federal Power Commission's regulatory activity, the arguments for and against regulation itself and the particular regulatory approaches adopted, the changing environment of natural gas supply and demand in which regulation has unfolded, the outlook for the future, and other pertinent questions are all covered in these pages. It is our hope that this publication will contribute to a better public understanding of the issues at stake and to an improved basis for policy in the future.

Sam H. Schurr
Director
Energy and Minerals Program
Resources for the Future, Inc.

November 1971

CONTENTS

FOREWORD v

INTRODUCTION, *by Keith C. Brown* 1

Factual Issues in Dispute 2
Competition and Rent in the Natural Gas Producing Industry 2
Costs and Elasticity of Supply 3
Demand Elasticity and Future Demand Growth 4
Current Quantities Supplied and Demanded 5
Transient Results from Removing Price Controls 6
Evaluating Regulatory Alternatives 6
Effects of Current FPC Price Regulations 7
Decontrol of New Gas Prices 8
Decontrol of All Gas Prices 9
The Market Value Standard 10
Proposals for Improving Industry Performance 10
Conclusions 11
A Note on Economics and Natural Gas Regulation 12
The Goal of Optimum Resource Allocation 12
Regulation, Optimum Resource Allocation, and Income Transfer 13
Economic Rent 13
Regulation and the Optimal Consumption Rate over Time 14
Other Costs of Regulation 14
Market Price, Regulation, and Resource Allocation 15
Analytical Difficulties Connected with the Natural Gas Producing Industry 16

1. NATURAL GAS IN THE FUTURE NATIONAL ENERGY PATTERN, *by John J. Schanz, Jr., and Helmut J. Frank* 18

Present Patterns of U.S. Production and Consumption of Natural Gas 18
Role of Pipeline-owned Gas in Meeting U.S. Gas Needs 20
Variations in Natural Gas Consumption Statistics 21
Trends in Natural Gas Consumption 24
Behavior of the Intrastate Market 28
Projections of U.S. Energy and Gas Requirements to 1990 under Customary Assumptions 31
Review of Individual Studies of U.S. Energy Consumption 32
The Consensus 35

Estimates of Future U.S. Gas Consumption 36
Trends by Use and by Region 37
Future U.S. Gas Demand under Alternative Assumptions 38
Econometric Studies 39
Residential and Small Commercial Uses 41
Industrial Markets 44
Impact of Environmental Policies 45
Policy Implications 46
Summary 48
Appendix: Reviews of Seven Gas Demand Studies Utilizing Econometric Techniques, *by Jean E. Draper* 49

2. INTERFUEL RELATIONS GOVERNING NATURAL GAS DEMAND AND SUPPLY, *by Richard J. Gonzalez* 56

Special Advantages of Natural Gas 56
Natural Gas in Competition with Other Fuels 56
Relation of Hydrocarbon Liquids and Gases 57
Displacement of Manufactured Gas by Natural Gas 58
Price Relation of Oil and Gas 59
Determining a Natural Gas Price Policy 59
Federal Regulation of Interstate Gas Prices 60
Balancing Gas Demand and Supply 61

3. PROJECTED COSTS OF ALTERNATE SOURCES OF GAS, *by R. Bruce Foster* 63

Canadian Pipeline Gas 63
Current Prices of Canadian Gas 63
Future Prices of Alberta Gas 63
Projected Quantities of Imported Canadian Gas 65
Current Canadian Gas Transmission Costs 66
Projected Canadian Gas Transmission Pipelines 67
Trans Alaskan Pipeline 68
Projected Gas Transmission Costs 69
LNG Pipelines 70
Importation of Pipeline Gas from Mexico 71
Nuclear Stimulation of Natural Gas Reserves 71
LNG from Alaska and Foreign Sources: General Considerations 72
Tanker Construction Costs 72
Availability of Overseas Natural Gas Reserves 72
Port Facilities 73
Shipping Distances 73
Relationship of Base Load Markets 73
Relationship to Peakshaving Markets 74
Transportation of LNG from Alaska 75
LNG Importation Projects 76
El Paso Project 76
Philadelphia Gas Works Project 76
Distrigas Project 76
Columbia Gas System-Standard Oil Company Project 77
Estimated Cost of Imported LNG 77

Synthetic Gas from Coal 78
- Bituminous Coal Research Two-Stage Superpressure Process 78
- Kellogg Molten-Salt Process 79
- Consolidation Coal Company's CO_2 Acceptor Process 79
- IGT Hydrogasification Process 79
- Char-Oil Energy Development Process 81

Comparative Economics of Synthetic Gas Processes 83
Availability of Coal Reserves 84
Projected Gas Costs in Various Coal Regions 86
- Southern Illinois 86
- Appalachia 86
- North Dakota 87
- Wyoming 87
- Colorado 87
- Four Corners Area 87

Projected Gas Costs from Heavy Oil 87
Conclusions 88

4. PRODUCER REGULATION: A COMMISSIONER'S VIEWPOINT, *by Charles R. Ross* 90

Regulation of Natural Gas Producers—Background 91
The Kennedy Commission 93
Problems Faced by the Kennedy and Johnson Commissions 93
Guidelines for In-Line Prices 95
Area Pricing 96
The Problem of Incomplete and Insufficient Data 97
Competitiveness of the Industry 99
Effect of Air Pollution on Demand 100
Pricing of Interruptible Gas 101
Jurisdiction over Direct Sales 102
Gas Shortage 103
Increased Price As an Incentive to Increased Production 106
The Future 108
Conclusion 111

5. CHANGING ELEMENTS IN THE NATURAL GAS PICTURE: IMPLICATIONS FOR THE REGULATORY SCHEME, *by Ralph S. Spritzer* 113

The Regulatory Scheme 114
- Pipelines 114
- Producers 116

The "Crisis" in Supply 121
The Regulatory Response: Pricing Policy 123
Other Aspects of Governmental Policy Bearing on the Supply of Natural Gas 125
- Federal Leasing Policy 126
- A Comment on Incentives 128
- Import Policy 129
- Liquid Natural Gas 131
- Synthetic Gas 135

Conclusions 136

6. STRUCTURE OF THE NATURAL GAS PRODUCING INDUSTRY, *by Clark A. Hawkins* 137

Concentration 138
Rent, Windfall, and Supply 161

7. THE REGULATION-INDUCED SHORTAGE OF NATURAL GAS, *by Paul W. MacAvoy* 169

An "Inventory" Approach to Excess Demand 172
Supply-Demand Models of Gas Field Markets 175
An Introductory "Engineering" Model of Supply and Demand 175
An Economic Model of Supply and Demand 179
Distribution of the Gas Shortage among Final Consumers 185
Appendix: Data Sources 190

8. PRICE, REGULATION, AND THE SUPPLY OF NATURAL GAS IN THE UNITED STATES, *by Edward W. Erickson and Robert M. Spann* 192

Joint Costs in Oil and Gas Expenditures 194
Effects of Regulation and Other Public Policy 199
Statistical Estimation of the Price Elasticity of Supply of Natural Gas 201
Interpretation of the Results for 1946-58/59 206
Estimated Supply Response, 1959-67 211
Market Clearance through Price Adjustment 214
Summary and Conclusions 217

9. PRODUCER REGULATION FOR THE 1970s, *by Milton Russell* 219

The Public Interest in Natural Gas Field Price Regulation 219
Regulation for Monopoly Control 220
Regulation to Alter Use Rate 221
Regulation to Redistribute Income 221
The Three Proximate Goals Summarized 222
FPC Field Price Regulation 222
Supply and Demand for Natural Gas 223
The Field Market Model: Jurisdictional and Nonjurisdictional Sales 224
The Effect of FPC Area Price Regulation on Economic Welfare 227
Welfare Effects of a Sub-Equilibrium Price 227
Area Price Regulation and Resource Misallocation 229
Regulation and Allocational Efficiency 230
Reform of Current Area Price Regulation 230
Restructuring of Producer Regulation 232
Interarea Differentials 232
The Base Price for Gas 233
Revisions of Area Prices 233
The Multi-Price System 233
Field Price Deregulation 234
Short-run Effects 234
Long-run Effects 235
Conclusions 235
Producer Regulation for the 1970s 236

10. THE FUTURE OF NATURAL GAS REGULATION WITH AN INADEQUATE SUPPLY OF DOMESTICALLY PRODUCED NATURAL GAS, *by James R. Nelson* 238

Environmental Issues and Demand for Natural Gas 238
Interrelationships of Natural Gas and Petroleum 239
Regulation of Sources of Natural Gas Supply 246
  Natural Gas Imports from Canada and Mexico 246
  Importation of Liquefied Natural Gas 251
  Synthetic Pipeline Gas 256
Monopoly versus Economic Rent 257
Synthetic, or Chronological, Marginalism 258
Public Utility Regulation and Comparative Advantage 259
Micro-economics and Macro-politics 260
Bygones Are Forever Bygones 260

INTRODUCTION

*Keith C. Brown**

Natural gas together with natural gas liquids supplies about one-third of the energy from fuel consumed in the United States. Since 1925 not only have absolute volumes of gas consumed increased continuously but also the percentage of total energy supplied by gas steadily increased until the 1960s. Emphasis on pollution control and price increases in other fuels have recently caused further increased demand for natural gas.

The Natural Gas Act of 1938 gave the Federal Power Commission (FPC) authority to regulate the "sale in interstate commerce of . . . gas for resale," but exempted from FPC jurisdiction the "production or gathering of natural gas." Until 1954 the FPC held that it had no jurisdiction over prices paid by interstate pipelines to independent producers. In 1954 the Supreme Court ruled that the commission did indeed have such jurisdiction. During the period 1954-60, the commission undertook to decide rate cases involving individual gas producers. This attempt led into such a morass of difficulties that in 1960 the commission adopted a new procedure under which it attempted to establish uniform ceiling prices for each producing area in the country. This policy is often referred to as "area pricing."

A very considerable fraction of the natural gas used in the United States is not subject to FPC producer regulation. In the 1958-68 decade, approximately 37-38 percent of total domestic marketed production was sold within the state in which it was produced;[1] most of this gas was not subject to FPC producer price regulation. Unless current regulatory policies change drastically, there seems little likelihood that this percentage of gas used intrastate will drop much in the next decade, and it could well increase. Further, over 3 percent of all the gas used in the United States in 1968 was imported by pipeline, mostly from Canada, and the Canadian Energy Board projects that this percentage may rise to nearly 7 by 1980.[2] Such imports must be licensed by the FPC. While the FPC will not approve an import license application if it deems the price at the entry point too high, it clearly cannot specify precisely the maximum prices Canadian producers may receive. Also, though current imports of liquefied natural gas (LNG) from foreign countries are negligible, four large LNG importation projects have been announced. It is possible that LNG imports could amount to several percent of the total domestic usage of natural gas by the end of the decade. Again, such imports apparently require an FPC license and thus approval of the price at the point of delivery, but the FPC cannot specify precise ceiling prices paid at the wellhead to foreign producers. Finally, manufactured gas may reappear in appreciable quantities, though perhaps not by 1980. Thus, if no changes are made in current FPC powers, the FPC will probably be able to apply direct producer price regulation to something less than 60 percent of the gas used in the United States in the 1970s, and this percentage may well decline throughout the decade.

*Associate Professor of Economics, School of Industrial Administration, Purdue University.

[1] See John Schanz and Helmut Frank's paper (chap. 1), especially table 1-8.

[2] See R. Bruce Foster's paper (chap. 3), table 3-2.

A regulatory commission that controls only part of a market usually has a difficult task. But the situation of the FPC is currently being made more difficult by the fact that, at present prices, customers of interstate pipelines would like to purchase more gas than they are able to obtain. There is widespread complaint that the current system is not working properly. Thus, the FPC, the executive branch of the federal government, and the Congress, in attempting to respond to the present situation, are all faced with difficult decisions about regulatory alternatives. Discordant views coexist because of divergent beliefs about the structure of the natural gas producing industry, the goals of regulation, and the appropriate trade-offs between goals when all cannot be attained simultaneously.

In the debate about natural gas regulatory policy, there are two sorts of issues about which disputes have arisen—factual issues and preferential issues. Factual issues can be investigated empirically if adequate data are available and findings can be discussed and evaluated. Preferential issues are matters of judgment and are not subject to empirical verification in the same way as factual issues. Factual issues are important largely as inputs to the economic analysis of regulatory alternatives; the preferential issues are important when one attempts to decide the preferable alternative given some analytic inputs.

FACTUAL ISSUES IN DISPUTE

Several of the most important factual issues that have been in dispute with respect to natural gas regulatory policy concern: (1) the existence of workable competition in the natural gas producing industry, (2) the responsiveness of natural gas supply to producer price changes, (3) the responsiveness of natural gas demand to consumer price changes, (4) the magnitude and likely duration of the deficit between the amounts of gas pipelines would like to purchase and producers would like to sell, and (5) the transient results of possible changes in regulatory policy. A discussion of these factual issues follows.

Competition and Rent in the Natural Gas Producing Industry

A firm is said to possess market power when it can behave persistently in a manner different from the behavior forced on it by a competitive market. Much study has been undertaken in an attempt to determine whether buyers or sellers typically possess market power in the natural gas producing industry. A summary of the more important published studies with some extensions is found in the first half of Clark Hawkins's paper (chapter 6). While it is clear that neither on the demand side nor the supply side does the market fit the economist's ideal of price competition, most economists feel the evidence warrants a verdict of effective or workable competition on the supply side (i.e., absence of market power on the part of producers) with perhaps some monopsony[3] (purchaser market power) in some special areas on the demand side.[4]

However, some have claimed that pipelines, because of their regulated status, often exhibit a type of behavior that, in effect, allows producers to possess some measure of market power:

> . . . pipelines pass their gas purchase outlays on to distributors, and ultimately on to consumers, through the pipelines' regulated cost-bases resale rates, which is a factor tending to reduce their bargaining motivations.[5]

In other words, some producers may obtain more for their gas than they would if the pipeline company did not feel that any cost increases could be easily passed on without decreasing the pipeline's

[3]Economic theory predicts that a firm that purchases an input in a monopsonistic market will purchase less of that input at a lower per-unit price than it would pay if the input market were competitive. The proper regulatory policy to achieve more efficient allocation of resources in this case is a minimum price set at the competitive level. There have been almost no advocates for such a minimum price of natural gas.

[4]In spite of the economists' studies, there are those who do not believe that the gas producing industry is competitive and who feel that it is in need of regulation because of this lack of competitiveness. It is clear that there are many incentives for cooperation among gas producers just as there are incentives for cooperation among producers in practically any industry. However, economists have been able to uncover little evidence of pervasive cooperation, either active or passive.

[5]FPC, Opinion no. 546 (*Southern Louisiana Area Rate Proceeding*), September 25, 1968, p. 18.

profit expectations. If the pipelines are really nonaggressive in their bargaining for gas, this must be the reason. For pipelines are under strong incentive to transport ever increasing volumes of gas; if they do not continually expand, accumulated depreciation may decrease the rate base and therefore the maximum dollar amount of profit. Thus, if pipelines do not have the strongest possible bargaining motivations, this lack is attributable to a consumer price that is below competitive equilibrium; that is, attributable to regulation and not to some necessary feature of the market structure.

If a firm sells in a market that is not workably competitive, it may have the opportunity to earn a rate of return on invested capital in excess of that which it could have earned had the industry been competitive. This differential return or "monopoly profit" is often called "excess" or "unearned" income. However, there is another sort of return that has been said by some to give rise to "unearned" income. This type of return is called economic rent[6] and can be earned even though the product is sold in a workably competitive market. The general public may tend to equate economic rents and monopoly profits, but economists regard these two types of returns in very different lights. Monopoly profits are symptoms of markets in which efficient resource allocations[7] are not being made. Efficient resource allocation can be attained by making the market competitive or by properly regulating the firm(s) in the noncompetitive market. In either case, monopoly profits will disappear. But economic rents are quite different. Such rents can result from the operations of highly competitive markets and not be a symptom of inefficient resource allocation.[8] Rents accrue to landowners who sell production rights if expectations change after the lease agreement is signed. A competitive market in natural gas production would produce a stream of rents for landowners and short-run rents (which can be negative as well as positive) for producers. Similar rents earned in other markets (for example, by the owners of farmland and by farm operators) seem to be accepted in our society today without demands that they be limited by price controls.

Costs and Elasticity of Supply

A great deal of testimony has been presented before the FPC concerning the cost of finding and developing natural gas deposits. Not much of this testimony is very useful to the economist trying to evaluate future regulatory alternatives. In the first place, there appears to be no way to compute a noncontroversial cost for natural gas. Oil and gas are joint products. They are often produced simultaneously from the same well. Even if a field produces only natural gas, prior to drilling into the formation there was a probability that it contained oil. Further, the geological information obtained from developing the field may be valuable for future liquid petroleum exploration. Any method for assigning some fraction of the costs to gas and the rest to oil is arbitrary. Very complex costing formulas have been developed and used by the FPC to determine average cost in an area, but the arbitrariness of such formulas cannot be denied.[9]

Second, most of the testimony has been directed toward finding average costs. The economist theorizes that firms will wish to expand output so long as the incremental revenue expected to be derived from an output expansion exceeds the incremental cost of the expansion. Thus, the economist assumes firm decisions about the level of output to be based on incremental costs rather than average costs.

Third, the calculated costs have usually been historical and not easily projectable into the future. It is especially difficult to assess incentives for, and the productivity of, exploration. Past exploration may have taken place under conditions much different from those that will prevail in the near future.

While allocating joint costs in any theoretically acceptable way appears hopeless, estimating the responsiveness of reserve additions to price does not. Edward Erickson (chapter 8) has found reason

[6] See note at end of this introduction for discussion of this term.

[7] Ibid.

[8] In fact, they play a role in efficient resource allocation. See Milton Russell's paper (chap. 9).

[9] This discussion of the joint cost problem was paraphrased from Russell. See also Clark A. Hawkins, *The Field Price Regulation of Natural Gas* (Florida State University Press, 1969) for a detailed study of this cost allocation problem.

to believe that the elasticity[10] of supply of new discoveries of nonassociated natural gas may be approximately +0.5. Paul MacAvoy (chapter 7) has estimated the elasticity of all new reserves to be +0.45.[11] It is important to remember that there is considerable uncertainty associated with these estimates. The authors' papers should be consulted for the assumptions underlying them and their expected accuracy.

The above elasticity estimates were based on yearly data and thus represent intermediate term responsiveness of quantity to price change. It is difficult to say much about long-run elasticity of supply except that probably it is somewhat higher. One can say, however, that the immediate response to a significant unexpected price increase would be relatively slight. Some additional development might take place in existing fields (the quickest way to get more gas), and some previously submarginal wildcat prospects might become supra-marginal. But the really important additions to reserves would likely come only as a result of new exploration projects elicited by the price increase (testing followed by development of successful projects would take longer).

There is no reason to expect that elasticity of supply will be constant at all price levels. It is quite possible that elasticities might be considerably higher at higher prices than have yet been attained. As gas prices rise, both incentives for finding gas and the ability to find it may increase more rapidly with respect to a given percentage price change.[12]

In the long run, the prices of energy sources alternative to natural gas produced in the forty-eight contiguous states will set an effective ceiling on the delivered price of natural gas. These alternative sources include Canadian and Alaskan pipeline gas, imported liquefied natural gas, and synthetic gas from coal or oil. R. Bruce Foster (chapter 3) concludes that at the lowest cost at which alternatives could be made available at least one of the alternative supplies of gas can be transported to the major natural gas markets in the United States at prices that are close to 70 cents per Mcf and most market areas will have a choice of more than one source of supply in this general price range.

Demand Elasticity and Future Demand Growth

While it seems reasonably safe to extrapolate past trends in total energy use into the future, forecasting the usage of gas over the next decade is another matter. New large-scale uses of gas might develop—for example, automobiles and trucks powered by liquefied natural gas. The quantity of gas demanded for industrial use will depend in part on the stringency of antipollution regulations and the relative costs of gas and substitute fuels. Assuming, however, that there will be no major changes in fuel competitive relationships and no dramatic new uses for gas, the Future Requirements Committee predicts total requirements of 34.7 trillion cubic feet (tcf) for natural gas by 1980.[13] This is over a 60 percent increase from the 1968 recorded requirements of 21.5 tcf.

The economist's concept of elasticity is also applied to the demand schedule for a commodity. Here it measures the responsiveness of quantity demanded with respect to price changes, all other variables affecting quantity demanded remaining constant. As price rises, ordinarily one expects the quantity demanded to decrease. The more elastic the demand, the greater will be the percentage decrease in quantity demanded caused by a given percentage price increase. A number of prior econometric studies have attempted to estimate the elasticity of demand for natural gas in various markets.

[10] Elasticity is defined as the ratio of percent change in quantity to percent change in price. Therefore, an elasticity of 0.5 implies that a 20 percent increase in price would result in approximately a 10 percent increase in quantity.

[11] J. Daniel Khazzoom, formerly of the FPC Office of Economics, has also made some detailed estimates of the response of supply to price changes. See FPC, Office of Economics, *An Econometric Model of U.S. Natural Gas Supply*, testimony of Dr. Khazzoom, Area Rate Proceeding, Southern Louisiana Area, Docket no. AR 69-1 (September 1970). It is difficult to summarize Dr. Khazzoom's estimates here, since they were based on a different and more complicated model than those of MacAvoy and Erickson.

[12] See Hawkins's paper for an elaboration of this argument.

[13] For further information about this and other projections see Schanz and Frank.

Unfortunately, these studies have inherent defects that do not lead to substantial confidence in their results.[14]

Qualitative evidence supplied by John Schanz and Helmut Frank (chapter 1) tends to support the view that demand for gas by residential and small commercial users in the vicinity of current delivered prices is rather inelastic. On the basis of British thermal unit (Btu) equivalence, gas has a substantial price advantage over distillate fuel oil in most regions of the country. In addition, gas is regarded as cleaner and more convenient than fuel oil. Those who have already installed gas appliances would be most reluctant to change fuels until the appliances wear out. Thus, Schanz and Frank conclude that price increases of up to 10 cents per thousand cubic feet (Mcf) would cause little change in the quantity of gas demanded by residential and small commercial users in the short run and that in the long run such a price increase might make itself felt only through a slowing of growth of quantity demanded.

The industrial market for gas[15] presents a somewhat different picture. Schanz and Frank believe that "gas prices to industrial users could rise substantially, without significant loss of sales volume, in locations that account for approximately half of total sales. In addition, the demand by some other industrial users is inelastic, regardless of location, because of the inherent advantages this fuel offers. . . . Certainly in the long run and probably also in the short run, this leaves a very sizable market, chiefly on the East and West Coasts, where gas demand is highly elastic." In the latter market, price of at least one alternate fuel is competitive, and many present gas users already have dual burning facilities installed. Environmental considerations, however, may inhibit even this elasticity. Therefore, in aggregate, the industrial demand for gas is perceived as somewhat inelastic also for price increases up to 10 cents per Mcf, though perhaps not so inelastic as residential and small commercial demand. If gas increases drastically in price relative to other fuels, however, elasticity of demand ought to increase as low-sulfur coal and fuel oil become more attractive.

Current Quantities Supplied and Demanded

Most observers of the gas industry agree that, at current producer prices, many interstate pipelines would like to purchase more gas than they are able to obtain. At the other ends of the pipelines, this deficiency of quantity supplied over quantity demanded is being felt by new consumers being refused all or part of the volumes they would like to purchase, old consumers not being able to purchase the incremental amounts of gas they desire, and interruptible customers being interrupted more often.

This situation results from a decline in large gas discoveries after 1952 from the rate that had prevailed earlier without a compensating increase in small discoveries and also from increased demand due to a number of factors, such as air pollution regulations. The economist's conjecture is that economic incentives[16] (particularly current producer ceiling prices) are not sufficient to elicit the quantity demanded. There has been some talk about producers "sitting on reserves" in hopes of temporarily magnifying the problem so that larger price increases might be obtained than otherwise. Sitting on reserves, however, can have several interpretations. One is that firms are delaying the sale of gas from already developed reserves. Another might be that operators are not developing known discoveries. A third is that firms are slackening exploratory efforts on possible prospects.

Holding gas off the market, either from developed reserves or from reserves discovered but not developed, might be infeasible under many lease agreements[17] and in any case is expensive, since it carries with it the opportunity cost of the delay in starting the revenue stream from gas sales. A risk involved in delaying development is that drilling costs might increase in the interim. Nonetheless, such

[14] For a review of these studies by Jean E. Draper, see app. of chap. 1. It should be emphasized that all consumers in an area do not pay uniform rates. Rate structures are often quite complex.

[15] This includes the market for gas used as fuel for electric power generation.

[16] Since gas is discovered in the search for oil and vice versa, the relevant incentives arise not only from possible revenues from gas but also from possible revenues from oil.

[17] Particularly in those fields that are being exploited by more than one operator.

holding back might appear profitable if the delay time were relatively short; with firm discount rates of approximately 15 percent, delays of more than two years would have to be expected to yield really sizable profit increases in order to render delay alluring. However, as both Ralph Spritzer and Charles Ross (chapters 4 and 5) point out, it is quite conceivable that the industry is currently "taking it easy" in exploration.

In any case, even if economic incentives were made stronger by immediately raising ceiling prices, substantial domestic gas supply increases should not be expected for several years because of the lag in supply response of natural gas to changes in economic incentives. Further, massive supplies of LNG from foreign countries or Alaska or gas from coal are not likely before the end of the 1970s.

As producer prices increase, presumably the quantity supplied (after the time lag mentioned above) would begin to rise and the quantity demanded would fall somewhat. How much would producer prices have to go up in order that quantity supplied would equal quantity demanded? Erickson and MacAvoy have provided tentative estimates. They both expect that approximately a 50 to 100 percent increase in wellhead prices would be required.[18]

Transient Results from Removing Price Controls

Much economic analysis is done in a so-called "comparative static" framework; that is, an initial equilibrium position is compared with a later equilibrium that is different because some specified change has occurred. Almost no attention is paid to the path followed by economic variables in the transition from one equilibrium to another.

However, especially in markets where long time lags exist between changes in price and the full resulting response, the time path followed by economic variables during the transition from one semi-equilibrium to another may be of considerable importance because the transition time may be very lengthy. Also, in regulated industries what happens in the interim may have policy consequences.

If producer price regulation were removed, it seems likely that prices might rise very steeply for a while until producers would be able to respond to the newly increased incentives[19] and some consumers would be able either to reduce their energy consumption or to switch to alternative fuels. Then, over time, prices could be expected to settle back to a point above the level at which they were restricted but below the maximum initial increase. The temporary appearance of higher than long-run equilibrium prices would perform the valuable function of allocating the presently limited supply to those who need it most (as shown by their willingness to pay more) but might be very alarming to consumers who would not realize such prices would be impermanent. It is very difficult to conjecture how large such an overshoot might be or how long it might persist.

EVALUATING REGULATORY ALTERNATIVES

In setting ceiling prices payable to producers of natural gas who fall under its jurisdiction, the FPC operates on the basis of estimated average cost.[20] The commission is using a multi-price system. For example, in the Permian Basin area, the "old" gas (gas sold under pre-1961 contracts and all oil-well gas) ceiling price was set on the basis of average historical cost of gas-well gas including a 12 percent rate of return on invested capital. New gas (sold under post-1961 contracts) was priced chiefly on the basis of the then current national average finding and producing cost. In South Louisiana the commission has established a similar system, except that three vintages of gas were defined instead of two.[21]

[18]With these results, also, there is associated considerable uncertainty, and the reader is referred to the respective authors' papers for their assumptions and methodology.

[19]See the papers by Russell and Schanz and Frank.

[20]For a criticism of FPC cost estimation procedures, see the previous section, "Costs and Elasticity of Supply."

[21]See Ralph Spritzer's paper (chap. 5).

And now (as of December 1970) it appears that the FPC will add another tier to this price structure with new higher prices allowed for all contracts dated after June 17, 1970.

Effects of Current FPC Price Regulations

What have been the effects of this current system? First, the quantity of gas demanded at current prices considerably exceeds the quantity that producers are willing to supply. It is not clear that the current FPC regulatory system is entirely responsible for this situation,[22] but it certainly seems to have contributed to its development. Whatever the reasons for its existence, if this situation continues the FPC will be forced into the uncomfortable role of rationing the available supplies, unless it wants to leave this sensitive task to the pipelines. Those who would like to purchase gas but cannot are forced to use more expensive or less desirable fuels. This leads to resource misallocation.

Also, it appears that the FPC ceilings are now holding down prices in most areas—that newly marketed gas would command higher prices if the ceilings did not exist. This has led to a considerable income transfer from producers and royalty owners to some gas consumers; those consumers who are able to purchase gas transported by interstate pipelines pay less for their gas than they otherwise would have paid, the losers being the producers, royalty owners, and consumers who are unable to purchase all the gas they would like.

The migration of firms (who otherwise would prefer to locate somewhere else) to locations where they can obtain nonjurisdictional gas because they need an assured supply also causes resource misallocation. As Milton Russell (chapter 9) shows, so long as nonjurisdictional purchasers are not forbidden to pay a market-clearing price for gas, it is the jurisdictional purchasers who may not be able to purchase as much gas as they would like.

The current system also produces resource misallocation because interarea value differentials are not reflected appropriately.[23] The value of a gas reserve to the ultimate consumer depends on the total cost to the consumer at the consuming point. Location is a particularly important determinant of this cost because of transportation expense.[24] The use of national cost patterns for new gas and historical costs for old and associated gas have created a pricing pattern that eliminates some of the quantity that might be supplied from high cost—but high value—reserves, and forces the jurisdictional customer into greater transportation and other costs.

Too, the multi-price system creates resource misallocation. Resources are correctly allocated in a competitive market when the market price for all volumes sold is equal to the incremental or marginal cost of the last units produced. A profit-maximizing firm selling in a competitive market will choose an output level at which marginal cost equals selling price. If some gas is sold at a price less than the market price for new gas (which roughly equals the marginal cost of new gas), the average price paid by the pipeline for gas of all vintages will be below current marginal cost. This means that consumers will want to purchase more gas in the present than if gas were sold at marginal cost. This "rolled-in" price may also be very misleading to consumers about to invest in capital equipment that will tie them into using gas for years to come. Even if producer ceiling prices never again change and pipeline and distribution expenses remain constant, the price of gas will rise as increasingly larger fractions of the pipelines' supply come from "newer," higher-priced gas. These effects will be present regardless of whether the cause of the lower price paid for some gas is a long-term contract entered into in the past or to an FPC ceiling price for "old" gas. But to the extent that the FPC multi-price system holds

[22]Part of the excess of quantity demanded over quantity supplied may be due to a sudden increase in demand coupled with the existent lag between increased producer incentives and increments to supplies. Under similar conditions in other markets, sellers have refused to raise prices to the market-clearing level, preferring a temporary shortage to a large price increase followed soon by a price decrease.

[23]The rest of this paragraph is abstracted from Russell.

[24]See James Nelson's paper (chap. 10) for a discussion of the importance of transportation costs in the natural gas industry.

down the price of "old" gas, it causes resource misallocation. On the other hand, this system clearly moderates price increases to existing users in times of rising prices.

Ceiling price regulation may have improved resource allocation where pockets of producer monopoly power have existed. But to the extent that the existence of such monopoly has been transitory or localized, the beneficial effects on resource allocation have been small.

As a conclusion to this section about the effects of the current system, it is interesting to consider the attitudes of various groups in the economy toward the current regulatory system. But attempting to identify appropriate groups and to ascertain whether each group would, on balance, prefer the present system or some alternative to it is a very complicated and speculative activity. The following conclusions can, however, be hazarded.[25] Consumers who can get all the gas they want probably favor the present system; those who cannot would favor some other alternative that would allow them to do so. Clearly, the gas producing industry does not like the current system. The coal industry, however, probably does. The coal producers likely believe that more gas would be available (and used) if higher prices were paid gas producers and that thus the demand for coal would decrease if the regulatory system were changed to allow these higher prices. Some may dislike the current system, believing it limits the quantity of gas available and thus forces the use of more pollutional fuels. Others, who belong to the "zero growth" school, may be glad to see a shortage of a fuel, a shortage that might in its small way discourage industrial growth.

If the current system of regulation is to be retained, one might achieve uniform treatment of all gas production by expanding FPC jurisdiction. There is little doubt that an act of Congress bringing all natural gas production under FPC jurisdiction would be ruled constitutional. In times like these when quantity demanded exceeds quantity supplied, to some it seems far more equitable for the FPC to control gas allocation in all gas markets through certification of sales than to have the intrastate markets cleared and interstate markets bear the full supply deficiency. Such a law might also help relieve locational resource misallocation. If the FPC controlled gas allocation and regulated gas prices in all markets, there would be no artificial incentive for firms to move to gas-producing states. However, most economists acquainted with the industry would not favor such an extension of FPC jurisdiction. They feel that the allocational problems involved would be too large for the FPC to handle.

There are at least two goals, however, toward which the commission might work in the framework of the present system of regulation—the goals of reducing producer uncertainty and of shortening regulatory lag. Uncertainty is a cost to an economic unit, and action that would, for example, foreclose future administrative reductions of already approved gas field prices would increase the actual price of gas to producers while leaving the transfer price unchanged. Similarly, publication of the criteria for future pricing decisions would likely increase the certainty equivalent of prospective return from given exploration and development expenditures.

The reduction of regulatory lag would result in more timely price increases in periods of secularly rising gas prices. Lag reduction would result not only in a higher effective price for producers but would also reduce producer and consumer uncertainty. Suggestions for the use of index-type cost adjustments to new gas prices have been offered as one possible means of reducing regulatory lag.

Decontrol of New Gas Prices

As an alternative to the current system, the FPC might abandon broad-gauged area price regulation for new gas contracts. Since this proposal is based on the supposition that workable competition exists in most markets at most times, this alternative does not imply that the FPC should relinquish all control over new gas markets. Rather, it means that the FPC should continuously survey natural gas markets for evidences of noncompetitive behavior. Only in the few markets in which a lack of workable

[25] See Russell's paper for a thorough discussion of this subject.

competition was observed would the FPC supervise contract terms.[26] In the long run, such a regulatory system would have several effects.

First, it would allow the various producer gas markets to clear. The prices of gas sold by pipelines would rise until the quantities the pipelines would supply would equal the quantities demanded.[27] Neither the pipelines nor the FPC would have to engage in rationing.

Second, it would improve resource allocation. All those who would be willing to pay current prices would be able to purchase gas. There no longer would be any artificial incentive for firms to migrate to obtain an assured gas supply. Pipelines would be allowed to pay more for gas that is more valuable to them because of locational or other reasons. Further, the resource misallocation associated with the FPC imposed multi-price system would disappear over time as the old gas with controlled prices was exhausted.

Since only new contracts would be decontrolled, the price rise to consumers would be slower than if all vintage ceilings were lifted.[28] This might be more socially acceptable than a more sudden consumer price increase. In such a case, however, it would seem to be the duty of the FPC to warn consumers of continuing price increases until a new equilibrium was attained.

The form of income redistribution caused by the current system would disappear over time. Additional economic rent would accrue to producers and landowners.

Decontrol of All Gas Prices

This option would be essentially the same as that discussed in the previous section except that the FPC would relinquish price control of all vintages of gas.[29] Producers with permissive clauses in their existing contracts could then demand or bargain for prices above the current ceilings. To the extent that producers of old gas were able to exact higher prices for it, average prices paid by pipelines and thus consumer prices would climb faster than if ceilings were lifted only on new contract prices.

Choosing this alternative would have effects similar to those of decontrolling only new gas prices with certain modifications. The resource misallocation caused by the multi-price system would be partially relieved rather quickly. Also, the more rapid rise in consumer prices might provide incentive for quicker development of substitute fuels.[30] Rents to producers and royalty owners of old gas would be substantially increased. Such producers would receive higher prices and landowners higher royalties for gas that otherwise would have been supplied at lower prices. On the other hand, rents to producers and royalty owners of new gas would be lower than with only decontrol of new gas because of the lower price for new gas that would be obtained. It is not obvious which of the two alternatives would yield the lowest aggregate rent to producers and royalty owners. The more sudden consumer price increases that would occur if this alternative were adopted might cause sharper social repercussions than the more gradual rise of consumer prices if only new gas were decontrolled. Substantial savings in FPC costs would result from adopting this proposal.

[26] How the FPC might monitor markets for competitiveness is discussed briefly by Russell. Since the FPC does not now have sufficient staff trained in this area, the monitoring responsibility could be assigned to the Federal Trade Commission or to the antitrust division of the Justice Department.

[27] In this discussion, it is assumed that the present methods of pipeline regulation remain unchanged.

[28] Many existing contracts between producers and pipelines contain provisions that would now entitle the producer to a price greater than the current FPC ceiling or would be subject to renegotiation if the FPC permitted. If the elasticity of supply of old gas is zero, the elasticity of new gas is not zero, and the elasticity of demand is not infinite, the initial price of new gas would rise higher under this alternative than under complete decontrol, but the initial average price of all gas would be less. Over time, the average price would rise to the long-run equilibrium price as the old gas was exhausted. The long-run equilibrium price would be the same under either alternative.

[29] Except, of course, in those areas where workable competition does not seem to exist. Surveillance to uncover any such areas would be required as in the previous alternative. (See fn. 26.)

[30] This assumes that pipelines would continue to price gas using average cost. Should they price incremental volumes using marginal costs, prices to incremental users would be higher if only new gas were decontrolled.

There would be, however, some serious questions about the future of FPC producer regulation if such a proposal were adopted. If ordinarily the commission really lets market forces determine prices, would such "do-nothing" regulation be viable? Or would irresistible consumer pressure arise for the FPC to reassert control in most markets in order to hold down consumer price increases? There is no consensus about the answers to these questions, but they should be considered by those attempting to evaluate this, and other, policy alternatives.

The Market Value Standard

Instead of decontrol of producer gas prices coupled with continued surveillance of markets for non-competitive behavior, the FPC might retain authority to fix ceiling prices but attempt to set them in a fashion somewhat different from the current cost-based criteria. Under a market value standard, field prices would reflect the value of gas to consumers.

The most important element in determining the value of a given reservoir to consumers is its location. The consumers must pay for transporting gas from the field to the burner tip. Therefore the value of a unit of field gas is inversely related to its distance to the consumer and also its distance from existing transportation facilities. Thus, gas found in West Virginia would be worth more to New York consumers than gas from a similar deposit located in Texas. Of course, gas produced at a given place could theoretically be used by consumers in a number of different locations. Nonetheless, Russell argues that the geographical distribution of major gas reserves and consumers is such that relatively unambiguous location value differentials could be determined for each producing area.

Once these differentials were settled, adjustments in price could be uniform nationally (ignoring basic industry shifts) and would follow from findings that the market was failing to clear appropriately. Essential to satisfactory operation of a market value system would be abandonment of the concept of vintage pricing for gas. Noncontractual differentials in prices for gas that has an equal value to consumers would violate the market value criterion.

Adoption of this pricing method would improve resource allocation by providing relatively larger incentives for the search for gas in high-value areas than in low-value areas. In fact, if the commission did its work well, prices in all the competitive markets would approximate those that would prevail in the absence of regulation.

However, the problems involved in implementing such a system are formidable. To get it started, the commission would have to estimate market-clearing consumer prices and to translate these into producing-area market values. And, as Spritzer points out, it is difficult to see what useful purpose such FPC regulation could serve in most markets, since decontrol of these markets would also result in a competitive price.

If convinced that the FPC would really attempt to set ceiling prices in this manner, producers might well favor this alternative to decontrol with continued contract price surveillance. The producing industry might feel it could at least partially defend itself from changes to too high prices by pointing out that producers were receiving no more than commission-determined maximum prices. A possible danger inherent in this alternative is that the FPC might become "captured" (as other regulatory bodies allegedly have been) and come to regard itself as the guardian of the producers' rate of return.

Proposals for Improving Industry Performance

Several proposals for improving the performance of the gas producing industry have been made. I shall list a few of them and state briefly what I consider their major advantages and disadvantages to be.

1. One set of proposals is aimed at providing increased competition in the industry by encouraging small independent producers. First, the FPC might allow larger advance payments by pipelines or other purchasers to producers. Such payments generate capital that may be very vital to a small producer in times of capital stringency. Second, the Department of the Interior might set a nominal fixed lease bonus and make percentage royalty the variable in bidding for outer continental shelf (OCS)

lands. Such easing of the initial capital outlay ought to increase the chances of smaller firms bidding without decreasing the present value of the net revenue stream to the government. Finally, perhaps consortiums of large companies ought to be forbidden to bid for OCS or other public land tracts except where testing is likely to be so expensive as to discourage even a single large firm. In this way the competition for tracts might be increased.

2. Several schemes for accelerating or yielding additional incentive for domestic exploration and development have been advocated. For example, the FPC might grant higher ceiling prices to those who guarantee to increase exploration activity in some specified way. This is an interesting idea, but it would seem very difficult to administer. How can one tell for sure that a company is doing exploration it otherwise would not have done? Others would make the depletion allowance available only to the extent new exploration expenses are incurred. This would have the advantage of tying the depletion allowance to the stimulation of exploration.[31] Of course, many economists feel the depletion allowance is an inefficient and resource-misallocating method for encouraging exploration.[32]

3. Another suggestion is to change the tax treatment of production of oil and gas on foreign soil by domestic firms to provide more of a relative advantage to domestic exploration and production. It is not clear, however, that such a procedure would be very effective. Increased taxes on foreign operations of domestic petroleum firms may well both increase the Treasury Department's tax take and cause diminished foreign investment, but it is difficult to see how such a move, per se, would much increase domestic investment.

4. Ross has suggested the formation of "a public corporation open to private participation to engage in the exploration and development of oil and gas. The public corporation should be limited to the public lands of the United States, including the outer continental shelf." Such a corporation might serve at least two major purposes. First, it might expedite the development of gas and other fuels on government lands. Second, such a corporation could go a long way toward plugging some of the information gaps that currently exist about gas supply. There are, however, other ways of attaining these goals, and the proposal also raises the troublesome question of the efficiency with which such a public corporation could be operated.

5. A final group of proposals deals with the availability of substitutes for natural gas.[33] One proposal is that synthetic gas prices not be regulated at the factory gate.[34] In view of the widespread nature of coal deposits, it seems quite likely that pipelines would be able to bargain vigorously for low prices. Indeed, if present cost projections for manufactured gas are accurate, pipelines that wish to purchase large quantities of such gas will likely have a strong incentive to bargain hard—the threat of losing customers to alternative fuels if their delivered price is too high. Similar arguments may be made for allowing LNG imports sans duty or other price regulation. Another suggestion is to modify the Jones Act to allow shipment of LNG between American posts by foreign flag vessels. Currently, Alaskan LNG is being transported to Japan but not to Hawaii, because it is cheaper to transport gas from Alaska to Japan on foreign flag vessels than to Hawaii on American ships.

Conclusions

FPC regulation of the maximum price that jurisdictional purchasers are allowed to pay producers of natural gas seems to have been at least partially responsible for the current unsatisfactory conditions existing in the natural gas producing and distributing industries. At current price levels, the quantity of gas demanded by consumers considerably exceeds that which producers are willing to supply.

[31] See Spritzer for a discussion of federal leasing policy and tax incentives.

[32] Also, such treatment might not be in line with other incentive payments. In a private communication, Richard Gonzalez has suggested that depletion, patents, and copyrights are rewards for contributions to society that may encourage desirable actions but are not granted on the basis that the recipient must continue to repeat past efforts.

[33] See Nelson.

[34] It would seem that under existing law synthetic gas is not natural gas and therefore not subject to regulation, but there apparently has been no definitive ruling on the question.

There is no evidence that large enough quantities of substitutes for natural gas produced in the forty-eight contiguous states (gas from coal or oil, Canadian imports by pipeline, or liquefied natural gas from Alaska or foreign countries) will be available in the next decade to substantially reduce the projected demand growth for natural gas, either at current or prospective equilibrium natural gas prices. Therefore, providing market conditions under which quantity of natural gas supplied is likely to increase to quantity demanded seems to be the only way to end the current problem.

Probably it would have been better not to have instituted blanket regulation at all, but rather to have confined producer regulation to those (apparently few) areas where workable competition among producers did not exist. However, there is no consensus about how the current system should now be modified. Suggestions range from some modification of the current policy to the abandonment of the federal attempts to influence most gas prices in the field. I have discussed the advantages and disadvantages of the current system and some proposed remedial policies in an effort to present to policy makers the important issues as an economist sees them.

A NOTE ON ECONOMICS AND NATURAL GAS REGULATION

None of the papers in this volume contains an inclusive summary of elementary economic theory of markets and regulation. Since some readers may not be completely familiar with the economic terminology and concepts used in the papers, I offer such a summary here.

Ultimately, the decisions about regulating a market in an economy rest on judgments about what will be best for society as a whole. However, criteria for evaluating social welfare differ among individuals. Even when one looks at only one aspect of human behavior—say, economic behavior—there is not likely to be unanimity of opinion about the relative desirability of alternative states. Nonetheless, economists have developed a method of analysis that tends to pervade their thinking about the desirability of economic alternatives. If one is to understand the recommendations of economists, it is useful to survey their economic theory of regulation.

The Goal of Optimum Resource Allocation

There are two types of economic theory.[35] Positive economic theory attempts to describe the behavior of economic units within an economy. Moral issues are outside its scope. Positive theory can be used to predict economic behavior in well-defined situations, but it does not address itself to the desirability or undesirability of the resultant behavior. Normative economic theory, however, attempts to deal with what ought to be. Normative theorists sometimes describe their goal as maximizing economic welfare (though there is no consensus on how to measure economic welfare). Thus, normative economics is often called welfare economics.

Many economists believe that the greater the total real (deflated) net value of goods and services produced in an economy, the greater the economy's potential want-satisfying power. They therefore think maximizing the real value of the goods and services produced in an economy, given the resources available for production, to be a valid sub-goal. If each resource is used in such a way that it contributes the maximum possible amount to the total value of goods and services produced, then these resources are said to be efficiently allocated. But, of course, the social welfare is affected by many things other than just the total value of goods and services produced; how these goods and services are distributed among the members of the society is of extreme importance also. In an economic system like that of the United States, where most resources are privately owned, many economists would

[35] Much of the discussion that follows is extremely simplified. In particular, no mention is made of external economies or "second-best" problems, and, unless mentioned, the analysis is static rather than dynamic and proceeds in a partial equilibrium rather than in a general equilibrium framework.

advocate the following scheme for achieving the ideal distribution of the fruits of economic activity: (1) let payments be made to the owners of resources in such a manner that resource owners will have the incentive to use their resources in the most efficient ways; and (2) if the resulting income distribution is deemed to be inequitable, correct the inequity through a system of taxes and subsidies that will not affect resource allocation. If this scheme is followed, the total value of the goods and services produced is maximized and the means to purchase these goods and services is "fairly" divided among the members of the economy.

Regulation, Optimum Resource Allocation, and Income Transfer

It can be shown that if all markets in an economy are purely competitive and if the economists' usual assumptions about the behavior of entrepreneurs and resource owners are correct, resources will be allocated efficiently and the maximum possible value of goods and services will be produced. But if in one market pure competition does not prevail, then the price in that market may diverge from what it would have been if the market had been purely competitive. Such a price distortion will cause inefficient resource allocation and the total value of the economy's output will be reduced. Further, economic units that possess "market power" in an imperfect market may be able to earn income in excess of what they would have earned had the market been competitive. This income is earned at the expense of other groups in the economy, including those consumers who must pay higher than competitive prices. Government regulation of imperfect markets thus may aim at the following goals: (1) establishing a price in the market that will cause resources to be allocated efficiently, and (2) preventing the holders of market power from using it to obtain "excess incomes." There is a method of regulation that, at least in theory, can accomplish both of the above goals. In essence, this method requires the regulatory authorities to set the market price at the level that would prevail if the market were purely competitive.

There are at least two reasons, however, why regulators may feel pressure to set ceiling prices at a lower than competitive level. Even in competitive markets, owners of scarce resources may earn rents. Those who feel rents earned in a market would be "excessive" if prices were at the competitive level may recommend lower ceiling prices in order to reduce the level of rents. Second, there may be some who feel that the existing distribution of income is not satisfactory and would like to see income (of whatever origin) transferred from producers to consumers. Since selective taxes and especially direct subsidies (other than those paid to the most needy cases) do not appear to be very popular politically, a ceiling price below competitive level may be perceived as a viable means of effecting this transfer. However, pursuing the goal of income redistribution by the indirect means of regulation involves a cost. If regulators operating in a market cause the price to deviate from the competitive level in order to effect a redistribution of income among the participants in that market, resources will be misallocated in this market and the total value of the output of goods and services in the economy will be reduced. Therefore, if regulators decrease some price other than the competitive one, the desirable (at least to the regulators) effects of income redistribution are achieved only at the cost of a reduced value of national output.

Economic Rent

Economic rent is a payment in excess of that which is necessary to bring a factor of production into use. Almost all payments to landowners can be classified as rent, since the inconvenience cost to the landowner is usually very small in comparison with the payments he may receive. Therefore, many owners would have offered their holdings at much lower prices had those lower prices been the highest available. They are often able to get more than their reservation price, as the next two paragraphs will show.

Those who explore for oil and gas have as a major incentive the hope of uncovering information that will enable them to lease potentially productive lands under such favorable terms that they will earn, on average, above normal returns on their investments. They hope to be able to do this because landowners in general have done no exploration themselves and thus are dependent on explorers for information about their lands. But if there is workable competition among producers, rivalry between the potential lease purchasers will tend to bid up the prices paid to landowners until, on the average, landowners are getting nearly all the certainty equivalent of the producers' return stream in excess of their normal profit, based on present expectations about the future.

However, the key phrase in the preceding paragraph is "based on present expectations about the future." If the present value of expected net revenue increases after the lease is signed, then the producing firm will be the prime beneficiary; the landowner's lease bonus has already been fixed and increased royalty payments will typically absorb only a relatively small fraction of increased revenues. But so long as expectations remain constant at their new level, the earning of rent by producers is only a one-time affair. Competitive bidding for new prospects will, on the average, squeeze out rent from producers, passing it on to landowners.[36]

Regulation and the Optimal Consumption Rate over Time

Many minerals are found only in limited supplies in local deposits and are apparently neither being created in, nor migrating into, these deposits at an appreciable rate. Such a resource, whose total underground stock can only be diminished over time as a flow of the resource is consumed, is said to be a nonreplenishable resource. There has been considerable discussion in the literature about whether supplies of certain nonreplenishable resources are being used too rapidly according to some criterion that takes into account the satisfaction of future as well as current consumers. Those who believe that specific resources are being used too fast often point to uses that they consider to be wasteful or "inferior." If these uses were curtailed, more of the resource could be conserved for the future. If prospective purchasers of gas for "inferior" uses are forbidden to buy gas and the supply schedule for gas is not infinitely elastic, then not only will some gas be saved for the future but the equilibrium price for gas for the remaining uses will be lower. But these effects were purchased only at the cost of reduced real value of goods and services in the economy in the present.

However, it is always debatable whether units of some resources are likely to be sufficiently more valuable in the future to justify forgoing consumption of them in the present by those who are willing to pay prevailing prices. In the past, technological improvements that enabled the exploitation of lower-grade deposits or the discovery of unsuspected deposits have often removed the threat of serious future shortages and consequent drastic price increases. Or, new substitutes may have lessened the demand for the resource with the same result. On the other hand, the advent of such fortuitous occurrences cannot usually be guaranteed. On this question, too, one's judgment is likely to be subjective and to depend on one's faith in technological progress.

Other Costs of Regulation

Some costs of regulation other than possible resource misallocation need to be mentioned. Regulation carries with it costs that can be directly allocated to the affected firms and the regulatory body. These costs include the expenses of the regulatory body and its staff and extra employees necessary in the regulated firms. Other costs result from regulation-caused inefficiencies, such as a long time delay between petitions by firms and regulatory decision or diversion of managerial talent from trying to make the firm operate more efficiently to trying to "put one over" on the regulators. This latter attempt may give rise to what Professor James McKie of the University of Texas has called the "tar-baby effect." Though regulators may control certain variables, such as price, firms may find that they can partially offset the resulting undesirable effects (to them) by making changes in uncontrolled

[36] See Hawkins's paper for an expanded discussion of this question.

variables, such as terms of sale. "Extension of control in response to perpetually escaping effects of earlier regulation . . . usually enmeshes the regulatory authority in a control effort of increasing complexity with little gain in efficiency, but a growing feeling of frustration."[37] In any case, the adversary atmosphere is probably not conducive to efficiency either on the part of the regulator or the regulated. For example, former FPC Commissioner Charles Ross complains of a record "developed more for the purpose of achieving some small victory over opposing counsel than for assisting the (regulatory) commission."

Market Price, Regulation, and Resource Allocation

The positive theory of price determination in markets can be illustrated with the "Marshallian Cross" graph that shows intersecting supply and demand schedules, that is, schedules of quantities per unit of time that would be demanded or supplied at various prices, other determinants of demand and supply remaining constant. Traditionally, as seen in the accompanying figure, price is the ordinate and quantity the abscissa on this diagram. The demand schedule is conventionally drawn sloping downward; the higher the price, the smaller the quantity demanded. Similarly, the supply curve slopes upward; the higher the price, the larger the quantity supplied. The price at which the demand and supply schedules intersect is called the equilibrium price; at that price the quantity demanded equals the quantity supplied and the market clears.

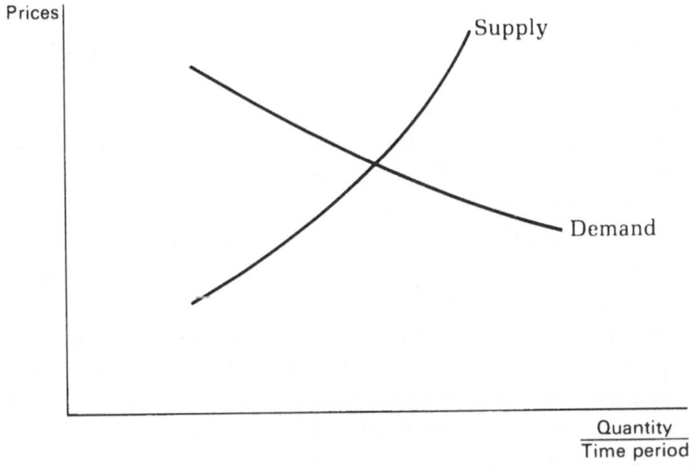

If the market price for some reason does not equal the equilibrium price in an unregulated market, forces will be generated that will tend to push the market price toward the equilibrium price. If the market price is below equilibrium, quantity demanded will exceed that supplied, and unsatisfied consumers will bid up the price toward equilibrium. If market price is above equilibrium, quantity supplied will exceed that demanded, and producers will have to cut prices in order to sell the surplus. Equilibrium will be attained after the surplus has been disposed of and price has fallen to the level at which quantity supplied equals quantity demanded.

In a competitive market, if regulatory officials decree a ceiling price below the equilibrium price in a producing area, they will induce a shortage (a larger quantity will be demanded than supplied), and the regulators must enforce some workable rationing plan if a black market is to be avoided. If a

[37] James W. McKie, "Regulation and the Free Market: The Problem of Boundaries," *The Bell Journal of Economics and Management Sciences*, vol. 1, no. 1 (Spring 1970), p. 9.

decreed minimum price is above the equilibrium level, a surplus of the commodity will be produced, and the regulators must find a way to prevent the surplus from depressing the actual price at which many transactions take place.

As has been pointed out, if the market is workably competitive, letting the market move to equilibrium price is a necessary condition for efficient resource allocation. On the other hand, if the market is not workably competitive the equilibrium price is not that necessary for efficient resource allocation. If the product sellers have monopoly power, an appropriate ceiling price below the equilibrium price will cause sellers to wish to sell more at the ceiling than they would in an unregulated market. If the problem is monopsony on the part of the purchaser in the market, then the appropriate remedy aimed at achieving efficient resource allocation is to set a suitable minimum price. At this minimum price, buyers will be willing to purchase more than their unregulated market demand schedule would call for. If, however, a market is competitive, no regulation of that market is needed to aid efficient resource allocation; indeed, any attempt at regulating a competitive market will likely hinder efficient resource allocation.

Analytical Difficulties Connected with the Natural Gas Producing Industry

A number of formidable problems arise when one tries to apply this analysis to the natural gas industry. It is almost never economically efficient to produce all the recoverable natural gas in a reservoir within such a short period as a year; reservoirs are usually developed to have productive lives of from five to over twenty years. Most gas is sold by producers under long-term contracts. The present value of such a contract to a producer is not a function of just the current price per Mcf that the producer receives; it also depends on such things as the specified time pattern for gas production, any take-or-pay provisions, possible contractual price escalations, etc. Further, gas is not of uniform quality from field to field, nor are all deposits of equal quality of equal value to a pipeline, because of locational differences. Nevertheless, it is possible to think of converting all actual prices into an equivalent price for some standard contract, quality, and location.

Another problem is that of defining a frame of reference for discussions about supply and demand. Some would define supply and demand in terms of annual production. Others have noted that current annual supplies are largely dependent on conditions that have existed in the past, that once contracts have been entered into and wells have been drilled, the quantity that producers wish to supply from a reservoir is not as responsive to change in present or expected future prices as it was before the capital was committed and the contracts signed. Therefore, these others say, a more appropriate quantity is the annual volume of new reserves. This quantity is certainly more sensitive to current conditions than is annual supply. Also, the current rate of reserve addition helps one to predict shifts in the supply schedule in the intermediate future. However, the relationship between reserves added in one year and the resulting yearly supply pattern from these reserves has varied over time; the desirable rate of exploitation at the time of development can vary because of changes in price, interest rates, etc. Also, the proving of reserves is not the same thing as development, though in many cases there is little incentive to prove reserves unless development is imminent. Nevertheless, since one of the prime concerns of regulators is what may happen to supply in the future, I prefer (with one exception) in this paper to work with reserve additions as the quantity for supply and demand schedules.

An inherent limitation in the above analysis is that it is static. A dynamic analysis is preferable for several reasons. It has already been pointed out that the production pattern over the life of a reservoir is heavily influenced by the economic conditions prevailing at the time the reservoir is developed. But before reservoirs can be developed, they must be discovered through exploration. The number and types of reservoirs available for development at any time are a function of past technology and expenditures on exploration. There are lags on the demand side, too. Once a consumer has installed capital equipment that uses gas, he may have strong incentives to continue using gas until such time as replacement of the capital equipment appears attractive, even in the face of substantial gas price increases.

Thus, there are long time lags built into the system. Not much is known, however, about the lag structure or magnitude, particularly for supply.

These lags can be the boon or the bane of the regulator. If the regulatory body makes a mistake but corrects it quickly, then the damage done may not be very obvious, since current quantities supplied and demanded depend so strongly on past conditions. But this same lag structure makes it difficult for a regulator to discern that he has made a mistake; conclusive evidence may be a long time in coming. And the results of misguided policy may continue to haunt the regulator for some time after the policy error has been corrected. During such a time, regulatory authorities may be under great stress. Faced with public clamor to remedy the situation as soon as possible and with uncertainty about the time pattern of effects generated by a rectifying action, regulators may have a tendency to overcorrect.

A somewhat related difficulty has its basis in a severe lack of information about the exploration process. Discovery and development take place each year on tracts selected from an inventory of prospects that have been uncovered by exploration in the past. If the inventory is not replenished by further exploration, then the supply schedule will change in future years even though drilling and the development costs have remained constant as the better prospects in the inventory are removed. Thus, the responsiveness of the quantity of new reserves supplied to price changes is a function of the current inventory that has been accumulated as a result of past exploration. Therefore, it is possible that current responsiveness of reserves supplied is not at all indicative of future responsiveness. For example, a considerable change in price might result in only a small change in reserve additions in the next year or two but might so change the incentive for exploration as to cause large changes in reserves during subsequent years. And, vice versa, any change in reserve additions caused by a price change will peter out unless the price change causes a change in exploratory activity.

The picture is further complicated by the fact that much gas is found as a by-product of the search for oil. Also, the demand for natural gas depends in part on the price of fuel oil that can be used in place of gas for certain purposes. Thus, supply of gas and demand for gas are dependent to some extent on the state of the oil industry. In economic jargon, not only are oil and gas joint products, they are also substitutes for each other in some uses.

The static analysis also makes it difficult to handle changes in producers' expectations about the future. The opportunity cost of delaying development of a reservoir is equal to the holding or carrying costs for the reservoir plus the diminution in the present value of the net revenue stream caused by delay in developing the project. If producers believe that a sufficient price rise may occur in the future to more than offset the opportunity costs of delay, then they may hold back their undeveloped reserves.[38] Thus, the quantity of new reserves supplied may vary from year to year even though costs of drilling and production remain constant. For the best available discussion of current knowledge about the responsiveness of supply to price changes, the reader is referred to the papers of Erickson and MacAvoy.

Other difficulties in applying the Marshallian Cross static analysis could be mentioned. Nonetheless, it is quite useful as an analytic tool if it is not used blindly. I have attempted here to discuss the important caveats necessary when applying this analysis to the natural gas producing industry.

[38] In all fairness, however, I should point out that situations in which holding back reserves is attractive are likely to be rare. With today's high interest rates opportunity costs of delay are likely to be so large that really substantial future price increases would be necessary to offset them. Further, unless the entire reservoir is controlled by one operator, delay without collusion will not be feasible, since leases almost always require development to prevent drainage by others producing from the reservoir.

1 NATURAL GAS IN THE FUTURE NATIONAL ENERGY PATTERN

John J. Schanz, Jr., and Helmut J. Frank***

The objective of this paper is to provide a statistical framework for the subsequent discussion of regulation of the U.S. natural gas industry by other authors. The initial section is a brief review of the regional pattern of gas supply and consumption, including some details concerning trends in types of use. Two special topics are also included in this first section—pipeline companies as suppliers of gas and the intrastate market as compared with the interstate market.

The second section is devoted to an examintion of future energy requirements predictions for the United States. A general outlook for the possible magnitude of energy consumption in 1990 is drawn from a number of contemporary studies. The gas component of future energy needs is then considered with particular emphasis on the assumptions and methodologies employed in recent studies. The section concludes with a résumé of the Future Requirements Committee's projections concerning future regional and use-category trends.

The final section is devoted to a discussion of the characteristics of gas demand and the difficulties of trying to assess future gas demand under alternative assumptions with respect to price regulation. The analytical problems that are likely to be encountered on departure from the customary simplifying assumptions concerning supply and relative price are discussed. The section concludes with brief examination of the policy alternatives available to regulators and policy makers.

The authors have relied considerably on the work of the Future Requirements Committee in discussing the future gas needs of the nation. However, the manner in which this and other statistical information has been employed is the responsibility of the authors. Interpretation of the statistical trends and the qualitative discussion of gas-demand forecasting represent the opinions of the authors and are not intended to reflect the viewpoint of the Future Requirements Committee or its individual members.

PRESENT PATTERNS OF U.S. PRODUCTION AND CONSUMPTION OF NATURAL GAS

At the present time the utilization of gas by the U.S. economy is almost totally dependent on natural gas provided by domestic oil and gas wells. Approximately 1 percent of total consumption is provided by manufactured gas, liquefied petroleum gas (LPG), or mixed gas. Although imports are expected to play a somewhat larger role in the not too distant future, they were only 3 percent of the total supply available to the United States in 1968 (see table 1-1). This, however, represents a marked increase over the 1 percent supplied as recently as 1958. Historically, most U.S. imports have originated in Canada and are brought into the United States by pipeline. A small amount of gas is imported

*Professor of Economics, University of Arizona.
**Professor of Natural Resources and Senior Research Economist, University of Denver.

Table 1-1. Salient Statistics of U.S. Natural Gas Industry, 1958-68

(trillion cubic feet)

Year	Gross withdrawals[a]	Marketed production[b]	Imports	Total supply[c]	Consumption[d]
1958	13.15	11.03	0.14	11.79	10.76
1959	14.23	12.05	0.13	12.85	11.82
1960	15.09	12.77	0.16	13.64	12.51
1961	15.46	13.25	0.22	14.09	13.01
1962	16.04	13.88	0.40	15.98	14.56
1963	16.97	14.75	0.40	15.98	14.56
1964	17.44	15.46	0.44	16.78	15.45
1965	17.96	16.04	0.46	17.46	16.03
1966	18.93	17.21	0.48	18.83	17.19
1967	20.25	18.17	0.56	19.87	18.17
1968	21.33	19.32	0.65	21.30	19.46
Average annual growth rate	4.4%	5.3%	14.4%	5.6%	5.6%

[a]Gross output of natural gas from domestic oil and gas wells.
[b]Gross withdrawals, less gas used for repressuring, vented, and flared.
[c]Marketed production, plus withdrawals from storage, plus imports.
[d]Marketed production corrected for interstate movement, change in storage, and transmission losses.

SOURCE: U.S. Bureau of Mines, *Minerals Yearbook*, for the years 1958-68.

from Mexico. The proposed importation of liquefied natural gas (LNG) from other areas outside of North America is not yet a reality.

During the decade 1958 to 1968, the utilization of domestic gas increased at an annual rate of 5.6 percent per year. However, this did not require a commensurate rate of increase in output at domestic wells, since gas returned to the reservoirs, flared, or vented has remained close to 2 trillion cubic feet (tcf) per year for the entire period. This, coupled with the increase in imports, has held gross withdrawals from the well as reported by the U.S. Bureau of Mines to an annual growth rate of 4.4 percent for the 1958-68 period.

By the end of the 1960s, the South Louisiana producing area had moved into a dominant position as a source of natural gas (see table 1-2). With output increasing by 13 percent per year for more than ten years, the area now produces almost twice as much gas as any other producing region. The three

Table 1-2. Production of Natural Gas by Producing Areas in the Continental United States, 1956, 1964, and 1969

Producing area	Volume[a]			Annual average growth rate		Percentage of continental U.S. production		
	1956	1964	1969	1956-69	1964-69	1956	1964	1969
	(----- *bil. cu. ft.* -----)			(---------------- *percent* ----------------)				
South Louisiana	1,393	3,513	6,789	13.0	14.1	12.8	22.9	32.9
Texas Gulf Coast	2,550	3,289	3,827	3.2	3.1	23.5	21.4	18.6
Hugoton-Anadarko	1,881	2,582	3,119[b]	4.0	3.9	17.3	16.8	15.1
Permian Basin	1,364	1,541	2,277	4.0	8.1	12.6	10.1	11.0
Other Southwest	2,210	2,424	2,359[b]	1.3	(0.5)	20.4	15.8	11.4
San Juan Basin, New Mexico	197	432	553	8.3	5.1	1.8	2.8	2.7
Rocky Mountain area	313	478	575	4.8	3.8	2.9	3.1	2.8
California	485	638	665	2.5	0.8	4.5	4.2	3.2
All other	456	443	472	0.7	1.3	4.2	2.9	2.3
Continental U.S.	10,849	15,341	20,636	5.1	6.1	100.0	100.0	100.0

[a]At 14.73 pounds per square inch absolute.
[b]Estimated.

SOURCE: Radford L. Schantz, Foster Associates, Inc., testimony in *Southern Louisiana Area Rate Proceedings*, FPC Docket no. AR 69-1, May 1, 1970.

other major gas-producing regions, which are roughly equal to one another in production and growth rates, are the Texas Gulf Coast, the Permian Basin, and the Hugoton-Andarko region. The latter two areas occupy a region that extends from western Texas northward into the Mid-Continent area. There is also production in the eastern portion of Oklahoma, in Arkansas, and in the area directly north of the Gulf Coast producing fields, but this region has declined in both importance and growth in recent years. The balance of U.S. gas production is found in an area that extends from New Mexico northward through the Rocky Mountains and in California, the Appalachian region, and the North Central states.

Since current production reflects very closely the proved reserves of gas, the distribution of known gas supplies mirrors the present regional pattern of production. Moreover, with one exception, the potential supply of gas yet to be discovered probably will be distributed in a pattern that rather closely resembles the regional picture of current proved reserves and contemporary production.[1] The one exception is the potential for gas to be discovered along the North Slope of Alaska. However, prospects for gas in this area are still considered to be speculative, and it is anticipated that the areas that provide the bulk of U.S. gas at the present time will probably continue to provide most of the domestic gas supply for a number of years to come. The relative importance of each area in the future will reflect the magnitude and success of further exploratory efforts.

The interstate movement of gas from the current major producing areas follows five main pathways (see figure 1-1). A cluster of interstate transmission lines moves large volumes of gas from the Gulf Coast fields northeastward along the western flank of the Appalachian mountains to the industrial and population concentrations of the Midwest and North Eastern Seaboard. A second branch moves gas from the Gulf eastward across the southern states. The gas then moves both north and south to serve Florida and the southeastern coastal states.

A third major route for gas to follow is the one that moves northward from the inland producing areas of the Southwest and Mid-Continent and then fans out to feed gas to the industrial states surrounding the Great Lakes. The two remaining pathways are not quite so striking in terms of volume or in the number of pipelines involved. One moves across the desert southwest from the Permian Basin and New Mexico to serve California. The other is the movement of gas northwestward from the Mid-Continent to the metropolitan areas of the Rocky Mountain region. In addition, there is also a certain amount of intermountain shipment.

Role of Pipeline-owned Gas in Meeting U.S. Gas Needs

Considerable data for the period 1946-65 have been assembled by the Federal Power Commission (FPC) in connection with RP66-24 concerning ownership of reserves and production of gas by the pipeline companies themselves, by affiliated companies that were "on-line" and providing gas to the parent pipeline company system, by affiliated companies owned by the pipeline but not providing gas to the pipeline system, and by interstate companies that did not own any gas themselves or through affiliated companies. The twenty-nine pipeline companies owning their own gas at the time of the original hearings were identified as Group I.

The data developed by the FPC in a special questionnaire for RP66-24 indicate that the gas owned and produced by the Group I interstate pipeline companies themselves and by their affiliates has never been more than a modest part of our total gas production in the post-World War II period. The reserves held by the Group I companies increased after 1946 and reached a peak in 1960. At that time they constituted 9.1 percent of the U.S. total proved gas reserves held at the end of 1960. The Group I reserves declined from 1960 to 1965, and FPC statistics since 1965 indicate that the reserves

[1] Potential Gas Committee, *Potential Supply of Natural Gas in the United States* (Colorado School of Mines, October 1969).

held by Group I have continued to decline through 1968. They were approximately 7 percent of U.S. reserves at the end of 1965.

The addition of both on-line and off-line reserves held by affiliates of the pipeline companies does not change the picture significantly since the holdings of the pipeline companies themselves are the larger share. Total reserves owned by the interstate pipeline companies and by affiliates reached a peak in 1961. It was 11 percent of the U.S. total at that time.

Despite some brief periods of decline, production of pipeline-owned gas, both internally and through affiliates, has climbed in the years 1946–68 to meet the growing need for gas. However, since the reserve additions have lagged behind the national trend, the pipelines' own production has gradually become a less important segment of our natural gas production.

Group I's own production was 10.1 percent of the national gas production in 1946 but declined to only 5.5 percent by 1965. Using FPC statistics to determine the trend since the RP66-24 questionnaire, it is indicated that the pipeline's own production had fallen to approximately 5 percent of the U.S. total production in 1968. Adding production from affiliated producing companies makes only a small change in this picture. Total production by pipeline companies and their affiliates was 13.1 percent of total U.S. natural gas production in 1946. This had declined to only 7.9 percent by 1965.

It is beyond the scope of this paper to discuss the variations among individual pipeline companies in developing their own reserves or the motivations that have led to an overall decline in this activity, but it is obvious that there has been little inclination on the part of interstate pipeline companies to develop their own sources of supply. What little incentive there may have been prior to 1960 seems to have lost much of its leverage in recent years, and the general pattern has been for pipeline companies as a group, but not necessarily in individual cases, to reduce further their involvement in reserve acquisition. This decline in reserve holdings has led to a decline in the relative importance of pipeline company reserves in the U.S. picture, a development that must ultimately cause pipeline company production to diminish also.

Variations in Natural Gas Consumption Statistics

Before examining the present pattern of natural gas consumption in the United States, it seems appropriate to review the similarities and differences in the sources of data concerning gas utilization. At present, there are three groups that assemble statistics on gas consumption—the U.S. Bureau of Mines (USBM), the American Gas Association (AGA), and the Future Requirements Committee (FRC) of the gas industry.

A considerable amount of confusion is encountered when one attempts to compare the statistical reports published by these three sources. For example, the FRC reports 1968 requirements as 21.5 tcf; the USBM reports 1968 consumption was 19.3 tcf; and the AGA reports natural gas sales in 1968 as 144.7 trillion therms (approximately equivalent to 14 tcf). Obviously, a difference of over 7 tcf between the top and bottom number is a matter of concern when trying to determine the gas needs of the United States.

Since three different agencies are involved in the collection of these statistics, and since their surveys are conducted at different times and by different methods, a perfect reconciliation of the differences in the statistics is not to be expected. Fortunately, the disparity in the numbers is not nearly as large as the absolute numbers seem to indicate once the differences in definitions, points of measurement, and units used are identified and understood.

The USBM natural gas statistics are basically designed to show the disposition of natural gas produced by the U.S. gas producing industry as it flows from well to marketplace (see figure 1-2). The FRC statistics, in contrast, are summations of the terminal utilization of gas by all users within the fifty states. On the other hand, the AGA numbers are restricted to arriving at total sales of gas by utility companies.

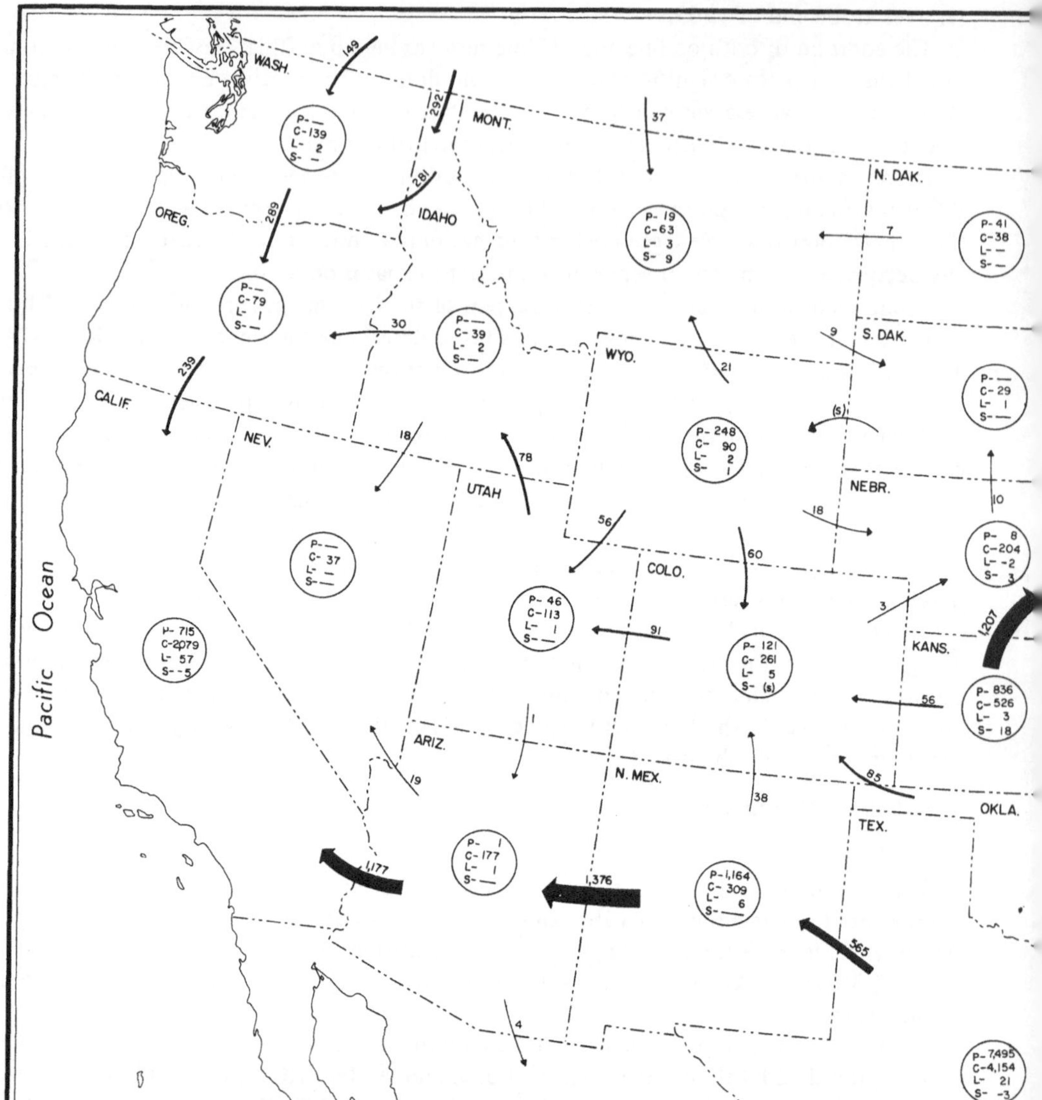

FIGURE 1-1. Production, consumption, and net interstate movements of natural gas in the United States, 1968.
Source: U.S. Bureau of Mines, "Natural Gas Production and Consumption: 1968," *Mineral Industry Surveys*, August 25, 1969.

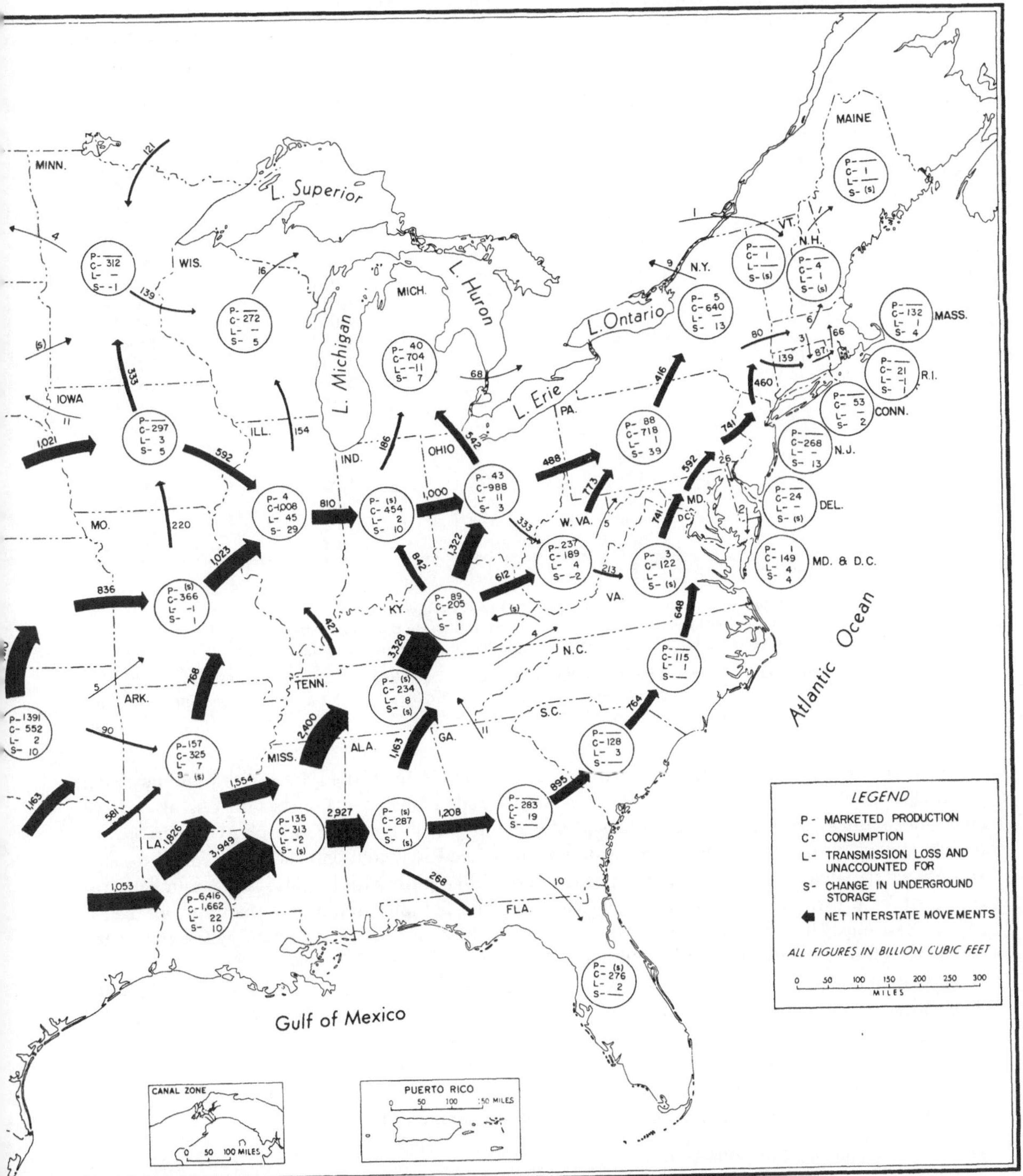
LEGEND
P - MARKETED PRODUCTION
C - CONSUMPTION
L - TRANSMISSION LOSS AND UNACCOUNTED FOR
S - CHANGE IN UNDERGROUND STORAGE
NET INTERSTATE MOVEMENTS
ALL FIGURES IN BILLION CUBIC FEET
0 50 100 150 200 250 300
MILES
MINN.
WIS.
MICH.
IOWA
ILL.
IND.
OHIO
PA.
N.Y.
VT.
N.H.
MAINE
MASS.
R.I.
CONN.
N.J.
DEL.
MD. & D.C.
W. VA.
VA.
KY.
MO.
TENN.
N.C.
S.C.
GA.
ALA.
MISS.
ARK.
LA.
FLA.
L. Superior
L. Michigan
L. Huron
L. Ontario
L. Erie
Atlantic Ocean
Gulf of Mexico
CANAL ZONE
0 50 100 MILES
PUERTO RICO
0 50 100 150 MILES

Table 1-3. Interstate Pipeline-Owned Natural Gas Production, 1946-68

(billion cubic feet)

Year	Group I[a]	III[b]	IV[c]	Total	U.S. production
1946	486	159	–	645	4,916
1947	520	152	–	672	5,599
1948	540	163	–	703	5,975
1949	520	163	1	684	6,211
1950	557	172	6	735	6,855
1951	652	206	4	862	7,924
1952	675	219	8	902	8,592
1953	577	271	11	859	9,188
1954	586	267	9	862	9,375
1955	640	278	27	945	10,063
1956	668	315	28	1,011	10,849
1957	689	313	31	1,033	11,440
1958	681	323	39	1,043	11,423
1959	753	319	42	1,114	12,373
1960	784	311	45	1,140	13,019
1961	779	308	46	1,133	13,379
1962	749	311	49	1,109	13,637
1963	801	330	47	1,178	14,546
1964	844	345	48	1,237	15,247
1965	888	353	45	1,286	16,252
1966	920	n.a.	n.a.	n.a.	17,491
1967	949	n.a.	n.a.	n.a.	18,381
1968	985	n.a.	n.a.	n.a.	19,374

NOTE: n.a.–not available.

[a]Group I–29 interstate pipeline companies that have produced their own gas.
[b]Group III–affiliates that deliver gas to the parent company interstate pipeline system.
[c]Group IV–off-line affiliates that do not deliver gas to the parent company interstate system.

SOURCES: For 1946-65, FPC, *Pipe Line Producing Area Rate Proceedings*, Docket no. RP 66-24, vol. 1, p. 214; vol. 3, p. 680; bol. 4, p. 1060. For 1966-68, FPC, *Statistics for Interstate Natural Gas Pipe Line Companies*, FPC S-183, July 1967; FPC S-189, July 1968; FPC S-197, July 1969.

Since the FRC is measuring the terminal utilization of gas in the United States, it should bear some physical relationship to the total amount of gas that is made available in any year for final use. However, this requires accounting for gas that is produced but goes into storage or back to a reservoir, thereby delaying final use to a later time. Consequently, the FRC requirements can be found to be approximately equal to either AGA net production or USBM gross withdrawals if these numbers are corrected for British thermal unit (Btu) content, shrinkage, net importation, net change in storage, and repressuring. These adjustments leave a residual statistical difference of something less than 1 tcf–part of which may be accounted for by the broader coverage of the FRC survey.

Numerical comparisons of USBM or FRC data with the AGA total sales data are virtually impossible, since the necessary detailed statistical data to make adjustments are not available. Since the AGA data are limited to utility sales, the numbers are not at all similar in magnitude to either the USBM's "gas delivered to consumers" or the FRC's "regional requirements." This is unavoidable since the AGA, in measuring utility sales, excludes field use, deliveries by nonutility gas suppliers, gas transfers that do not involve a sale, and all kinds of internal company use or losses.

Trends in Natural Gas Consumption

USBM data on U.S. energy consumption indicate that dry natural gas increased from 27 percent of the U.S. energy market to 31 percent in 1958-68.[2] This resulted from a growth rate of 5.6 percent

[2]USBM consumption data are employed to obtain growth rates for the past decade since only historical data for the period 1964-68 are included in FRC, *Future Natural Gas Requirements of the United States*, vol. 3 (Denver Research Institute, University of Denver, September 1969).

Table 1-4. Interstate Pipeline-Owned Natural Gas Reserves, 1946-68

(billion cubic feet)

Year	Group I[a]	Group III[b]	Group IV[c]	Group total	Proved U.S. reserves
1946	12,562	3,511	2	16,075	159,704
1947	13,756	3,527	1	17,284	165,026
1948	12,637	3,663	1	16,301	172,925
1949	11,652	3,896	222	15,770	179,402
1950	11,441	3,946	237	15,624	184,585
1951	12,066	3,801	233	16,100	192,759
1952	15,554	4,136	258	19,948	198,632
1953	15,908	4,679	251	20,856	210,299
1954	15,679	5,315	250	21,244	210,561
1955	16,583	5,479	518	22,580	222,483
1956	20,342	5,600	490	26,432	236,483
1957	21,166	5,624	466	27,526	245,230
1958	21,628	5,485	521	27,634	252,762
1959	22,420	4,735	894	28,049	261,170
1960	23,860	4,914	886	29,660	262,326
1961	23,719	5,318	892	29,929	266,273
1962	23,021	5,323	898	29,242	272,279
1963	22,349	5,134	890	28,373	276,151
1964	21,444	5,144	856	27,444	281,251
1965	21,503	5,237	840	27,580	286,469
1966	20,506	n.a.	n.a.	n.a.	289,333
1967	20,345	n.a.	n.a.	n.a.	292,908
1968	20,085	n.a.	n.a.	n.a.	289,350

NOTE: n.a.–not available.

[a]Group I–29 interstate pipeline companies that have produced their own gas.
[b]Group III–affiliates that deliver gas to the parent company interstate pipeline system.
[c]Group IV–off-line affiliates that do not deliver gas to the parent company interstate system.

SOURCES: For 1946-65, FPC, *Pipe Line Producing Area Rate Proceedings*, Docket no. RP 66-24, vol. 1, p. 214, vol. 3, p. 680, vol. 4, p. 1060. For 1966-68, FPC, *Statistics for Interstate Natural Gas Pipe Line Companies*, FPC S-183, July 1967; FPC S-189, July 1968; FPC S-197, July 1969.

per year for gas compared with slightly over 4 percent for total energy (see table 1-1). The FRC survey of actual requirements for the 1964-68 period shows an even higher rate of growth of 6.7 percent. USBM data indicate a more modest growth of 5.7 percent for these same years.

USBM's state consumption data have been aggregated into totals for the eleven regions utilized by the FRC to permit comparison between the two series. The areas of most striking growth in gas consumption in 1958-68 were the Great Lakes, the Pacific Northwest, and the Pacific area (Alaska and Hawaii).

Other areas with less spectacular growth but still above the national average were New England, the Southeast, and the Northern Plains. The remaining five regions fell below the national average (see table 1-5). FRC data on growth rates in regional requirements for the period 1964-68 correspond closely for most regions with bureau data. The major differences in the FRC and USBM regional series are found in the Southeast, the Gulf Coast, and the Pacific Northwest where the FRC growth rates are noticeably higher than those of the bureau.

The growth rate in natural gas consumption in all regions is at least equal to or considerably greater than the overall growth in national energy utilization. This nationwide competitive strength of natural gas in the marketplace has resulted in gas gaining a greater share of the total market almost everywhere during the last decade. In those areas where natural gas service has become available in recent years, such as in New England and the Pacific Northwest, the opportunity for market penetration has been greatest. Some regions that have been favorable for gas are those where industrial growth or population migration have stimulated rapid growth in gas use.[3] The areas that have displayed the slowest

[3]There are also areas where pipeline capacity inadequacies in the previous decade built up large unsatisfied demand backlogs. These supply problems were cured about the turn of the decade.

FIGURE 1-2. Schematic Diagrams of U.S. Gas Usage Statistics.

1. U.S. Bureau of Mines

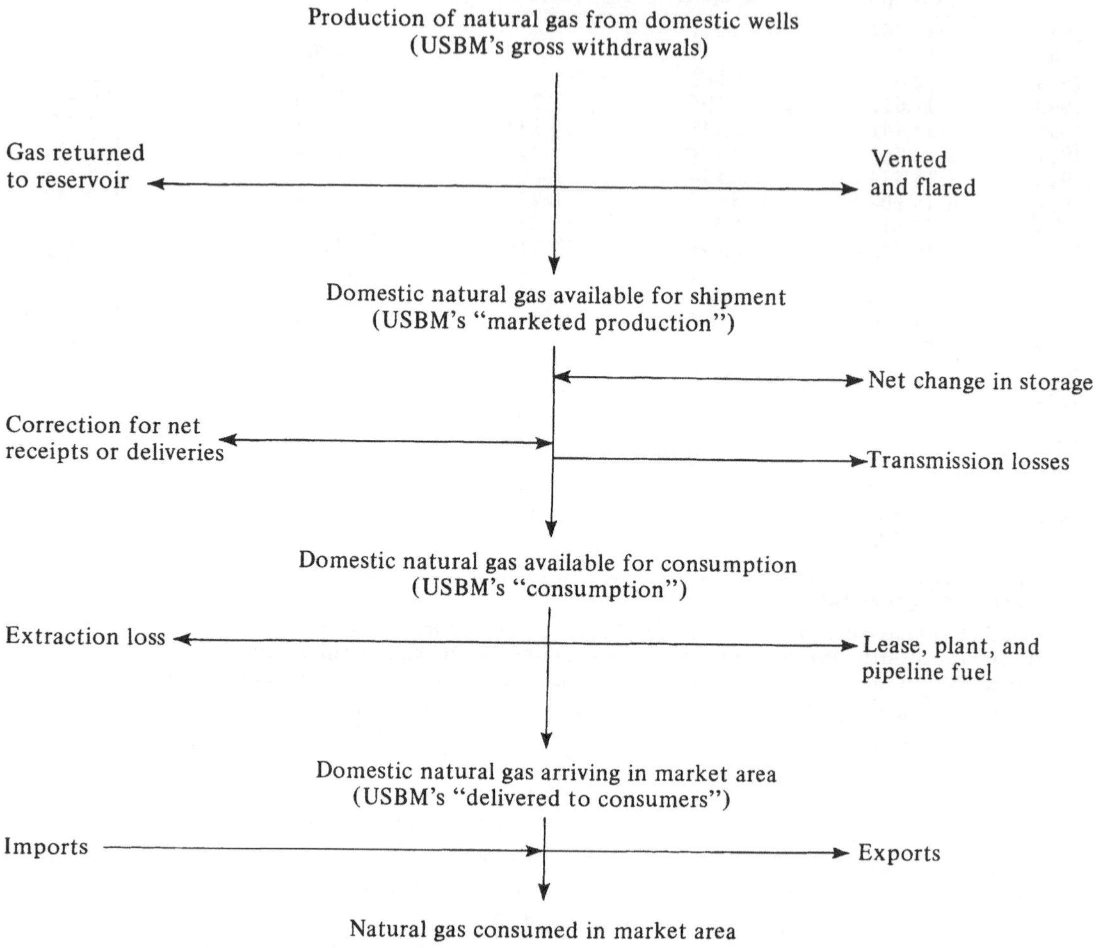

2. Future Requirements Committee

All consumption of domestic and imported gas for residential, commercial, industrial, and electric generation uses

plus

Pipeline use, company use, main line extraction plant use, shrinkage, other uses, losses and unaccounted for

plus

Field pumping and drilling fuel, field plant uses and losses, vented and flared gas, shrinkage, and other losses

equals

U.S. gas requirements

FIGURE 1-2 (Continued)

3. American Gas Association

Direct sales of gas by utilities and pipelines to residential, commercial, industrial, and electric utility customers

equals

total gas utility sales

4. Relationships between USBM, FRC, and AGA Statistics

FRC gas requirements ≅ USBM gross withdrawals – gas for repressuring – exports + imports ± net change in storage

FRC gas requirements ≅ AGA net production + shrinkage factor – exports + imports ± net change in storage.

AGA gas utility sales ≅ USBM natural gas delivered to consumers + consumption of other gases – exports + imports – direct deliveries to consumers by producers or nonutility firms – transfers by combination companies

AGA gas utility sales = FRC gas requirements – field use – direct deliveries by nonutilities – transfers – company use – pipeline use – extraction plant use – losses – shrinkage – unaccounted for

NOTE: These diagrams explain how the reported statistics can vary because of differences in points of measurement, in what is included or not included in the measurement, and in the units of measurement.

AGA – Utilizes therms of 100,000 Btu/therm.
USBM – Utilizes cubic feet of 1,032 Btu/cubic foot.
FRC – Utilizes cubic feet of 1,000 Btu/cubic foot except for field use, which is in original volumes.

Table 1-5. U.S. Regional Consumption of Natural Gas, 1958-68

Region	1958	1963	1968	Average annual growth rate 1958-68[a]
	(bil. cu. ft.		)	(percent)
1. New England	108	152	212	7.1
2. Appalachian	1,970	2,625	3,304	4.9
3. Southeast	587	878	1,322	7.2
4. Great Lakes	972	1,572	2,438	9.5
5. Northern Plains	459	636	881	5.9
6. Mid-Continent	946	1,279	1,443	4.3
7. Gulf Coast	3,846	4,863	6,453	4.8
8. Rocky Mountain	319	449	528	4.2
9. Pacific Southwest	1,444	2,007	2,602	5.4
10. Pacific Northwest	92	157	258	9.9
11. Pacific	0	4	18	30.4

[a]Growth rate was derived from a least squares trend line fitted to the regional consumption for the ten-year period.

SOURCE: U.S. Bureau of Mines, *Minerals Yearbook*, for the years 1958-68.

growth are primarily those where gas has been available over a long period and where there is less opportunity for market capture to provide an extra impetus to the growth rate.

The long-term trends in individual categories of gas use are difficult to determine because of the lack of usable statistical information. FRC and USBM data are continuously comparable for only the last few years. The AGA series covers a longer period but understates total use because it only includes utility sales (see tables 1-6 and 1-7).

Despite the sparsity of statistical information, the available data reveal that the distribution of gas consumption among the various categories of use is remarkably stable. Both AGA and FRC data indicate that residential use has encountered a modest decline in its share of the total gas market in recent years. This is almost matched by a small increase in the percentage of commercial use. AGA statistics show virtually no change in the relative importance of the industrial, electrical utility, and other categories of the gas market over the past ten years. FRC statistics, however, show an increase of 4 percent in the past five years in the share of the gas market credited to firm industrial use.

Based on the statistical evidence available, it must be concluded that all categories of gas use have participated almost equally in the strong growth of gas consumption in the immediate past. There are no major trends that would indicate that any change in the pattern of gas utilization had become apparent by 1968.

Behavior of the Intrastate Market

There has been considerable interest in the growth of the FPC-regulated interstate market compared with that found in the market for intrastate gas. An immediate problem involved in studying these two markets is the lack of intrastate statistics. Federal sources of gas statistics only report on the interstate markets. An inquiry made of the appropriate agencies in seven large gas-producing states reveals that information on intrastate sales is either not gathered by these state agencies or no attempt has been made to extract this information from the raw data received from industry so that it would be readily accessible.

As a result, there has been no recourse other than to approximate intrastate sales through statistical manipulation of total production and interstate sales data. For the purposes of the discussion here, intrastate sales for individual states and for the nation were estimated by taking the total marketed production reported by the USBM and subtracting sales by producers to interstate pipelines as re-

Table 1-6. U.S. Natural Gas Requirements, by Category of Use, 1964-68

	Residential		Commercial		Industrial	
Year	Volume (*bil. cu. ft.*)	Percent of total	Volume (*bil. cu. ft.*)	Percent of total	Volume (*bil. cu. ft.*)	Percent of total
1964	3,914	23.6	1,256	7.6	4,292	25.8
1965	4,053	23.4	1,343	7.7	4,794	27.6
1966	4,248	22.9	1,460	7.9	5,378	29.0
1967	4,442	22.6	1,573	8.0	5,773	29.3
1968	4,642	21.6	1,707	7.9	6,386	29.7

	Interruptible		Other and region 11[a]		Field use		
Year	Volume (*bil. cu. ft.*)	Percent of total	Volume (*bil. cu. ft.*)	Percent of total	Volume (*bil. cu. ft.*)	Percent of total	Total
1964	3,616	21.8	1,083	6.5	2,448	14.7	16,609
1965	3,711	21.4	1,170	6.7	2,268	13.1	17,339
1966	4,041	21.8	1,283	6.9	2,113	11.4	18,523
1967	4,120	20.9	1,332	6.8	2,452	12.5	19,692
1968	4,525	21.1	1,434	6.7	2,797	13.0	21,491

[a]Region 11 included with "other" since regional data by class of service are not available.

SOURCE: Future Requirements Committee, *Future Natural Gas Requirements of the United States*, vol. 3 (Denver Research Institute, University of Denver, September 1969).

Table 1-7. U.S. Natural Gas End-Use Consumption, by Category of Use, 1958–68

	Residential		Commercial		Industrial		Electrical utility		Electrical and industrial		Other		
Year	Amount (*bil. therms*)	Percent of total	Amount (*bil. therms*)	Percent of total	Amount (*bil. therms*)	Percent of total	Amount (*bil. therms*)	Percent of total	Amount (*bil. therms*)	Percent of total	Amount (*bil. therms*)	Percent of total	Total
1958	26.3	29.2	7.3	8.2	40.4	44.8	12.3	13.7	52.7	58.5	3.7	4.1	90.1
1959	28.0	28.3	7.9	8.1	45.3	45.7	13.6	13.7	58.8	59.4	4.3	4.3	99.1
1960	30.2	28.9	8.9	8.5	46.7	44.6	14.2	13.6	60.9	58.2	4.6	4.5	104.7
1961	31.6	29.2	9.6	8.9	47.5	43.9	14.5	13.4	62.0	58.3	4.9	4.6	108.1
1962	33.9	29.3	10.7	9.2	50.6	43.8	15.4	13.3	66.0	57.1	5.0	4.4	115.6
1963	35.3	28.9	11.1	9.1	54.0	44.2	16.6	13.6	70.6	57.7	5.2	4.3	122.2
1964	37.7	28.3	12.5	9.4	58.7	44.2	18.7	14.1	77.5	58.2	5.3	4.0	133.0
1965	39.2	28.9	13.3	9.8	61.4	45.2	17.3	12.8	78.7	58.0	4.9	3.6	135.8
1966	40.9	28.0	14.4	10.2	66.4	45.4	19.0	13.0	85.5	58.5	5.7	3.8	146.2
1967	42.8	28.0	15.6	10.2	69.8	45.6	19.6	12.8	89.3	58.4	5.3	3.5	153.0
1968	44.7	27.1	16.8	10.2	75.5	45.9	21.4	13.0	97.0	58.9	6.1	3.7	164.6

SOURCE: American Gas Association, *Gas Facts*, for the years 1958–68.

ported by the FPC. In this approach, it is assumed that any gas not shipped out of the state must be consumed within the state. This probably understates intrastate consumption, since some of the gas purchased by interstate pipelines is shipped to destinations within the state. In recent years the FPC has begun to report these intrastate deliveries by interstate pipelines, and they have amounted to about 6-7 percent of the total marketed production.

As a cross-check on this method of estimation, intrastate sales for 1965 were also estimated by subtracting shipments into individual states from that state's consumption. Here the assumption is that the amount of gas consumed in excess of the amount of gas from outside the state must have originated from intrastate production. This particular year was used because of the availability of USBM fuel balances for each state. This second approach produced an intrastate sales estimate for the United States that was only 5 percent different from an estimate using the customary method. The more important producing states also showed, in most cases, very close checks between the two estimates. Rather large differences were observed, however, in the states with little production. Fortunately, this did not have a very great effect on the U.S. total.

It is frequently conjectured that restrictive price regulation for interstate sales may be prompting producers to show preference for developing their intrastate markets and, in the process, denying gas to the interstate market and possibly restraining its growth. Obviously, it is not possible to determine what interstate sales might have been without interstate price regulation in recent years, but the estimates of intrastate sales do not seem to substantiate any contention that there was marked growth in them through 1968 (see table 1-8). In the period 1958-68 intrastate sales appear to have grown at 4.9 percent per year, which is less than the 5.5 percent per year growth in producer sales to interstate pipelines. Further, that portion of marketed production that appears to have been retained within states has held very close to 37-38 percent of total marketed production for the entire decade.

Table 1-8. Interstate and Intrastate Natural Gas Sales, 1958-68

Year	Marketed production (*tcf*)	Interstate sales (*tcf*)	Interstate sales as % of production	Estimated intrastate sales (*tcf*)	Intrastate sales as % of production
1958	11.03	6.74	61.1	4.29	38.9
1959	12.05	7.40	61.4	4.65	38.6
1960	12.77	7.97	62.4	4.80	37.6
1961	13.25	8.14	61.4	5.11	38.6
1962	13.88	8.59	61.9	5.29	38.1
1963	14.75	9.04	61.3	5.71	38.7
1964	15.46	9.65	62.4	5.81	37.6
1965	16.04	9.89	61.7	6.15	38.3
1966	17.21	10.69	62.1	6.52	37.9
1967	18.17	11.40	62.7	6.77	37.3
1968	19.32	12.21	63.2	7.11	36.8
Annual growth rate, 1958-68		5.5%		4.9%	

SOURCES: U.S. Bureau of Mines, *Minerals Yearbook*, for the years 1958-68; and FPC, *Sales by Producers of Natural Gas to Interstate Pipeline Companies, 1968*, FPC S-202 (January 1970).

It has been shown that natural gas occupied a strong competitive position throughout this ten-year period. Since the supplies of natural gas were adequate both to meet intrastate demand and to ship gas into interstate commerce, both markets were able to grow. Not too surprisingly, new gas markets in those forty-two states that are not major producers appeared more rapidly than in the eight states that are major producers and have always relied on attractively priced gas from local sources. The statistical record for 1958-68 does not indicate that the intrastate market was so attractive that significantly more gas flowed to intrastate markets and that this retarded the growth of interstate gas markets.

It should be recognized that the growth in intrastate sales is concentrated in approximately eight states. These eight states at the present time account for over 90 percent of both U.S. production and U.S. intrastate sales (see table 1-9). Consequently, the growth in intrastate sales tends to be limited by

Table 1-9. Natural Gas Production, Consumption, and Intrastate Sales, Eight States, 1968

State	Marketed production	Consumption	Estimated intrastate sales	Intrastate sales as % of production	Intrastate sales as % of consumption
	(----------	---------- bil. cu. ft.	----------	----------)	
California	715	2,079	715	100.0	34.4
Kansas	836	526	234	28.0	44.5
Louisiana	6,416	1,692	1,431	22.3	86.1
New Mexico	1,164	309	407	35.0	131.7[a]
Oklahoma	1,391	552	331	23.8	60.0
Texas	7,495	4,154	3,480	46.4	83.8
West Virginia	237	189	115	48.5	48.5
Wyoming	248	90	52	21.0	57.8

[a]Obviously, intrastate sales cannot exceed consumption. This reflects a statistical error resulting from the intrastate estimation method.

SOURCES: U.S. Bureau of Mines, Minerals Yearbook, for the years 1958-68; and FPC, *Sales by Producers of Natural Gas to Interstate Pipeline Companies, 1968*, FPC S-202 (January 1970).

what these eight states can consume. At present, even they do not depend wholly on intrastate gas. Concerning the future, based on FRC data, these eight states will represent 47 percent of total requirements in 1990. This would appear to be the approximate limit to which intrastate sales might go under conditions of adequate supply and competitive price levels and assuming that the producing states would rely totally on their own production.

Between 1958 and 1968, the producing states as a group appeared to be shifting toward using less intrastate gas rather than more. In seven of the eight major producing states, Texas being the exception, sales to interstate pipelines represented a greater percentage of the states' marketed production in 1968 than in 1958. In six of the eight states, estimated intrastate sales relative to total state consumption were less in 1958 than in 1968. Texas and Louisiana were the two exceptions.

Texas represents the major exception to the general rule that until 1968 growth in intrastate sales did not appear to be at the expense of opportunities for more interstate gas utilization. However, with Texas representing close to 40 percent of the national marketed production and nearly 50 percent of all intrastate sales, this is not a minor exception. But it must be recognized that intrastate sales in Texas are now estimated to be nearly 85 percent of her consumption. At such a time as her intrastate sales approach 100 percent, then, the growth rate in intrastate sales in Texas could be no faster than that of her own internal consumption.

One final observation about interstate sales versus intrastate should be made. The statistical record we have been examining reflects a period when presumably there was a reasonably adequate supply of gas. Under these circumstances, even though the producers would find a higher intrastate market attractive and might be inclined to seek out these markets, the consumers would be attracted by the interstate prices. It was to be expected that any available and uncommitted supplies of gas would tend to move to satisfy this demand. A change in the supply picture would mean that the behavior of the past is not indicative of what will happen in the future. Preliminary information for the period 1968-70 seems to indicate that under restricted supply conditions, available gas is tending to gravitate toward the more attractive prices in the intrastate market at the expense of growth in the lower-priced interstate markets. It is likely that the intrastate market may now be maintaining a stronger growth pattern than the interstate market.

PROJECTIONS OF U.S. ENERGY AND GAS REQUIREMENTS TO 1990 UNDER CUSTOMARY ASSUMPTIONS

In several studies released during the past three years an attempt has been made to assess total U.S. primary energy requirements at the point of production through the next ten to thirty years. Five studies are reviewed in this paper with particular emphasis being given to the assumptions, methodology, and conclusions. Since none of the studies includes a specific projection on energy consumption for 1990, extrapolation to that year was done where necessary. The five studies reviewed are as follows:

Study	Year released	Year of farthest projection
Ebasco	1970	1985
Chase Manhattan Bank	1968	1980
Texas Eastern Transmission Corp.	1968	1985
U.S. Bureau of Mines	1968	2000
U.S. Bureau of Mines	1970	2000

The most recently released studies are reviewed since it is assumed that accessibility to more current information and statistics improves the reliability of the forecasts. Only the Bureau of Mines studies make projections on high, medium, and low energy outlook bases. The other three studies are specific in their final forecasts of energy consumption. Where the individual forecast is made to a year preceding 1990, we have projected the preceding trend to that year. If the study goes to 2000 without a specific 1990 forecast, interpolation is employed. For example, if the study includes a projection to 1985 and if a 4 percent compounded growth rate has been used, the same percentage rate has been applied to the period 1985-90. Obviously, the group that made the original forecast to 1985 assumes no responsibility for the added five-year estimate. In one case this has meant extrapolating ten years, and for the other two studies, we have extrapolated five years.

Review of Individual Studies of U.S. Energy Consumption

The first energy study reviewed has been prepared by Ebasco Services Incorporated. Ebasco is a firm of engineers, constructors, and utility consultants that periodically releases information on energy in the United States. In addition to forecasting total future requirements, this firm analyzes the individual components of the energy sector.

Ebasco uses the latest USBM reports of actual energy consumption as a basis for making forecasts. Past trends appear to determine the Ebasco forecasts of future total energy consumption. No mention is made of what specific statistical techniques may have been used, such as multiple regression or correlation analysis. The Ebasco studies are updated periodically.

A summary of the Ebasco projections on total energy consumption follows. The 1985 estimate has been extrapolated to 1990 based on an anticipated 4 percent annual increase. This percentage of annual increase is the figure used by Ebasco for each of its five-year interval forecasts for 1970 to 1985:

	1970	1975	1980	1985	1990
Total energy consumption (*quadrillion Btu*)	68.2	83.3	102.0	124.0	(150.9)

In October 1968 the Energy Economics Department of the Chase Manhattan Bank released its study on future energy consumption entitled *Outlook for Energy in the United States*. The Chase study forecasts, among other things, total energy use by the United States to 1980. All data are reported in terms of million barrels daily of oil equivalents. For comparative purposes, the energy equivalent in barrels of oil has been converted to Btu. The factor used in the Chase study was 5.8 million Btu per barrel of oil.

The basis for energy use projections in this study is, essentially, anticipated population growth. In addition to population growth, changes in the age-group structure that have been caused by changes in the birthrate are used as a basis for projections. The focus is on the population included in the

20-to-35 age group. Marriage and family formation particularly characterize individuals included within this group. Consequently, the Chase analysts felt that for most people in this age bracket, demand for goods and services reaches a peak. Economic activity, therefore, is considerably influenced by the rate of growth of this group.

By analyzing future population growth projections, the study anticipated an annual increase in the nation's energy requirements of 3.8 percent in 1968-80. This translates to an energy consumption figure of 46 million barrels of oil per day by 1980 or 97.4 quadrillion Btu for the entire year. We have projected this figure to 1990, assuming a continuous 3.8 percent annual increase, which results in an anticipated energy consumption figure of 141.4 quadrillion Btu for that year.

In the study, population growth trends are also used to forecast the individual shares for each of the primary energy sources. Information on end-use markets and the geography of U.S. population projections are relied on to make these forecasts.

The Chase study on anticipated energy consumption to 1980, with our own 1990 projection based on the Chase 1980 estimate and a 3.8 percent continued annual growth, can be summarized as follows:

	1970	1980	1990
Total energy consumption (*quadrillion Btu*)	66.3	97.4	(141.4)

The third study, prepared by the Texas Eastern Transmission Corporation (TETCO), is entitled *Competition and Growth in American Energy Markets, 1947-1985.* The unique characteristic of the TETCO study, which distinguishes it from the other four, is the approach to analyzing the end uses of energy. From such an analysis and the resulting projections, the total fuel consumption by the entire economy is estimated. The following are the assumptions of this study:

1. The 1961-65 economic growth rate is more indicative of the near future growth rate than that which characterized the period 1947-65.
2. Energy price trends will continue in a manner similar to that of the 1960-65 period.
3. There will be no extraordinary technological developments.
4. There will be an adequate supply of all fuels to meet consumption demands without upsetting price relationships.

Initially, end-use requirements were projected. This established a basis for forecasting total energy requirements by summation. Finally, the supply sources were forced to fit into this framework.

Basically, regression analysis and judgment estimates were used to forecast requirements in each of the four major sectors: residential, commercial, industrial, and transportation. The information from the TETCO study on total energy requirements for the United States to 1985 is summarized below:

	1965	1975	1985	1990
Total energy requirements (*quadrillion Btu*)	52.4	79.6	119.6	(146.2)

As has been done previously, the 1990 projection was extrapolated from TETCO's projection for 1985, and the assumption was that there will be a continued annual growth rate of 4.1 percent for the five-year period 1985-90.

The fourth study to be discussed, *An Energy Model for the United States*, was prepared for the Bureau of Mines by W. Morrison and Charles Reading (Information Circular 8384). One essential point of difference between it and the three previous studies is the bureau's long-range forecast after 1980, based on a different technique than that used for the earlier years. Forecasting total energy demand for the 1960-80 period relies primarily on the ability of the forecasters to quantify the major components of energy demand by source, form, and sector. The historical trends in these particular components are then projected through such techniques as single or multiple correlation. Consequently, the trends resulting from such projections are fitted to complex growth curves on a conditional or probability basis.

Various growth rates are used during this period to forecast total consumption. Three projections are made to 1980: low (2.2 percent annual growth), medium (3.3 percent annual growth), and high (4.5 percent annual growth).

Beyond this period, after 1980, the forecasters recognize that major shifts may occur in the structure of energy demand and supply. Such shifts, if they occur, will definitely alter the projected energy pattern. In order to accommodate such structural changes, the bureau's forecasts to 2000 are based on a new method of quantitative forecasting variously known as technological or contingency forecasting.

Contingency forecasting has been developed as a method for accommodating in projections significant technological innovations that, in this case, would probably displace the conventional energy system. The bureau uses technological or contingency forecasting for its long-term estimates of energy consumption to the year 2000. The particular contingencies decided on were chosen from a number of possible energy forms or conversion systems viewed as having future applicability: fuel cells, magnetohydrodynamics, fusion, thermoelectricity, and solar energy, to name a few.

The results of the two forecasts for the periods 1968–80 and 1980–2000, with the authors' interpolation of the medium forecast for 1990, are as follows:

Year	Range	Domestic demand (*quadrillion Btu*)	Percentage of average annual change
1965	Actual	53.8	–
1980	Medium	88.1	3.3
	Low	73.9	2.2
	High	104.7	4.5
2000	Medium	168.6	3.3
	Low	106.5	2.0
	High	209.4	4.0
(1990)	(Medium)	(122.0)	3.3

The most recent USBM estimate on future energy requirements is contained in the 1970 *Mineral Facts and Problems*. In the section on energy consumption projections, high and low ranges were estimated on the basis of contingency forecasts. Such forecasts attempt to assess the impact of technological, environmental, economic, social, and related factors on the demand for the respective resources, with their major end uses.

By the year 2000, total energy consumption is expected to increase from the 1968 level of 62 quadrillion Btu to a range of from 166 quadrillion Btu to 239 quadrillion Btu. The Bureau of Mines anticipates that the growth in demand for electricity generation by utilities will be the major phenomenon characterizing the future energy market. It is estimated that the total gross energy inputs into electric power generation will increase approximately 400–500 percent in the thirty-year period between 1970 and 2000.

Since the forecasts are made to the year 2000, the estimates for 1990 have been interpolated. The following table summarizes the forecasts and shows the respective growth rates used to make the high and low forecasts (in trillion Btu):

1968 actual	62,424
1990 low (at 3.1%)	122,189
1990 high (at 4.3%)	157,614
2000 low (at 3.1%)	165,991
2000 high (at 4.3%)	239,109

The Consensus

An examination of these five contemporary forecasts of future U.S. energy consumption shows that despite different assumptions and methodology, there is a definite clustering of 1990 forecasts in the 140 quadrillion Btu to 150 quadrillion Btu range. This represents an energy growth rate of nearly 4 percent per year. Assuming that energy continues to grow at approximately 80 percent of the gross national product (GNP) rate—as it has in the past—this 4 percent growth in energy use implies a GNP growth rate of close to 5 percent per year. Perhaps some would feel more comfortable with economic growth rates at less than 5 percent but more than 4 percent. Others might wish to pursue refinements in the assumed pace of efficiency, the independent variables employed, or the manner in which the components of energy utilization are analyzed. However, it must be recognized that unless some rather dramatic contingencies are introduced into the model, these refinements cause a swing of only 10 percent in the expected consumption of energy in 1990. It would seem safe to say that these kinds of refinements do not seem really too significant when one recognizes that the probable error in the forecast is of similar magnitude.

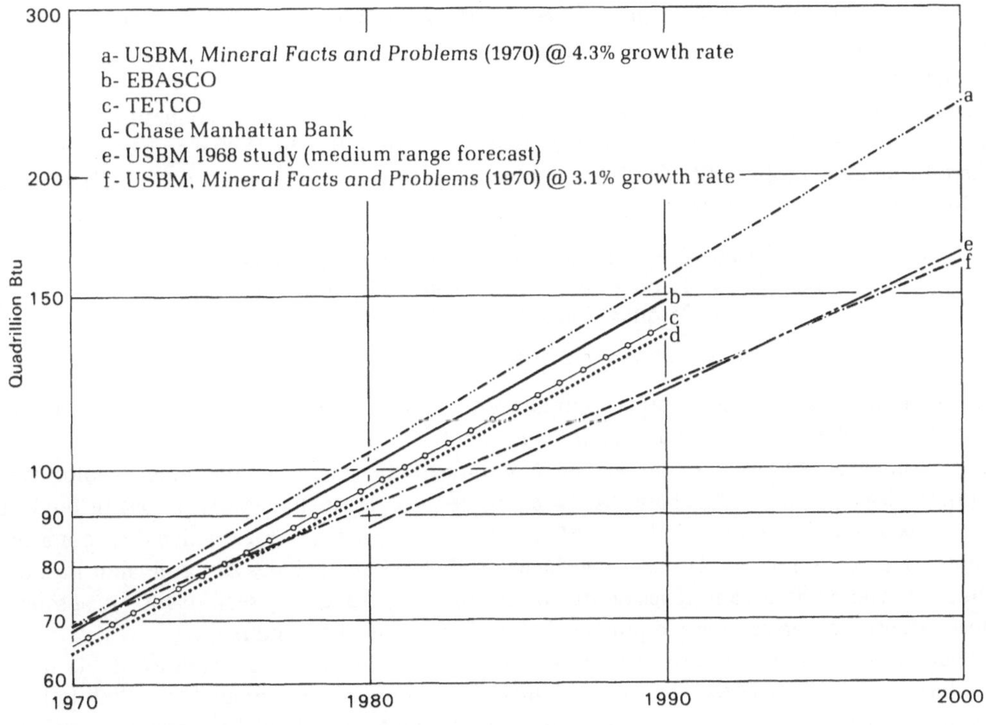

FIGURE 1-3. Comparative results of total energy forecasts.

We accept the assumption that U.S. primary energy consumption will be of the order of 140 quadrillion Btu to 150 quadrillion Btu by 1990, figures probably quite adequate for the purposes of this study. This acceptance carries with it a recognition that no dramatic or rapid changes are assumed for the United States in the technology of energy utilization, the national ethic concerning consumption of material wealth, population growth, the ability to avoid any greater military conflict than the nation is now engaged in, and the successful avoidance of serious economic dislocations of long duration. If any one of these should not hold or if some other major contingencies were to be assumed, then the primary energy pattern from now until 1990 could take on a markedly different shape.

Estimates of Future U.S. Gas Consumption

While there is a noticeable variation in recent forecasts of future U.S. energy consumption, the degree of variation among most studies is quite small when compared with the range of judgment concerning the future consumption of natural gas. In addition to the differences in method and in the basic assumptions concerning economics and technology that are encountered in most energy forecasts, there are several additional judgments that have to be introduced when the gas proportion of the total energy market is derived. The two most significant decisions made by the various forecasters are how they will handle the future price or competitive relationships and what assumptions they will make concerning supply.

Most forecasters have traditionally avoided this issue by assuming that there would be no marked changes in the competitive relationships and that adequate supplies would be available. Chase Manhattan's recent studies, in which limited gas supplies are assumed, are the main exception.[4] Some forecasts are for gas alone and no attempt has been made to treat explicitly the derivation of the gas share of the total energy market. The rate of growth derived for gas in these studies may suggest what competitive position gas will hold, but it is not identified in any precise manner.

In addition to the energy studies discussed previously that show the gas sector separately, we can add two forecasts that deal with gas alone. These are the sales forecast of the American Gas Association and the estimate of future U.S. requirements made by gas industry personnel through the Future Requirements Committee.

By employing the same methods of extrapolation and interpolation that were used for energy to permit comparisons in 1990, it can be observed that gas consumption in 1990 may be expected to fall anywhere from 30 quadrillion Btu (USBM 1968 low range) to 47 quadrillion Btu (TETCO and the FRC). In between these two extremes can be found:

Ebasco–41 quadrillion Btu, USBM 1968 high range
–41 quadrillion Btu, USBM 1970 low range
–35 quadrillion Btu, USBM 1968 middle range

Chase Manhattan–32 quadrillion Btu

Thus the highest forecast is almost 60 percent above the lowest. The AGA forecast of 43 quadrillion Btu is for utility sales only and is not directly comparable.

These comparisons can be quite misleading. Not only are the assumptions and methodologies employed quite diverse, but the data bases used are not identical. Most studies rely on one of the USBM series. The AGA forecast is a projection of AGA's own utility gas sales series, which does not attempt to include all gas consumption. On the other hand, the FRC study includes all terminal utilization of gas within the United States and relies on its own independent survey of users. Its historical series has the most extensive coverage of gas requirements by all segments of the industry.

It is perhaps more meaningful to compare the growth rates shown in these various studies of future gas use than to look solely at the absolute amounts of consumption they predict for 1990. In its two studies, the USBM indicates gas consumption may grow from a low rate of 1.9 percent per year to a high of 3.4 percent. Chase Manhattan, which assumes a limitation in gas supplies between now and 1990, has a declining energy share for gas and a growth rate of 2.6 percent.

The growth rates of Ebasco and the FRC fall in the intermediate range at 3.2 percent and 3.6 percent, respectively. The FRC rates between 1968 and 1990 show more rapid growth in the early years, implying market capture, and slower growth in later years, indicating a possible small loss of market share. Under its assumption of a GNP growth rate of nearly 5 percent, TETCO derives a 4 percent growth rate for gas. This produces an almost constant energy market share for gas. The AGA takes the most aggressive position with gas sales growing at 5.1 percent per year compared with its assumption

[4]Testimony of John G. Winger and John D. Emerson, *Southern Louisiana Area Rate Proceedings*, Docket no. AR 69-1 (July 15, 1970).

of a GNP growth of 4.3 percent. This indicates that in its latest forecast the AGA accepted a continuation of the recent historical growth rates in gas use with market capture taking place throughout the period.

The FRC forecast has the greatest detail in terms of categories of use and regional data. Based on a company-by-company survey, it is a composite of the entire industry's subjective evaluation of what each system would require to meet customer and industry needs if gas were to hold its current competitive position with regard to price and if adequate gas from some source is made available to the systems to meet these needs. As such, it is not a forecast of future gas consumption but an appraisal of the amount of gas that would have to be obtained from domestic reservoirs, imports, or synthetic sources to satisfy the demands that would be made under present-day price relationships (see figure 1-4).

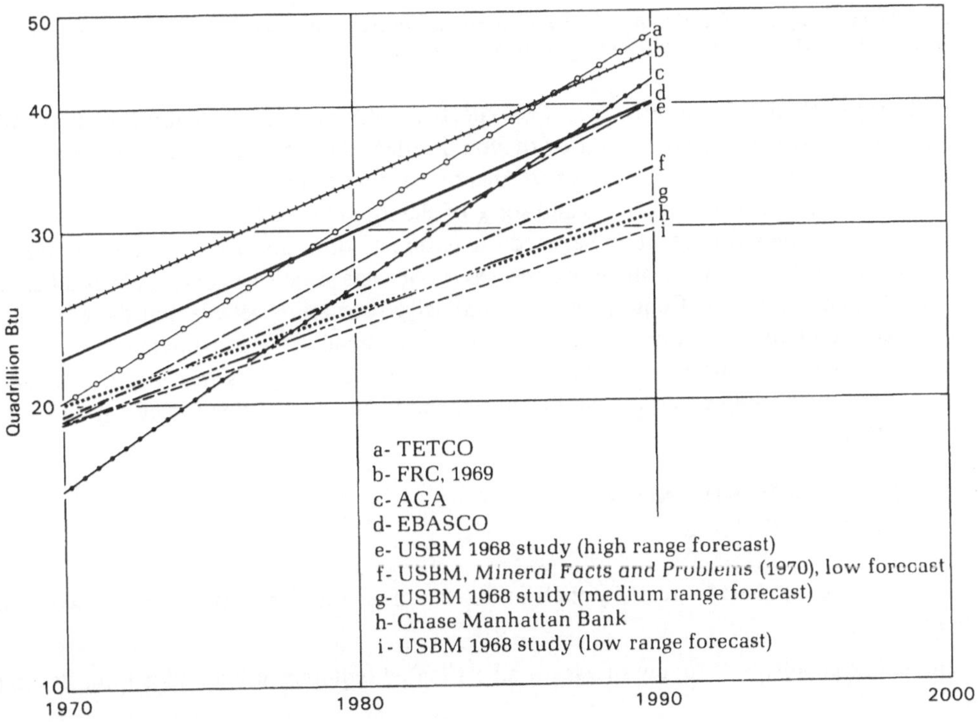

FIGURE 1-4. Comparative results of gas energy forecasts.

Trends by Use and by Region

In the FRC 1969 study, it is indicated that residential gas use will grow less rapidly between 1968 and 1990 than the other markets. This to a great extent reflects past market successes. In those areas where gas has been available, the distributors are reaching saturation points in residential use where market capture can no longer inflate the growth rate. Future residential use will more closely mirror new household formation and new types of residential gas use. The firm industrial segment of gas use is shown by the committee to be the area of fastest growth–4.7 percent for the period 1969-90. Commercial use will also grow, exceeding the overall growth rate for all gas uses. Interruptible uses of all types will perhaps decline slightly in relative importance in the committee's estimation, although there are marked variations in the trend of interruptible use between regions. Field use, which reflects the amount of activity in the production sector, will rise or fall depending on what one assumes will be the future production in oil and gas (see table 1-10).

Table 1-10. U.S. Natural Gas Requirements, by Class of Service, 1968, 1980, 1990

Class	1968 Volume (*tcf*)	1968 % of total	1980 Volume (*tcf*)	1980 % of total	1990 Volume (*tcf*)	1990 % of total	Annual average % rate of change, 1968-90
Residential	4.7	21.9	7.0	20.2	9.2	19.7	3.1
Commercial	1.7	7.9	3.0	8.7	4.5	9.7	4.5
Industrial	6.4	29.8	11.8	34.0	17.4	37.2	4.7
Interruptible	4.5	20.9	7.3	21.0	9.7	20.8	3.5
Other[a]	1.4	6.5	2.3	6.6	3.0	6.4	3.5
Field use	2.8	13.0	3.3	9.5	2.9	6.2	0.2
Total	21.5	100.0	34.7	100.0	46.7	100.0	3.6

NOTE: 1,000 Btu per cubic foot except for field use.

[a]Includes requirements for Region 11.

SOURCE: Fugure Requirements Committee, *Future Natural Gas Requirements of the United States*, vol. 3 (Denver Research Institute, University of Denver, September 1969).

Turning to regional growth patterns, the FRC expects the Pacific Northwest and New England to have the greatest percentage growth. This growth would reflect the more recent arrival of natural gas service in these areas, which makes it possible for market capture to take place for a number of years in the future. Although the southern states and some of the mountain states are expected to have economic and population growth that will exceed the national norm, this may not be reflected in gas consumption in all of these areas. In these regions, gas service has always been available and has always been the cheapest energy source. Consequently, the industry forecasters believe that these areas have reached the greatest amount of saturation and that their growth will be approximately on a par with the national pace. Areas of concentrated industrial activity—the Great Lakes area, the central East Coast and adjacent inland states, and the Gulf Coast, are all indicated as doing somewhat better than the national average.

FUTURE U.S. GAS DEMAND UNDER ALTERNATIVE ASSUMPTIONS

The validity of the numbers forecast by the Future Requirements Committee rests on two basic assumptions:

1. There will be an adequate supply of gas for all estimated requirements for all periods covered by the survey.
2. The present-day relationship of the cost of gas to the competing fuels will remain the same in the future. However, individual companies should consider known changes in these relationships in their market area.[5]

Granted these assumptions, the resulting projections appear to be quite reasonable, if not conservative. The growth potential of the American economy, accompanying geometric rates of increase in energy consumption, the general appeal of gas and, recently, environmental factors, all underscore such a conclusion. For some FRC regions where the regional committees were not in a position to measure fully the potential impact of all on-going developments, the projections could turn out to be on the low side.

The real question, however, is whether the industry will be able to supply the quantities of gas that consumers would like to purchase at the assumed prices. Even if the resource base were available, could the industry convert *potential* into *actual* supplies at the "presently prevailing cost relationship"? The problem is hardly solved by broadening the supply concept from "U.S. natural gas" to "gas of

[5]FRC, *Future Natural Gas Requirements*, p. 16.

comparable quality from whatever source." In that case, the potential supply may become greatly expanded, but it is unlikely that substantially greater quantities will become available *at present cost relationships* in most regions of the country, nor could this occur soon enough to meet projected requirements during the 1970s.

It must always be recalled in using the FRC forecast that the committee has operated under the constraint of measuring requirements without introducing the supply side. Even if the committee had not operated under this limitation, it is far easier to question the relevance of the assumptions that were actually used than to suggest replacement assumptions that are both valid and usable. For one thing, it would be necessary to know which of at least two broad alternative courses of policy action—rationing of the limited supply at fixed prices or unfreezing of field prices until supply and demand balanced out—might be followed if and when supply shortages become serious. Finally, even if one assumed that the second alternative would be adopted, it would then be necessary to estimate how high field prices would have to rise before a significant response on the supply side set in and also the volume of sales that would be lost because of customers shifting to other energy sources at successively higher prices.

The FRC did not (and does not now) have information available that would assist in answering the elasticity question on the demand side, nor is its counterpart, the Potential Gas Committee, in a position to produce the data needed to arrive at alternative price-quantity combinations on the supply side. The first hesitant steps are just now being contemplated for experimentation with possible procedures for estimating demand elasticities within individual distributor's systems. Any ensuing results will be tentative and illustrative at best.

Econometric Studies

A number of attempts have been made by others, chiefly academic investigators, to determine with the aid of econometric techniques the responsiveness of gas demand to changes in gas prices. We have examined the results of a half-dozen such studies that represent the sum total of the work published on gas demand to date. These employ multiple regression techniques based on time series or cross-sectional data. Our examination found that not one of them has yielded results in which one could have a substantial degree of confidence. Omitting most of the technical details here,[6] the major reasons for this conclusion are:

1. The geographical units chosen are states, regions, or entire industries, nationwide. This means that the gas quantities, prices, incomes, and other data all represent very broad aggregates and not specific observations for one location and supplier. These aggregates and the prices derived from them are therefore a composite of many rate structures, often of a large number of supplying companies. Even the average price per thousand cubic feet (Mcf) realized on sales to, say, all commercial users of various types and sizes is hardly a meaningful figure. Such an average is a calculated mean, derived from hundreds or even thousands of billings to customers who are using different quantities, possibly under different rate classifications, each of which contains a basic rate level and a number of blocks with successively lower rates for larger quantities purchased. Also, management decisions may reflect different company strategies with respect to expanding interruptible versus firm loads or to seeking boiler fuel markets. Availability of market area storage will frequently dictate what markets are developed. Pollution has become the most recent nonprice factor influencing the market as consumers seek to substitute gas for high-sulfur oil or coal.

2. No account is taken, in most cases, of differences in the cost of installing alternative fuel-burning equipment (gas versus oil or electricity, single versus multiple burners) or of the differences in conditions facing old versus new users. These costs, of course, are not identical, and the net cost to the customer is further affected by assistance provided by suppliers to varying degrees, especially by the electric companies.

[6] See appendix on p. 49.

3. In several cases the number of variables used was large relative to the total number of observations. This tends to reduce the number of degrees of freedom and may result in a "forced fit"; that is, an impressively high but completely meaningless coefficient of determination.

4. The interpretation of the price elasticity coefficient obtained is not at all clear. These elasticities are estimated as the regression coefficients for data transformed logarithmically. This procedure gives meaningful estimates of the elasticities only if the original data are hyperbolic. None of the studies examines the form of the data, so it is not possible to assess the usefulness of the estimates based on logarithmic transformation. Also, time series data frequently involve autocorrelated errors. In some cases no test for autocorrelation was presented. In other cases the Durbin-Watson test was performed, often indicating the presence of substantial autocorrelation.

5. In numerous instances the relationships that result between prices and gas quantities sold or between income and gas sales are not significant, or their response is even in the wrong direction. That is, gas consumption is shown as increasing with higher prices or as decreasing with higher incomes. Such results do not indicate that gas demand fails to follow the normal "law of demand" or that gas is an "inferior good." They merely confirm the serious flaws in procedures listed above.

6. The results that are perverse or insignificant might be ignored if criticism were limited to them alone. However, considering the techniques employed and the faultiness of the data, one cannot even have confidence in those results that appear to be reasonable. For example, the widely cited price elasticity coefficient obtained by Balestra, -1.3, is subject to the same criticisms as most of the other findings. It is impossible to tell from the studies, therefore, whether the demand for gas is price elastic, inelastic, or possibly unit elastic.

7. In addition to these specific criticisms applied to the studies of gas demand, there are the general assumptions involved in projecting by a mathematical relationship any set of past data into the future. For example, reasonable stability of the coefficients maintaining these assumptions for comparatively short periods, say two or three years, may not be unduly risky. However, as the forecasting period is extended the validity of these assumptions becomes increasingly tenuous, so that the usefulness of econometrics for long-term forecasts is severely limited.

These criticisms are limited to the studies, reviewed in the appendix, that have been completed and were available to the authors of this paper. It is hoped that some of the more serious objections do not apply to other investigations currently under way or planned, that new studies not subject to such deficiencies will be undertaken in the near future, and that the results will be made available to policy makers. Clearly, in order to overcome the aggregation problem, these studies will have to be confined to individual markets, probably to single companies or portions of companies. They will have to deal with homogeneous groups of users subject to similar economic characteristics and operating under the same rate schedule, and they will have to take into account differences in local conditions, including the availability of alternative energy sources and pressures for pollution abatement. If such studies can be designed, it may be possible to obtain meaningful results by the application of quantitative analysis.

In the absence of such studies, the investigator's work necessarily has to be limited to a qualitative analysis of major gas markets, based on more or less scattered information available from a large variety of sources. In the following discussion, we will draw three major distinctions. First, markets will be separated between small users (residential and small commercial) and very large users (industrial customers and electric power companies). Second, for each type of user, major geographical distinctions will be noted, including differences in the price of gas, the availability and prices of competing energy sources, and market penetration by various sources. Third, where appropriate, the importance of economic factors other than competitive price, as well as of noneconomic factors, will be stressed. Because of the incompleteness and other imperfections in the data, only broad impressions rather than precise quantitative conclusions can be drawn from the analysis. Nevertheless, the factual information appears to be sufficient to serve as a basis for an attempt at sketching the broad general pattern of gas competition today and for the immediate future.

Residential and Small Commercial Uses

Residential customers consumed 4.642 billion cubic feet of gas in 1968, or slightly over one-fifth of total gas requirement as measured by the FRC (see table 1-11). If small commercial users were added, the total might come closer to one-fourth of national consumption.[7] The gas is consumed by a variety of appliances, including furnaces, hot water heaters, ranges, clothes dryers, gas lights, and other novelty items. Separate information on the total number of each of these appliances installed

Table 1-11. Regional Distribution of Gas Requirements, 1968

(billion cubic feet)

Region	Residential (1)	Commercial (2)	Industrial: Firm (3)	Industrial: Interruptible (4)	Industrial: Total (3 + 4 = 5)
1	120	37	26	28	54
2	1,473	515	951	347	1,298
3	253	131	383	542	925
4	943	336	609	411	1,020
5	254	103	103	374	477
6	399	114	309	435	744
7	386	158	3,759	712	4,471
8	145	82	33	197	230
9	620	202	177	1,336	1,513
10	49	30	36	143	179
U.S.	4,642	1,707	6,386	4,525	10,911

Percent

Region	Residential	Commercial	Firm and interruptible industrial
1	2	2	a
2	32	30	12
3	6	7	8
4	20	20	9
5	6	6	5
6	9	7	7
7	8	9	41
8	3	5	2
9	13	12	14
10	1	2	2
Total	100	100	100

[a]Less than 0.5 percent.

SOURCE: Future Requirements Committee, *Future Natural Gas Requirements of the United States*, vol. 3 (Denver Research Institute, University of Denver, September 1969).

and their average consumption is not available, although shipments by manufacturers are reported for most items. It is clear, however, that the amounts used for heating homes, apartments, businesses, etc., lead all other uses combined by a wide margin, with hot water heaters and ranges following, in that order. (Gas-fired air conditioners have been used for many years both in large commercial-sized units and for residential purposes; residential units, although growing in numbers, are still less common than electrical units.)

On the basis of Btu equivalence, gas has a substantial price advantage over its nearest competitor, distillate fuel oil, in most regions of the country. The major geographical exception is the Atlantic

[7]While the AGA draws a very clear dividing line between residential and commercial uses (see its *Glossary for the Gas Industry* [1967], pp. 10-11, and *Gas Facts* [1969], p. 210), in practice the distinction is considerably more blurred. According to the FRC, in some cases commercial sales include those to smaller multiple dwelling units, public institutions, and small industrial customers. See FRC, *Future Natural Gas Requirements*, p. 47.

Seaboard where gas generally is as costly, and in some locations it is somewhat more costly (see table 1-12). This is substantiated by the high, and in most cases rising, saturation ratios achieved by gas in domestic heating (see table 1-13). Apart from the East Coast, only the Great Lakes area (East North Central) has a saturation ratio below 90 percent, and there it has been rising rapidly (to 82.4 percent in 1968). In the Middle Atlantic States, where the saturation ratio averaged 47 percent in 1968, gas and distillate fuel oil prices have been near parity in recent years. In New England, where the saturation ratio was 43.6 percent in 1968, gas heating is slightly more expensive than heating with distillate fuel oil (see table 1-13).

Price relationships in Btu equivalent, of course, do not tell the entire competitive story. On one hand, gas lays claim to advantages over fuel oil in terms of cleanliness, reliability, saving of storage

Table 1-12. Average Prices of Gas, No. 2 Fuel Oil, and Electricity in Twelve Cities, 1969

(dollars per million Btu)

City	Gas for residential heating[a]	No. 2 fuel oil[b]	Electricity[c]
Baltimore	1.70	1.68	3.58
Boston	1.96	1.71	4.72
Buffalo	1.25	1.81	4.43
Chicago	1.20	1.60	3.28
Detroit	1.13	1.65	3.66
Milwaukee	1.50	1.62	4.16[d]
Minneapolis–St. Paul	1.17	1.60	3.87[e]
New York	1.74	1.72	4.04
Philadelphia	1.84	1.70	3.81
St. Louis	1.10	1.66	3.84
Seattle	1.53	1.92	2.20
Washington, D.C.	1.76	1.68	4.02

NOTE: Annual averages of monthly data are unweighted for gas and fuel oil. Electricity data derived from annual bill.

[a] Based on "heating blocks" for 100 therms, assuming 75 percent efficiency.

[b] Retail prices, including taxes. Converted at 5,825,000 Btu per barrel (138,690 Btu per gallon) and assuming 75 percent efficiency.

[c] Based on incremental rates in the range of 15,000 kwh per year to 25,000 kwh per year. Converted at the "standard factor" of 1 kwh = 3,412 Btu (implies 100 percent efficiency).

[d] Effective March 5, 1969.

[e] Minneapolis rate.

SOURCES: Gas and fuel oil data from U.S. Department of Labor, Bureau of Labor Statistics, *Retail Prices and Indexes of Fuels and Electricity*, various issues, 1969. Electricity data from FPC, *All Electric Homes in the United States, Annual Bills* (January 1, 1969).

Table 1-13. Saturation Ratios for Gas in Domestic Heating, by Census Regions, 1950, 1960, 1968

(percent)

	1950	1960	1968
New England	4.6	25.3	43.6
Middle Atlantic	11.7	35.1	46.9
East North Central	26.1	64.3	82.4
West North Central	49.7	82.6	90.5
South Atlantic	30.1	66.7	78.8
East South Central	62.3	87.4	94.4
West South Central	99.4	99.7	99.1
Mountain	82.6	95.5	96.6
Pacific	87.9	97.3	97.3
U.S.	40.7	68.2	79.3

NOTE: Data represent proportion of total residential gas customers using gas for heating.

SOURCE: American Gas Association, *Gas Facts* (1969), table 110, p. 121.

cost, and so on. As a consequence it has been making substantial inroads on fuel oil, even in areas where it is more expensive. On the other hand, gas itself has been losing markets to all-electric homes, even though heating with electricity may be 2.5 to 3.5 times more expensive than heating with gas[8] (see table 1-12, col. 3). These losses have been concentrated in the southeastern United States where electric power rates have been particularly low, although electric heating is making inroads on gas markets in wide locations elsewhere in the country. In some degree, the peculiarities of the block rate schedule (used by both gas and electric companies) offer an incentive to large-volume users. Electric power companies are able to offer particularly attractive rates to heating customers since these represent off-peak users whose rates can be determined on the basis of incremental costs, whereas for the gas industry the demand for heating coincides with the seasonal peak of industry sales. (Note that for air conditioning the conditions are reversed.) Additionally, the selection of heating equipment is heavily influenced by first costs of installing equipment, with little regard given to subsequent operating expenses. In many cases selection is determined largely or entirely by the builder and his supplier rather than by the owner-occupant.

Economic factors appear to predominate in the selection of fuel for hot water heaters, although there may be some basis for choice on performance characteristics. Very probably, although not inevitably, a home heated with gas will have a gas-fired hot water heater. A home heated with fuel oil may have a gas hot water heater, while one heated with electricity probably will not. For other appliances, particularly ranges and clothes dryers, noneconomic considerations appear decisive in the selection of fuels. Thus, a housewife accustomed to cooking with electricity generally will be most reluctant to shift over to gas because of differences in timing and other quality characteristics, and vice versa. In these cases the cost of fuel is small compared with the cost of acquiring the appliance and is relative to the family's income. The price of fuel appears to play a subordinate role.[9]

In assessing the implications of these demand characteristics, two basic distinctions must be kept in mind: the short run versus the long run and the availability versus the absence of close substitutes for gas. Even where an alternative fuel is available at comparable costs, owners of gas appliances cannot be expected to shift from gas as long as the appliance is not being considered for replacement on other grounds. Short-run demand thus is extremely inelastic within the relevant range.[10] An increase within this range, assuming it were carried directly into the residential rate structure, would be equivalent to an average increase of 5 to 10 percent above the 1968 price, nationwide. On the East Coast, where gas prices are substantially above the national average, the percentage increase would, of course, be less—in New England, not much more than half the national average.

Over a period of years, increased gas prices naturally would affect the demand by new customers and the replacement demand of existing users.[11] The effect would make itself felt only very gradually since household appliances typically have a life of around eight to ten years and in the case of heating equipment even longer. Considering the price advantage that gas currently enjoys over most of the country, a comparatively moderate increase in gas prices should not be expected to have a severe impact on the future growth of gas markets in most regions. On the East Coast, however, such a price rise either would make gas more expensive than distillate fuel oil, or widen the existing disadvantage. For uses where the decision is primarily based on price, the future growth of gas markets would thus tend to slow down, assuming that fuel oil prices maintained their present levels. This assumption can-

[8]TETCO, *Competition and Growth in American Energy Markets, 1947-1980* (Houston, Texas: TETCO, 1968), p. 31.

[9]The present value of total fuel purchases over the life of the equipment might bulk large, even as compared with the acquisition cost, but residential users, in contrast to businessmen, are not likely to make a comparison of the alternatives in present value terms.

[10]Here and in subsequent discussion, "relevant range" is considered to be an increase in field prices (reflected directly in city gate prices) of up to 10 cents/Mcf above those currently in effect. Some additional increases may be expected from factors other than changes in field prices, e.g., rising capital costs resulting from high interest rates.

[11]For a comparison of the relationship between short-term and long-run demand function, see Merton J. Peck, *Competition in the Aluminum Industry, 1945-1958* (Harvard University Press, 1961), pp. 52-56.

not be taken for granted, of course; an increased demand for distillate would tend to drive up its price, and a larger supply would require an inducement to oil refiners to shift their yields away from gasoline. In that event, therefore, the existing gas-fuel oil price ratio may be restored, with both selling at a higher level. In areas where electricity is the closest competitor of gas, such as the Southeast, any narrowing of the price differential would further tend to accelerate the loss of gas markets. Here the long-run gas demand is elastic. Even in this case, however, one cannot conclude that the result will be an overall reduction in the volumes of gas required; it is possible that the demand for gas will be transferred from direct use to indirect (generating electric power).[12] This possibility will require an examination of the sources of electric power generation.

Industrial Markets

The competitive position of gas for large-volume users (chiefly industrial customers and electric-power plants) differs substantially from that for small-volume users. Here the chief competitor of gas is coal. For conventional electric power plants, coal accounts for nearly two-thirds of the total fuel consumed; gas is second with over one-quarter, and heavy fuel oil accounts for the remainder (see table 1-14). (Discussion of power generation by nuclear plants will be postponed.) Again, there are

Table 1-14. Fuel Consumption by Steam-Electric Power Plants, by Census Regions, 1968

(*trillion Btu's*)

Census region	Coal	Oil	Gas	Total
New England	163	283	9	455
Middle Atlantic	1,143	425	127	1,695
East North Central	2,520	6	108	2,634
West North Central	392	6	348	747
South Atlantic	1,530	224	201	1,956
East South Central	961	1	118	1,079
West South Central	–	1	1,363	1,364
Mountain	181	11	160	352
Pacific	–	120	742	862
Total U.S.	6,891	1,076	3,177	11,144
		Percent		
	61.8	9.7	28.5	100.0

SOURCE: National Coal Association, *Steam-Electric Plant Factors* (1969), table 2.

substantial regional differences. Gas has a decisive price advantage over its competitors in and close to the major gas-producing states (the Gulf Coast and Mid-Continent regions); on the West Coast where coal is not available gas is priced competitively with heavy fuel oil. Along much of the Atlantic Seaboard, coal had a price advantage over both gas and fuel oil until very recently, but the gap has been closing as a result of increased coal prices and weak prices of heavy fuel oil, without sulfur restriction (see table 1-15).

Abstracting from environmental considerations, which will be discussed later, indications are that gas prices to industrial users could rise substantially, without significant loss of sales volume, in locations that account for approximately half of total sales. In addition, the demand by some other industrial users is inelastic, regardless of location, because of the inherent advantages this fuel offers, for example, as raw material for petrochemicals and in certain other processes. Certainly in the long run and probably also in the short run, this leaves a very sizable market, chiefly on the East and West Coasts where gas demand is highly elastic. Many present gas users, particularly electric power plants, have dual burning facilities installed and are free to shift—ignoring environmental constraints— with any significant rise in gas prices not accompanied by comparable increases in other fuels.

[12]Such a transfer could raise the demand for gas by a factor of three or more because of the loss of efficiency during conversion and transmission.

Table 1-15. Cost of Fuel for Steam-Electric Power Plants, by Census Regions, 1967

(cents per million Btu burned)

	Coal	Oil	Gas
New England	34.3	29.4[a]	32.0
Middle Atlantic	28.3[a]	35.0	35.8
East North Central	25.2[a]	64.6	28.0
West North Central	25.1	52.6	24.5[a]
South Atlantic	27.0[a]	32.3	31.6
East South Central	20.1[a]	55.2	23.9
West South Central	21.5	38.2	20.1[a]
Mountain	20.4[a]	26.8	25.9
Pacific	–	32.0	30.7[a]
Total U.S.	25.5	32.8	25.1[a]

[a]Denotes lowest-cost fuel.

SOURCE: National Coal Association, *Steam-Electric Plant Factors* (1969), table 2.

Any substantial increases in field prices, therefore, would confront the distributing companies in the coastal areas with an extremely nasty dilemma. First, they could attempt to maintain prices to users with elastic demand at existing levels and concentrate the entire increase on users with inelastic demand. (This assumes, of course, that even at the higher prices, rates to the elastic sector would still exceed incremental costs so that they could not be considered "unduly discriminatory" by regulatory agencies.) Alternatively, distributing companies could forfeit these markets to the competition and concentrate their efforts on supplying users with inelastic demand. In most cases, however, such a policy would result in a reduction of the load factor, thus raising unit costs to the remaining customers. These customers could, therefore, end up paying even more for gas than under the first alternative.

This conclusion rests on two major assumptions, in addition to ignoring environmental factors. First, in contrast to residential markets where gas is mainly in competition with light fuel oil, it is assumed that substantially larger quantities of heavy fuel oil suitable for use in industrial installations could be made available by the oil industry at little or no increase in prices. For the East Coast, this is probably a reasonable assumption if quality considerations are ignored, since supplies from the Caribbean and other sources could be substantially expanded, and since heavy fuel oil has free access to the U.S. market. On the West Coast, a sharp expansion in the demand for fuel oil probably would result in some price increases because of the higher transportation costs involved in obtaining additional quantities from overseas sources, but these would probably turn out to be modest over a period of time.

Second, the alternative of generating power from nuclear plants has been ignored in the discussion. This is not an immediate alternative to companies planning to install additional fossil fuel units–for one thing, because suppliers of nuclear reactors are fully booked for a period of from five to seven years ahead. Given enough time to construct new facilities, however, nuclear generating costs would impose a ceiling on the price at which gas (and other fossil fuels) could be sold to electric power plants, even in areas where gas currently has no close competitor.

Impact of Environmental Policies

The above conclusions are subject to major modification because of the impact of pollution regulations now being enacted. The coal industry is currently operating very close to capacity. The quantity of coal that could be readily made available contains objectionably high amounts of sulfur that cannot be removed with presently known technology, regardless of cost. The output of low-sulfur coal mines is fully committed to existing customers (electric power companies, steel mills, and foreign buyers). Large deposits of low-sulfur coal exist in the United States, particularly in the West, but an expansion of production would require opening of new mines. Coal companies now are requiring customers to agree to long-term contracts and to provide substantial financial support. Delivered costs for most users are likely to be substantially higher than in the past, in some cases as much as 50 percent.[13] And even the use of low-sulfur coal does not solve the problem of pollution from stack gases, whose removal requires expensive installations, such as electrostatic precipitators. The full effectiveness of this equipment has not been demonstrated, especially for very large units. Coal, it is apparent, currently does not represent a viable alternative to gas except in remote locations away from centers of population, and even in these areas growing problems are being encountered.[14]

The problem with heavy fuel oil is similar. Ordinary No. 6 (or Bunker C) fuel oil, which is available in the coastal areas, has a sulfur content of 2.5 percent to 3 percent or more. Following major shifts from coal to fuel oil on the East Coast in the fall of 1969,[15] supplies have become scarce and prices have risen as much as 50 percent. Low-sulfur fuel oil (1 percent or less) has been virtually unavailable in the market and carries a premium of 30–50 cents a barrel. Further expansion of heavy fuel supplies depends almost exclusively on imports, since production by domestic refineries (even at higher prices) is uneconomical and has been persistently reduced.[16] Additional supplies of low-sulfur fuel are becoming available from Caribbean refineries, where expensive sulfur removal equipment is being installed. There is also a possibility for larger imports from Eastern Hemisphere sources, but the quantities that can be obtained from these areas are not large, since there is a strong worldwide demand for fuel oil. Moreover, foreign oil will be of little or no help to industrial users in many locations.[17] The situation could be eased most readily by the processing of larger volumes of low-sulfur crudes, chiefly from Africa, but this presupposes a relaxation of import quotas on crude oil.

Even nuclear power plants are affected adversely by the pollution issue. Thus, objections to increased water temperatures and fear of contamination are severely limiting the number of available sites. Given sufficient time, many of these problems will no doubt be overcome as technology finds answers to the pollution riddles and as economic incentives for applying the new technologies are offered. For the foreseeable future, however, all indications point to gas as the only wholly unobjectionable energy source. Even where gas demand historically has been elastic, therefore, environmental considerations cause demand, under current and foreseeable conditions, to become highly inelastic.

Policy Implications

The preceding review of demand characteristics for gas and competing fuels has indicated that, for small users, gas demand is extremely inelastic in the short run. Over longer periods, and within the relevant range for field prices specified, a similar conclusion holds for most parts of the country. One exception is a shift to fuel oil by small users in the Northeast, but only on the questionable assumption that supplies of light fuel oil could be expanded substantially at constant prices. Another possibility is that rising gas prices, to the extent that they might narrow the spread between heating with gas and the cost of operating all-electric homes, could offer some further inducement to choose electric over gas heating. This would apply almost exclusively to new homes, since the cost of conversion to all-electric is prohibitive for existing homes. However, owners of new houses will probably face rising electricity rates. It should also be noted that even if there should be a shift from gas to electricity, the use of gas as a boiler fuel might more than offset any volume lost in the residential market.

For large-volume gas users, demand has been elastic in some locations and for some uses even in the short run, and in most cases (except special-process uses) demand has been elastic in the long run.

[13] *The Wall Street Journal*, January 3, April 6, April 27, and June 2, 1970; *Business Week*, July 5 and August 23, 1969.

[14] See, for example, "The Navajo Power Project and Environmental Protection Plans" (Salt River Project, Phoenix, Arizona, no date).

[15] *The Oil and Gas Journal*, February 9, 1970, p. 23.

[16] U.S. refiners, however, are temporarily increasing their output of heavy fuel oil in response to government appeals to avoid threatened shortages during the 1970–71 heating season.

[17] Note, however, that Commonwealth Edison Company has applied for a license to import low-sulfur fuel oil to Chicago, although the company, along with many electric power companies, would like to purchase much larger quantities of gas than can be made available.

This was true only while users were free from official restraints to choose their fuel on the basis of cost and convenience. Since the enactment of pollution regulations, the cost of acceptable alternatives has risen sharply and adequate supplies of other fuels of acceptable quality cannot be obtained in many instances. As long as these conditions continue—and the end is not foreseeable—industrial users will have little choice but to continue to rely on gas to the extent they can obtain it, regardless of any likely price increase.

With demand being highly inelastic, the crucial factor in the determination of equilibrium is the shape of the supply curve. This is the subject of other papers in this volume and has not been examined by the present authors. For purposes of discussion, it will suffice to assume that supply is likely to prove relatively inelastic for at least a time, in that the supply response to higher field prices will be neither immediate nor overwhelming, regardless of what may happen in the more distant future when new supply sources and technologies become available.

Under this assumption, regulatory policy can pursue one of two alternative courses. First, prices could continue to be held at levels substantially below those that would obtain if field price regulations were removed. In that event, substantial shortages would result and scarce supplies would have to be allocated among competing consumers by administrative action. Second, field prices, in effect if not in appearance, could be substantially decontrolled, permitting market prices to work toward balancing supply and demand.

Under present conditions, the latter course runs the danger that prices may suddenly rise very steeply, at least for a time, until suppliers are able to respond to the newly established incentives. This alternative cannot be recommended wholeheartedly or without reservations. Nevertheless, there are strong grounds on which it may be preferable to the other. From a negative point of view, deregulation would obviate the need for deciding administratively which customers and markets should enjoy adequate gas supplies at attractive prices and which should be excluded. In our view, any attempt to make such a determination would result in the probability of a misallocation of resources; it would saddle the FPC with responsibilities that neither it nor any other governmental body should be asked to shoulder, at least in peace time; and it would redirect the decision-making process from the realm of economics into the political arena, where extraneous pressures would inevitably overwhelm rational decision criteria. In short, any such effort would be foredoomed to failure. Moreover, adverse repercussions on all segments of the gas industry could not be avoided. For example, a significant slowdown in the growth rate—compared with historical rates and the growth of competing industries—should tend to make it more difficult and costly to raise capital and to attract top-level managerial, technical, and other talent.[18]

More positively, substantial increases in gas prices would accomplish several highly important objectives. For energy consumers (and local authorities concerned with setting pollution standards), a clear choice would be offered: Is a cleaner environment worth the added cost of burning a premium fuel? Continuing to pretend that ample gas supplies will be available at costs equal to, or even lower than, the "dirty" hydrocarbons seems objectionable as well as inherently self-defeating. For producers of natural gas, substantially higher prices would offer the challenge of responding to adequate incentives and of substantiating their claim that reduced incentives under regulation have been the major obstacle to providing consumers with ample supplies. For potential suppliers of supplementary gas sources (imports, LNG, and synthetic gas), higher prices would represent a clear signal to proceed with rapid development in response to the prospect that these new supplies could find a ready market as they become available. For American industry, science, and technology, higher gas prices would present a chance to search for new ways to utilize the fuels with which this country is amply blessed—especially coal—without polluting our cities. Past performance gives grounds for optimism if the rewards justify the necessary expenditure in money, time, and effort.

[18]See William J. Baumol, *Business Behavior, Value and Growth* (Harcourt, Brace & World, rev. ed., 1967), pp. 44–45.

Our argument against increased regulation and for allowing a freer rein to market forces rests on broad conceptual grounds, as well as on what we believe to be a sound economic basis. To be realistic, the choice of policies may well have to be tempered by practical considerations. Thus, immediate and complete decontrol of field prices might cause the cost of gas to consumers to rise so sharply that it would set off shock waves throughout the body politic. After all, in times of shortage, prices in the most perfect of competitive markets will be far above the "normal supply price," as defined by Alfred Marshall.[19] We do not advocate running the risks of the possible "backlash" to such a drastic policy change. But to pursue a course that would paralyze, rather than take advantage of, healthy economic forces would appear wholly ill-advised.[20]

SUMMARY

Most of the natural gas supplies of the United States originate in approximately ten producing basins and are augmented by imports from Canada and to a minor extent from Mexico. The Gulf Coast area of Texas and Louisiana has become by far the most important source in recent years with the inland area surrounding the Gulf fields second in importance. Geologic studies of our potential for undiscovered gas indicate that this pattern will continue to persist into the future, except for the probable emergence of Alaska as a new source of domestic gas. Modern technology will expand the opportunity for importation of LNG from more distant sources. The exploration and development of gas is dependent primarily on companies that are interested in both oil and gas. The role of the pipeline companies in acquiring their own reserves and producing gas has always been relatively minor and has diminished in recent years. The present reserve position of pipeline companies indicates that their production will continue to diminish both absolutely and relatively unless there is renewed activity in exploration and reserve acquisition.

U.S. natural gas consumption during the last decade has grown at a rate well above the national energy pace. The regions that have displayed the greatest growth have been the Great Lakes, the Pacific Northwest, New England, the Southeast, and the Northern Plains. There have been no noticeable shifts in the relative importance of the major classes of use, with all categories participating in the growth of gas utilization. The intrastate market is estimated to have remained at an almost constant proportion of the U.S. total gas market from 1958 to 1968. Only Texas seems to have had a tendency toward favoring intrastate market growth over expansion of interstate shipments.

Most energy forecasters utilize rather similar forecasting methodology and assumptions. As a result, there is a noticeable clustering of the projections in forecasts of recent vintage. If one accepts continued economic growth, adequate overall energy supplies, and no marked change in public attitudes toward seeking material well-being, then energy consumption should be of the order of 140 quadrillion Btu to 150 quadrillion Btu by 1990, slightly more than double the present level.

A much more diverse range of judgment is found in recent gas projections. These vary from 30 quadrillion Btu (21 percent of expected U.S. energy demand) to 47 quadrillion Btu (32 percent of U.S. energy). The lower projections usually reflect some assumed limiting factor in gas supplies, while the higher estimates are based on adequate supplies and constant fuel-price relationships. Under the

[19]See Marshall's *Principles of Economics* (Macmillan, 8th ed., 1947), bk. 5, chap. 5, pp. 363–80. It should be noted, however, that because of the "rolled-in" rate policy followed by the FPC the impact of phasing new higher-cost sources into the supply stream would be dampened and delayed in any event. Charles H. Frazier has suggested as an alternative that the commodity-charge component of the rate structure be geared to the incremental cost of supplies (plus an appropriate contribution to fixed costs). This would ensure the adherence of distributors to an economically sound policy in determining rates to off-peak customers. See Frazier's testimony in Virginia Electric and Power Company, Case no. 18759, filed March 9, 1970, pp. 38–39.

[20]Some degree of rationing is bound to prove inevitable during any transition period and has in effect already appeared. During recent months, a growing number of distributors in the East and Middle West have been forced to refuse service to new industrial customers, to limit sales expansion to present users and, in some instances, to cut supplies below previous quantities. See the *Oil and Gas Journal*, April 20, 1970, p. 83, and May 4, 1970, pp. 114–15.

latter assumptions, gas either holds its present energy market share or increases it slightly. The Future Requirements Committee expects that the Pacific Northwest, New England, the Great Lakes area, the Gulf Coast, and the industrial areas of the East will experience the greatest gas growth between now and 1990. As for individual markets, the commercial and firm industrial categories will show the greatest strength.

Attempting to forecast future supply and demand under varying price assumptions or trying to predict the impact of price changes under regulatory action is most difficult. Efforts to use econometric techniques to determine the response of gas demand to price changes, even for limited future periods, have not produced results in which we can place much confidence. It is hoped that further studies will refine these analytical tools and enhance their predictive capabilities.

A qualitative appraisal of small users of gas indicates that their demand is extremely inelastic in the short run in response to field price changes of 10 cents per Mcf or less. Over longer periods in regions where alternative fuels are available, the small user when influenced by price may gradually shift to the alternative sources, thereby slowing the increase in gas consumption. In these areas, the gas demand by the small user is elastic in the long term.

In the case of the large user (industry and electric utilities), demand in some regional industrial markets is price inelastic because of the large price or quality advantage enjoyed by the gas industry. However, in other parts of the United States, especially the East Coast area, the large-user gas market is highly elastic in the long run and probably in the short run in some types of use.

These possible responses of gas consumers to price change have been without consideration of the effect of new pollution regulations. Since gas appears to be the most unobjectionable energy source, in those regions and types of use where gas would have been expected to behave elastically, we may find that these markets have also become highly inelastic.

In view of the existing and expected behavior of gas demand and the probable configuration of the gas supply curve, there appears to be a necessity to choose between two basic positions: (1) control prices as in the past and allocate scarce supplies; or (2) permit gas prices to increase until supplies become adequate to meet demand at those prices. Neither course appears to be free from danger. The fundamental question is which alternative is the most workable and in harmony with basic economic forces. On balance, a heavier reliance on market forces would be clearly preferable to further extension of controls, even though such a course is bound to result in substantially higher gas prices to most users.

APPENDIX: REVIEWS OF SEVEN GAS DEMAND STUDIES UTILIZING ECONOMETRIC TECHNIQUES

*Jean E. Draper**

1.

J. B. Vermetten and Joh. Plantinga, "The Elasticity of Substitution of Gas with Respect to Other Fuels in the United States," Review of Economics and Statistics, *vol. 35 (May 1953)*–In this study the elasticity of substitution was calculated by determining a relationship between the ratios of gas prices and other fuels and the ratios of the respective quantities consumed. Data analyzed were for the year 1947 and were aggregated by state for eleven different industries. Not all states were included in every calculation–some states did not have industries in every category and in some states there was no possibility of substitution for certain fuels.

*Professor of Economics and Management, University of Arizona.

The industries chosen were fairly evenly distributed among the states and used gas, coal and coke, electricity, and fuel oil. Elasticities of substitution of gas with respect to coal and coke, electricity, and fuel oil were computed for each of the eleven industries. These elasticities ranged from -0.33 to -4.99 and were more homogeneous with respect to industries than with respect to fuels.

The intensity of energy—the percentage of the energy consumed in the value added by the process of production—was determined for each industry. In general, in industries with a high intensity of energy, the substitution elasticities of gas with respect to other fuels tended to be higher than in industries utilizing less energy.

Although it was not clearly stated, elasticities were apparently computed by regression analysis. The number of observations used in estimating elasticities ranged from six to forty-seven. The values of the correlation coefficients ranged from 0.38 to 0.98.

CRITICISM AND LIMITATIONS

Computations were based on data for one year (1947) aggregated by state and industry. It is difficult to interpret the results as elasticities in the usual sense because of the extensive aggregation. Some of the computations involved rather small samples and in some cases the correlation coefficients were quite small. The correlation coefficients ranged from 0.38 to 0.98; this means that the equivalent of R^2 ranged from 0.14 to 0.96, with a number of the values less than 0.50. In any event, results of a study based on 1947 data are by now badly out of date. Availability and cost of gas as well as of competing energy sources are no longer representative of current conditions.

2.

Harold Wein, "Analysis and Projections of Natural Gas Supply and Demand" (Prepared testimony, FPC Docket no. AR 61-1 Permian Basin Area)—In this study Harold Wein analyzed the interrelationships between prices, supply, demand, reserves, and the effect of these and other economic factors on natural gas supply and demand. The purpose of this study was to determine the economic consequences of setting a particular wellhead price for gas: how much exploratory activity and what level of reserves could be expected, what amount of residential, commercial, and industrial consumption would result, what reserve/production ratio would ensue, and what the effects would be of the wellhead price on burner tip prices.

Multiple regression analysis was used to estimate twelve linear or logarithmic equations that involved factors thought to determine exploration, determinants of residential-commercial demand, and determinants of industrial demand.

The demand models were concerned with factors that influence total national gas consumption. Cross-section analysis was used for the year 1961 with data aggregated by state; the forty-four states having natural gas service in this year were included in the analysis. Independent variables for the residential-commercial demand equation were burner tip price of gas, prices of competing fuels, degree days, and number of residential and commercial consumers having natural gas service. Independent variables for the industrial demand equation were industrial gas price, industrial oil price, coal price, industrial employment, number of kilowatt-hours, and degree days. When the demand equations had been estimated, they were used to determine the levels of demand in the years 1962–70 that would occur for different burner tip prices in conjunction with projected values of the other independent variables.

The results of the demand analyses indicated that the elasticity of industrial consumption with respect to price was almost four times as large as the elasticity of residential and commercial consumption with respect to price. Industrial demand was highly elastic, -2.5, while residential and commercial demand had an elasticity of -0.8. Residential demand by itself, had it been calculated, would have been even more inelastic, probably below -0.5.

The study concludes that burner tip prices of gas have substantial effects on total gas consumption and that, if the average residential-commercial burner tip prices increased by about 3 percent per year over the next decade, the growth in residential-commercial demand would virtually cease.

CRITICISM AND LIMITATIONS

Numerous criticisms have been made of the independent variables used in the explorations and demand equations of this study. Two of the independent variables used to predict demand in the residential and commercial sector of the demand models were the price of gas and the number of customers. Particularly when the equation is to be used for projection, number of customers and consumption should be simultaneously determined. Similarly, industrial demand and industrial employment should have been determined simultaneously, as should total demand and supply. Since estimation was not simultaneous, the assumptions of causality implicit in the projections are open to serious doubt, as are the resulting estimates.

Aggregation of the data by state makes interpretation of the results difficult. There is also the question of whether results obtained would hold for years other than 1961. The industrial demand equations were estimated also for the years 1955-60 and there was considerable variation in the estimated coefficients. Particularly when projections are involved, estimated coefficients must be quite stable or inferences based on them are unreliable.

Autocorrelation of the residuals was probably not a problem, since cross-section analysis was used. Multicollinearity probably was a problem and might well have been one cause of the instability of the estimated coefficients. No information was given concerning either autocorrelation or multicollinearity.

There was no check made on the assumptions involved in estimating elasticities by logarithmic transformation of the data, so interpretation of the estimates is difficult.

3.

Robert Stephen Villanueva, "An Econometric Estimation of User's Demand for Natural Gas" (Ph.D. dissertation, University of Pittsburgh, 1964)—This study was concerned with the estimation of the parameters of the natural gas demand schedules of the major sectors of the consumer markets and with the effect of the gas utility rate structure on the distribution of the fuel between the consumer groups. Multiple regression analyses were performed using time series and cross-section data for small volume consumers, classified as the residential-commercial market, and for large-volume consumers such as industrial and electric generation plants and large commercial establishments.

In the residential-commercial market for natural gas, regressions were estimated for total demand for natural gas, the demand for gas service by new customers, and demand per customer. Similar regressions were estimated for total utility gas. In the industrial category, models were developed for total demand and demand of individual industry categories. Time series regressions were performed for the same periods employing the same unit of aggregation, to the extent that the data permitted.

Independent variables included prices of natural gas and competitive fuels, income, temperature, housing starts, and measures of consuming-industry activity.

All time series data were aggregated by census region, and five of the nine census regions were used for analysis: East South Central, East North Central, West North Central, Middle Atlantic, and Mountain States. In the case of residential-commercial demand, the cross-section regressions were calculated for observations disaggregated by state; for industrial demand, cross-section analysis was applied to industry categories and thermal electric utility stations. The period used was 1950-60.

For the purpose of linearizing the data, reducing multicollinearity, and removing any serial correlation in the residuals, first differences were taken of the logarithms rather than of the original observations for time series analyses.

The regressions for regional natural gas demand of residential-commercial consumers resulted in estimated price elasticities of -1.83, -1.39, and -0.51 in three of the five regions analyzed; estimates for the other two regions were positive but nonsignificant.

The regressions for regional demand by industrial customers resulted in estimated price elasticities that ranged from -1.34 to -1.64 for three of the five regions; the other two estimates, one positive and one negative, were nonsignificant.

CRITICISM AND LIMITATIONS

The major criticism of this study concerns the extent of aggregation of the data—multiple regression equations were estimated for data aggregated by states, by regions, and by industry. In some cases the residual degrees of freedom were as few as nine; there were four or five independent variables; and the multiple R^2 was approximately 0.50, so that interpretation of the statistically significant coefficients is difficult. The author discusses the aggregation problem with respect to the rate structure and in some cases uses aggregation as an explanation for unexpected signs of estimated elasticities, but he does not solve the problem of accurately estimating the demand elasticities.

The data used for analysis are first differences of logarithms. This transformation was supposedly used in order to linearize the data and to reduce autocorrelation and multicollinearity. The appropriateness of the transformation is in doubt: there is no direct information given concerning linearity of the transformed data, but the values of R^2 are seldom above 0.50, which is certainly not strong evidence of linearity; in the cases for which it is possible to evaluate multicollinearity from the information given, there is evidence that the transformed independent variables are quite highly intercorrelated; many of the values of the Durbin-Watson statistic indicate the presence of autocorrelation. In addition to these statistical problems, there is the logical problem of interpreting the regression coefficients in terms of the transformation.

In several cases results for stepwise regression analyses were given, but in most cases only a final equation including both significant and nonsignificant independent variables was reported, and in many cases there were more nonsignificant than significant independent variables in the regression equation. Such a regression equation is difficult to interpret: not only is the value of R^2 artificially inflated, but the coefficients of the significant independent variables may be affected by inclusion of nonsignificant independent variables that are correlated with them.

4.

Pietro Balestra, The Demand for Natural Gas in the United States: A Dynamic Approach for the Residential and Commercial Market (*Amsterdam: North-Holland Publishing Co., 1967*)—The purpose of this study was to develop a dynamic demand equation for natural gas in the residential and commercial sector of the economy.

A static short-run demand function for gas relating the consumption of gas to the real price of gas and the real per capita income was estimated for time series data for the period 1950-62 for thirty-six states. Of the thirty-six estimates of price elasticity, thirteen were statistically significant, but seven of these had positive signs, contrary to theory. On this basis, the static model was judged to be inadequate to represent consumer choice in the gas market. The same data were then used to estimate a quasi-dynamic short-run demand function relating demand for natural gas to the availability of gas and the stock of gas-using appliances, as well as to the real price of gas and the real per capita income. Of the thirty-six estimates of price elasticity for this function, twenty-six were negative (ranging from -0.00002 to -7.75) and only three of the ten positive estimates exceeded their standard errors.

The conclusions from these short-run analyses were as follows: (1) the total demand for fuels is fairly inelastic, at least in the short run; (2) given the consumer's stock of major appliances, substitution of different types of fuels must be rather small in the short run; and (3) there is some evidence of competition between the different types of fuels, but the short-run analysis is inadequate to measure this effect.

A number of models (eight basic models with several versions of some models) were estimated for data for the period 1950-62 for forty-seven states. The independent variables included price of gas, price of substitutes, population, per capita income, percentage of urban population, and degree days. In some cases the models were estimated for cross-section data and in some cases for pooled cross-section and time series data. The pooled models were estimated for the separate periods 1950-56 and 1957-62, as well as for the entire period 1950-62. This was done in order to contrast the innovating and mature stages in gas consumption.

The conclusions drawn from the estimation of the dynamic models were as follows: (1) the results suggest that the time period studied can be separated into two subperiods corresponding to two technologically different stages of development; (2) the incremental demand for gas appears to be fairly elastic and the price elasticity is higher in the last part of the period; (3) the estimated average price elasticity for the period 1957-62 is approximately -0.7 for the incremental demand for gas, which is approximately twenty times as large as the corresponding price elasticity of the total demand for gas; and (4) an upward trend in the price elasticity is indicated by the yearly estimates as well as by the estimates for the two subperiods; in particular, the estimates for the last year (1962) show a price elasticity of -1.32. This upward trend in price elasticity suggests that competition from alternative sources of energy may become stronger in future years.

CRITICISM AND LIMITATIONS

Multiple regression equations were estimated for data aggregated yearly by states, and interpretation of the results is complicated by this extensive aggregation.

Several hundred multiple regression equations were computed using data for the period 1950-62. These equations differed with respect to the number of variables included; the period for which data were included; whether the analysis was time series, cross-section, or pooled time series and cross-section; and whether the equations were linear, logarithmic, double logarithmic, or involved first differences or first differences of logarithms. In some cases the results for these different regressions were in agreement, in other cases they were discrepant. In some cases values of R^2 were high, in other cases values of R^2 were low. In some cases estimates had values consistent with theory, in other cases estimates were not theoretically acceptable. It is difficult to interpret the results when so many analyses were performed, their results are not in agreement, and there is no clear theoretical basis for choosing among the different models.

No check is made on the assumptions necessary for transforming the data. No information is given on the basis on which the extent of multicollinearity can be determined, but the occurrence of many perverse signs suggests that multicollinearity may be fairly serious. No information is given concerning autocorrelation of the residuals, though this is probably a problem for the time-series analyses.

5.

Venkata Madhava Rao Tummala, "Alternative Methods of Estimation in the Demand for Natural Gas" (Ph.D dissertation, Michigan State University, 1968)—The purpose of this study was to develop dynamic models to explain the demand for natural gas in the residential, commercial, and industrial sectors. A simultaneous equations model was formulated to explain the influence of the burner tip price of natural gas and other relevant variables on the consumption in each of the sectors. A distributed-lag model was also developed, on the basis of the expected-income hypothesis, to explain the reaction or adjustment mechanism in the behavior of consumption of natural gas. Several alternative estimation procedures were used to estimate the parameters of the two models: ordinary least squares (OLS), two-stage least squares (2SLS), unbiased Nagar K-class (UNK), limited-information single equation (LISE), and three-stage least squares (3SLS).

The equations of the two models were estimated using annual time series data for the state of Michigan for a sample period from 1946 to 1964. The independent variables were: index of burner tip price in the appropriate sector, consumption of natural gas in the appropriate sector, index of price of elasticity in the appropriate sector, index of price of fuel oil in the apporpriate sector, disposable personal income (for the residential sector), number of degree days (for the residential sector), industrial employment (for the industrial sector), and index of price of coal (for the industrial sector).

In general, the OLS, 2SLS and UNK methods yielded comparable estimates; results of LISE estimation were erratic. The 3SLS estimates were theoretically plausible in most cases, and, since 3SLS estimation has some theoretically preferable asymptotic properties, the author concentrated his discussion on the results of this method of estimation.

From the simultaneous equations model, the burner tip price elasticities of demand in the residential, commercial, and industrial sectors were estimated as -0.4386, -0.6048, and -1.3344, respectively. From the distributed-lag model, the burner tip price elasticities in the residential and commercial sectors were estimated as -1.6320 and -1.4378, respectively. Estimation of the model for the industrial sector produced unsatisfactory results—the estimate of elasticity was -0.6983, indicating that price was inelastic, a result not in accord with economic theory. Cross-elasticities were estimated, but in many cases the sign was the opposite of that expected and these results seemed unreliable.

CRITICISM AND LIMITATIONS

The data are yearly averages for the state of Michigan (for the years 1946-64) and thus are aggregated to such an extent that they are inappropriate for an analysis of demand and the results are difficult to interpret.

There are nineteen sets of observations, but there are thirty variables with as many as seven or eight variables used in one equation. Consequently, values of R^2 are high, but few coefficients are significant, and many coefficients have unexpected signs.

Only a small part of the intercorrelation matrix is given, but this part indicates considerable multicollinearity and the presence of so many perverse signs leads to the suspicion of extensive multicollinearity and unreliable parameter estimation.

The results of estimation for the different models are in some cases very discrepant, more so than would be expected on the basis of the differences in the methods. This also supports the suspicion that the data are inadequate for the estimation being attempted.

Log linear transformations were used, with no check on the relevant assumptions, to estimate elasticities; this is likely to lead to erroneous results.

6.

Joseph Lee Steele, "A Critical Evaluation of the Uses of an Econometric Model in Cases Involving Natural Gas Rates" (Ph.D. dissertation, University of Texas at Austin, 1969)—This study evaluates the Wein model (see above) and does not contribute any additional information or findings.

7.

Damodar Gujarati, "Residential Electric Space Heating Market, 1963-1967" (National Economic Research Associates, Inc.)[1] —The purpose of this study was to determine the relationship between

[1] See testimony of Charles H. Frazier in Virginia Electric and Power Company, Case no. 18759, filed March 9, 1970, p. 19.

the customer saturation ratio—that is, the ratio of residential electric space-heating customers to total residential customers—and selected economic and noneconomic variables. The independent variables included price of electricity, price of gas, personal income of the consumer, degree days, and number of housing units of various types. Data were obtained for the period 1963-67 for twenty-seven privately owned utilities meeting the following criteria on the basis of 1966 data: electric space-heating customers numbered at least 2,000, average consumption per customer for space heating was at least 9,000 kilowatt-hours, and the saturation ratio was at least 1 percent.

Ordinary least squares multiple regression analysis was performed for data for each of the five years, 1963-67, and for data for the entire period 1963-67. The estimates for the cross-price elasticity—that is, the elasticity of the saturation ratio with respect to the price of gas—ranged from 0.4094 to 0.9262; all values were positive, as expected. There were three utilities whose saturation ratios far exceeded those of the other twenty-four utilities. When the regression equation was computed omitting these three utilities the estimates for the cross-price elasticity ranged from 0.0342 to 0.5414; all values were still positive, as expected.

The values of R^2 were quite low, ranging from 0.3359 to 0.6960 for the analysis of twenty-seven utilities and from 0.1772 to 0.3818 for the analysis of twenty-four utilities.

CRITICISM AND LIMITATIONS

Multiple regression equations were estimated for data aggregated yearly for each utility; interpretation is difficult because of the extent of aggregation. The values of R^2 were quite low, which also makes interpretation difficult.

No information was given concerning multicollinearity of the independent variables or autocorrelation of the residuals. Autocorrelation is probably not a problem, since the analysis is not of time series data. Multicollinearity may not be a problem, since perverse signs did not occur, but a priori it seems reasonable to suspect that personal income and number of housing units of various types, for example, might be quite highly correlated.

2 INTERFUEL RELATIONS GOVERNING NATURAL GAS DEMAND AND SUPPLY

*Richard J. Gonzalez**

A comprehensive analysis of the role of natural gas in meeting energy requirements must take into account several basic points:

1. Natural gas is an extremely useful and desirable fuel.
2. Natural gas competes with other fuels.
3. Natural gas and liquid hydrocarbons (such as crude oil, butane, and propane) are developed and produced jointly.
4. Natural gas virtually displaced manufactured gas because it was cheaper and better.
5. Changing demands affect the relative prices of joint products.

SPECIAL ADVANTAGES OF NATURAL GAS

Natural gas is often called a premium fuel. This designation arises from several advantages gas possesses that lead users to prefer it over other fuels even at some higher cost per therm as delivered to them. Cleanliness, efficient combustion, absence of air pollution, and lower costs for installation of furnaces and storage tanks and for fuel handling by the buyer are factors that make gas worth more than other fuels.

For residential and commercial customers, the extent of the use of natural gas throughout most of the United States reflects the advantages they find in this fuel relative to others. For large industrial users of energy, such as electric utilities, the choice of fuels is a matter of economics, taking into account differences in capital investment, in handling costs, and in meeting air pollution regulations, as well as the delivered cost of the alternate fuels available over the anticipated life of the plant. Total industrial use of natural gas in the United States doubled between 1956 and 1968, while the use of heavy industrial fuel oil increased 20 percent, and the use of coal and lignite gained only 11 percent.

NATURAL GAS IN COMPETITION WITH OTHER FUELS

Despite its many advantages for fuel users, the role of natural gas in meeting energy requirements was limited for many years by relatively high transportation costs. Until 1946 gas was used principally close to its source, and there were few long-distance interstate gas pipelines, mainly from the Panhandle-Hugoton field to the North Central states.

Technological developments in the construction of large-diameter welded pipelines and the rise in price of other fuels after World War II led to the creation of a vast system of interstate pipelines to carry gas from the southwestern producing areas to all major energy markets in the United States.

*Petroleum economics consultant.

Interstate shipments of natural gas increased about tenfold between 1946 and 1969 while total gas use increased about fivefold and total output of mineral fuels increased 90 percent. Table 2-1 shows the dramatic change in the role of wet natural gas (unprocessed) in the mineral fuel output of the United States.

Table 2-1. U.S. Production of Mineral Energy, 1946 and 1969

(trillion Btu)

Year	Coal and lignite	Crude oil	Wet natural gas	Total
1946	15,526	10,057	4,550	30,133
1949[a]	14,843	19,510	22,772	57,125
Change	-683	+9,453	+18,222	+26,992

[a]Preliminary.

SOURCE: U.S. Bureau of Mines.

Natural gas supplied two-thirds of the increase in production of mineral fuels in the United States between 1946 and 1969. Its share of this output increased from 15 percent in 1946 to almost 40 percent in 1969. By 1969 energy from U.S. natural gas exceeded that of coal and lignite by more than 50 percent and that of crude oil by more than 16 percent.

The manner in which users of energy have met their additional demands since World War II indicates clearly that they have found natural gas attractively priced in relation to other alternates. Under normal market conditions, without any interference by government, the price of gas would reach a competitive equilibrium with other fuels at a level that would balance long-run supply with demand.

RELATION OF HYDROCARBON LIQUIDS AND GASES

Liquid and gaseous petroleum hydrocarbons are found, developed, and produced together in varying relations in different geological provinces and different fields. A few fields have principally oil with little remaining associated gas, and a few fields have dry gas with little recoverable liquids, but most fields have substantial quantities of both liquids and gases. In the United States, recoverable reserves at the beginning of 1970 are estimated at 29.6 billion barrels of crude oil, 8.1 billion barrels of natural gas liquids, 59.6 trillion cubic feet (tcf) of associated dissolved gas, 211.9 tcf of nonassociated gas, and 3.6 tcf in underground storage.

For oil fields, the average ratio of gas to oil is about 2 thousand cubic feet (Mcf) per barrel. The energy content of a barrel of average crude oil in the United States is estimated at 5,800,000 British thermal units (Btu) and that of wet natural gas at 1,075 Btu per cubic foot. On this basis, about 73 percent of the energy recoverable from oil fields is crude oil and 27 percent wet natural gas.

For gas fields and gas condensate fields, the relation of liquids to gases varies widely. In some large fields, the gas is rich enough to warrant operation of processing plants to extract the liquids and re-inject the gas for future recovery. In other cases, the liquids and gases are both marketed as produced. U.S. Bureau of Mines (USBM) statistics indicate that natural gas liquids consumed in the United States have about 11 percent of the energy attributed to wet natural gas production. (The USBM *Monthly Petroleum Statement* for December 1969 reports 2,459 trillion Btu for natural gas liquids consumption and 22,772 trillion Btu for wet natural gas production in 1969.)

Economic considerations may favor efforts to direct the search for new petroleum supplies primarily for oil, as has been the case historically for many operators, or primarily for gas in some areas. However, some of the best gas fields, such as Panhandle-Hugoton and Katy, were found by operators looking for oil. More recently, many wells drilled in hopes of finding gas have produced oil instead.

Considering total new supplies (discoveries, extensions, and revisions) for oil and gas by five-year periods since 1960, there was a distinct upward shift in the relation of total new gas to new oil from

Table 2-2. Relation of New Crude Oil and New Natural Gas, for Five-Year Periods, 1950-1969

Period	New crude oil reserves (*bil. barrels*)	New natural gas reserves (*tcf*)	Mcf of new gas per barrel of new crude oil
1950-54	15.9	72.0	4.5
1955-59	14.5	106.1	7.5
1960-64	12.0	89.2	7.4
1965-69	13.4	85.4	6.4

NOTE: New supplies include discoveries, extensions, and revisions.

SOURCE: American Gas Association annual reports on gas reserves.

4.5 Mcf per barrel in 1950-54 to about 7.5 for 1955-64 and a decline to 6.4 for 1965-69. These changes suggest that gas prices rising relative to oil can exert a favorable influence on gas supply. The most favorable condition for additional gas supply occurs when the search for both oil and gas receives economic stimulus.

On the basis of the average gas/oil ratio for oil fields of 2 Mcf per barrel, oil discoveries in the 1960s may account, in the ultimate figures for this decade, for almost 30 percent of total new gas (25.4 billion barrels X 2 Mcf/barrel divided by 174.6 tcf of total new gas). This figure may be somewhat lower if part of the associated gas cannot be marketed economically because of distance from pipelines.

In addition, gas is sometimes discovered while wells are being drilled in the search for oil. In view of the large number of exploratory wells drilled by small operators who are interested primarily in oil production, the number of chance discoveries of gas must represent a substantial fraction of total gas discoveries. Taking the reserves of these gas discoveries along with the associated gas from oil fields, the search for oil may account for about half of total new gas supply in the United States.

DISPLACEMENT OF MANUFACTURED GAS BY NATURAL GAS

In 1945 utilities had about 19 million customers, of which less than 45 percent used natural gas. Other gas users relied either on manufactured gas or on a mixture of manufactured and natural gas. At that time the major use of natural gas was near producing areas and in the North Central states served by early interstate pipelines. After World War II, ability to transport gas at reasonable cost by large-diameter welded pipelines led to construction of a very extensive system of interstate gas pipelines serving all major energy markets. By 1965 natural gas had virtually displaced manufactured gas. The extent of this shift is shown in table 2-3.

Table 2-3. Natural Gas Operations of Utilities, 1945 and 1965

Type of gas	No. of customers (*thousands*)		Gas sales (*mil. therms*)	
	1945	1965	1945	1965
Natural	8,914	36,537	22,563	118,748
Manufactured	8,582	128	2,088	101
Mixed	2,400	618	1,194	1,256
LPG	81	55	23	26
Total	19,977	37,338	25,868	120,131

SOURCE: American Gas Association, *Gas Facts* (1966).

The prospective shortage of natural gas in the 1970s has led to consideration of using manufactured gas again. The main problems would be those of costs and of increasing coal output by the large amounts required to meet a major part of the increase in gas demand. The president of United Gas Improvement Company of Philadelphia stated recently that the cost of manufactured gas would run about 60 cents per million Btu, compared with about 45 cents per million Btu being paid by his company. For gas distributors closer to the point of production, the margin of added cost for manufactured gas would be even greater.

Panhandle Eastern Pipeline has estimated that 6 million tons of coal will be required to manufacture 300 million cubic feet of gas per day, or about 109.5 billion cubic feet a year. At that rate, 60 million tons of coal a year would be required to meet the rate of increase in demand of about 1 tcf a year estimated for the 1970s. The coal needed to supply 10 tcf of manufactured gas would be 600 million tons a year, or more than the total U.S. output of coal in 1969. Such volume of manufactured gas may never be economically feasible and would take decades to achieve in any case.

PRICE RELATION OF OIL AND GAS

For joint products (different cuts of beef, cotton and cotton seed, hydrocarbon liquids and gases, etc.), changing price relations are designed to achieve a balance between supply and demand for the entire output. (For example, the price of liver relative to steak increased sharply when recognition of the value of liver as a source of vitamin B brought about a great increase in demand.) Unless such balance is achieved, part of the joint output will be inadequate, while another part may be in balance with demand or in surplus supply.

The limited markets for gas relative to available supply prior to development of the extensive system of interstate gas pipelines drove natural gas prices down to about 5 cents Mcf at the point of production in the period 1937–47. During most of that time, the average price of crude oil in the United States was between $1.04 and $1.25 a barrel.

While gas prices were low, most casinghead gas associated with oil was flared because its value would not cover the cost of gathering. Gas wells were drilled only in the best fields with very large reserves per well. In other words, gas prices did not warrant development of any but the lowest-cost reserves, and gas made very little contribution toward covering joint costs.

As numerous new interstate pipelines were built after World War II, and contracted for at least a twenty-year supply to support their applications for certificates of necessity, gas prices responded to the increased demand. Between 1948 and 1958, while crude oil prices increased only 11 percent, the average value of natural gas at wells increased from 6.5 cents to 11.9 cents per Mcf. Since the average included gas sold under old long-term contracts at low fixed prices, prices paid on new purchases probably increased more than 100 percent in this ten-year period.

This change in the price of gas relative to oil was entirely predictable and in keeping with basic economic principles. At a symposium in 1949, I estimated that field prices of natural gas would rise to 15–20 cents per Mcf on the Gulf Coast. This conclusion was based on an analysis of interfuel competition that showed that the demand for gas would prove insatiable until a competitive equilibrium was reached with coal and fuel oil near that level. The price of natural gas had to find a new competitive equilibrium with oil and coal following developments that enabled it to reach new markets with large demands.

The problem of balancing demand and supply for oil and gas is complicated by the traditional twenty-year period for gas contracts. The price provisions in these contracts may be in line with other fuels initially, but they become out of line later, thereby creating demands different from those that would prevail if gas prices responded promptly and fully to all factors affecting the relation of gas to competing fuels.

DETERMINING A NATURAL GAS PRICE POLICY

The preceding analysis of the characteristics of natural gas and of the manner in which gas has increased its role in energy supply makes clear not only that bargaining among buyers and sellers of gas would be effective in balancing demand and supply, but also that regulation of gas prices on the basis of principles applicable to utilities would be very complex and unsatisfactory.

The main points that favor reliance on competition rather than regulation to determine the price of gas are as follows:

1. The business of locating, developing, and producing oil and gas has none of the characteristics of public utilities.
2. The joint-product nature of oil and gas presents insurmountable problems of cost allocation calling for constantly changing allocation between liquids and gases.
3. The nature and extent of interfuel competition means that any regulation reasonably satisfactory for consumers would have to be dynamic in nature, taking into account all forces affecting the relation of gas to other fuels, and not a static concept of cost determination suitable for a long period of years.

The main problem with regulation arises from the changing price level required to balance gas demand and supply. If it is assumed that, by chance, the regulatory process determined the right level of gas prices to balance demand and supply initially, the balance might quickly be upset by developments affecting other fuels, such as a rise in the price of coal, air pollution regulations that add to the cost of reducing emissions from other fuels, and higher interest rates.

These considerations are real and very important. The price of industrial coal at mines increased from \$5.217 per ton in 1967 to \$6.052 in 1969 and from \$6.068 to \$6.44 between the third and fourth quarters of 1969. These increases are equivalent to more than 3 cents per Mcf of gas in three years and to 1.5 cents per Mcf of gas in the final quarter of 1969. Air pollution regulations have had a marked effect on the demand for gas relative to other fuels. Higher interest rates have affected the relation of nuclear power plants to gas-fired electric generators because of the much higher capital costs of nuclear plants. The preceding considerations are entirely apart from those affecting the cost of gas and oil and illustrate why reasonable regulation of gas prices would be much more difficult than for utilities.

FEDERAL REGULATION OF INTERSTATE GAS PRICES

The preceding analysis makes it seem strange that the federal government should be regulating the prices at which producers sell gas to interstate pipelines. The sequence of events leading to this situation constitutes a classic case history of the mistakes that result from failure to understand basic economic principles and forces.

The Natural Gas Act was passed in 1938 in order to prevent interstate pipelines from making excessive profits from transporting natural gas. Previously, gas-distributing companies using manufactured gas had begun to switch to natural gas because it was much cheaper. Interstate lines could buy gas at low prices in the field, undersell manufactured gas, and make much higher profits than was customary for utilities.

If gas pipelines had been common carriers, as oil pipelines are, the regulation would have applied to the rates charged for transportation and not to the price of gas. However, the lines were engaged in buying gas in the field and reselling it after transportation. The wording of the Natural Gas Act dealt with the sale of gas in interstate commerce for resale, and provided opportunity for interpretation that the provisions designed for pipelines could also be applied to producers selling gas to interstate pipelines.

When gas prices began to increase in response to rising demand, representatives of consuming areas began to press for regulation of the price realized by producers on their interstate sales on the ground that a "regulatory gap" existed, which needed to be closed in order to protect consumers. The Supreme Court upheld this point of view in a decision in 1954. Since then, the Federal Power Commission (FPC) has been endeavoring to find a reasonable basis for regulation of gas prices.

A basic principle of our economy brushed aside in the 1954 decision is that government regulation should be applied only in very special circumstances under which competition cannot operate effectively to promote the public interest. There was ample evidence that competition among producers had kept gas prices at very attractive levels. For example, electric utilities on the Gulf Coast have paid much less for gas than they would have to pay for any other fuel, a situation that could not exist in the absence of effective competition among sellers of gas. Shortages of gas may now force prices on new contracts up to competitive equilibrium with other fuels as necessary to balance supply and demand, but even price increases of that amount will reflect basic market forces and not any lack of competition among numerous sellers.

BALANCING GAS DEMAND AND SUPPLY

The role of the federal government in regulating the price paid by interstate pipelines for natural gas does not change the price level required to balance demand and supply. It merely shifts responsibility for the correct decisions from numerous buyers and sellers to an administrative agency faced with evaluating conflicting testimony. The FPC must seek the economic truth that will serve the best interests of energy consumers. It will not be enough to provide cheap gas currently if the result of such action will be to cause shortages that force consumers to use more expensive and less desirable alternates.

The period of grace granted by past development of reserves and the time lag between a change in level of exploration-development and the ability to meet demands has now been exhausted. Some gas distributors have already announced that they can no longer take on increased demands or new business. The ability to deliver gas to meet peak demands will be a problem of increasing severity until a better balance is achieved between demand and supply for the long run.

Demand for gas has been stimulated by air pollution regulations and the rise in the price of coal. Profit prospects affecting investments in new wells have been reduced by the lower prices set in the Permian Basin and Southern Louisiana decisions; by higher costs, including income tax increases due to the Tax Reform Act of 1969; and by the outlook for changes in oil import controls that will discourage the search for oil.

The rate of development of new gas reserves has now fallen below the rate of use, and it appears that gas supply cannot be stimulated soon enough to preclude some shortages. Therefore, distributors are looking to alternates, such as imported liquefied natural gas and manufactured gas, at costs likely to exceed their current cost of natural gas by at least 15 cents per Mcf along the Atlantic Coast and in the Midwest.

The FPC must recognize that decisions made in the past on the basis of historical data do not provide the right answer as to what price will balance demand and supply for the future. It must evaluate the prospective impact of the prices it sets on supply and decide what action will serve the public interest.

From the standpoint of consumers, the key question will be whether additional supplies of domestic gas can be made available for less than the alternates they will be forced to use in case of a shortage of gas. Consumers are not concerned with theoretical allocations of cost between oil and gas or what rate of return is allowed on investment. Their interest lies in a practical answer that will provide them energy under the conditions most advantageous to them.

The need for the use of common sense can be illustrated simply. First, the supply of gas cannot be stimulated by cutting area prices, as has been done by decisions of the FPC, regardless of the estimates of costs determined by theoretical calculations based on assumptions as to allocation of joint costs. Second, regardless of what such calculations indicate as to the rate of return on gas investments, consumers will benefit from prices that stimulate additional gas supply so long as the final cost to them results in savings compared with alternates.

The alternates that will keep the price of gas reasonable to consumers include heating oil, fuel oil, coal, and nuclear power, as well as manufactured gas and imported fuels (gas from Canada, fuel oil, and liquefied natural gas). Gas from Alaska may also be a factor, but the distance from U.S. markets makes it inevitable that the cost of supplies from this source will be high in relation to existing price levels unless some way can be found to move gas at much lower cost.

From the standpoint of investors in oil and gas production, they cannot be coerced into developing additional gas but must be attracted into making new investments in the face of alternate opportunities. For example, an oil firm with alternate investment opportunities at 15 percent may not want to spend much on gas at a 12 percent rate of return but might be led to spend a great deal more if the prospective return were 18 percent. Using the Southern Louisiana decision as a bench mark, without any comment on its merits or deficiencies, it must be recognized that if 5.33 cents per Mcf represents a 12 percent return, an increase of 2.66 cents per Mcf in profit would mean an 18 percent return.

Regardless of what the correct figure on profits per Mcf of gas may be, the inescapable point is that returns on gas below the competitive level with alternate opportunities will discourage investments designed to increase gas supply. Little influence on gas supply can be expected until gas prices provide a return that investors consider commensurate with the risks.

Higher gas prices operate to balance the market through their effect on both demand and supply. On the demand side, limited supplies of gas are directed to the users who gain most from having gas rather than switching to alternates. On the supply side, operators are led to tackle additional prospects and well locations, which they will not drill at lower prices. Because the results of exploration are uncertain, some of the discoveries due to additional search could be large new gas fields.

The offshore continental shelf area of the Gulf of Mexico in particular appears to provide opportunity for development of large new gas supplies if investors are encouraged to undertake vigorous exploration. Additions to new investment and supply could easily be more closely related to the percentage change in profit margin than to the percentage change in price, provided the change in price caused the rate of return to cross from below to above the rate attainable on alternate investment opportunities.

The concluding point to be stressed about gas prices relates to the necessity of balancing supply with demand under complex conditions affecting not only gas but other fuels as well. The following factors constantly operate to determine the right economic price for natural gas: changing relations in the prices and cost of using competing fuels, changing costs in bringing forth new supplies of oil and gas, and the effect of a steadily declining value of the dollar on the willingness of risk takers interested in capital appreciation to invest in development of new gas if its price continues to be regulated under principles applicable to public utilities.

3 PROJECTED COSTS OF ALTERNATE SOURCES OF GAS

*R. Bruce Foster**

This paper is a discussion of the possible ranges of costs of pipeline gas from Canada and Alaska; liquefied natural gas (LNG) from Alaska and foreign countries; and synthetic pipeline gas produced from coal and oil at likely plant sites in the United States. It is hoped that these comparisons will provide the reader with a better understanding of the role that each alternative source may play in supplementing our domestic supplies of natural gas.

CANADIAN PIPELINE GAS

Since imported pipeline gas from Canada offers the most immediate prospects for additional gas supplies, let us consider first the possible range of cost of new gas from our northern neighbors.

Current Prices of Canadian Gas

Before attempting to project future cost of imported Canadian natural gas, it may be well to review 1968 and 1969 prices of Canadian imported gas (see table 3-1). It can be seen that the price of gas sold to transmission companies engaged in interstate commerce in 1969 ranged from 22.5 cents to about 37.2 cents per thousand cubic feet (Mcf). At the same time, gas sold to intrastate companies ranged from about 16.2 cents to 57.9 cents per Mcf. In all, about 680 billion cubic feet (bcf) of gas was imported at an average price of 24.52 cents per Mcf, a slight increase from the 23.82 cents per Mcf reported in 1968. This amounts to about 3.5 percent of the total amount of natural gas consumed in the United States in that year.

The flow of natural gas through Canada and into the United States is shown graphically in figure 3-1, which was prepared by the Dominion Bureau of Statistics. Although the data presented are for 1966, the pattern of flow has not changed markedly. It can be seen that virtually all Canadian gas is produced in the Western Canada Sedimentary Basin in Alberta and British Columbia. Notice that a larger quantity flows south into the western United States than flows east to Ontario and Quebec. At the present time there are no pipelines in Canada extending north of the lower tier of provinces.

Future Prices of Alberta Gas

Existing and proposed contracts for furture importation of natural gas from Alberta and British Columbia through existing pipelines provide for escalation of prices during the period of the contract. These increases will raise the 100 percent load factor price of gas imported at Sumas, Washington, from its present 25 cents to about 30.3 cents by 1974 and 31.2 cents by 1980. Similarly, gas imported at Emerson, Manitoba, will escalate about 1.5 cents per Mcf every five years from a 100 percent load factor price of 32 cents per Mcf in 1970-71 to a maximum of 37 cents per Mcf.

*Manager of Market Planning, Institute of Gas Technology.

Table 3-1. U.S. Natural Gas Imports from Canada, by Volume, Value, and Unit Cost, 1968 and 1969

Point of entry	Volume: Amount (*Mcf at 14.73 psia*[a]) 1968	Volume: Amount (*Mcf at 14.73 psia*[a]) 1969	Volume: % change	Value (*dollars*) 1968	Value (*dollars*) 1969	Unit cost (*cents/Mcf*) 1968	Unit cost (*cents/Mcf*) 1969
To interstate companies							
Sumas, Washington	148,842,855	171,938,412	15.5	35,201,258	43,055,299	23.65	25.04
Eastport, Idaho	50,437,458	47,987,474	-4.9	11,776,575	11,326,592	23.35	23.60
Noyes, Minnesota	3,665,152	19,894,859	442.8	1,010,079	5,428,293	27.56	27.28
Noyes, Minnesota	117,196,636	117,792,058	0.5	32,020,852	32,176,341	27.32	27.32
Eastport, Idaho	241,363,664	264,108,769	9.4	58,904,487	59,386,726	22.33	22.49
Niagara Falls, New York	454,728	5,416,638	1091.2	153,779	1,824,552	33.82	33.68
Niagara Falls, New York	–	5,682,312	–	–	2,116,606	–	37.25
Total	561,960,493	632,820,522	12.6	134,067,030	155,314,409	23.86	24.54
To intrastate companies							
Whitlash, Montana	17,344,490	17,175,917	-1.0	2,767,171	2,783,391	15.95	16.21
Babb, Montana	19,290,695	22,654,059	17.4	4,028,085	4,718,833	20.88	20.83
Massena, New York	4,414,324	5,361,252	21.5	2,267,271	2,722,039	51.36	50.77
Highgate Falls, Vermont	1,452,024	2,094,857	44.3	827,025	1,212,272	56.96	57.87
Total	42,501,533	47,286,085	11.3	9,889,552	11,436,535	23.27	24.19
Total imports	604,462,026	680,106,607	12.5	143,956,582	166,750,944	23.82	24.52

[a]Pounds per square inch absolute.

SOURCE: FPC, *FPC News*, vol. 3, no. 21 (May 22, 1970), pp. 3-6.

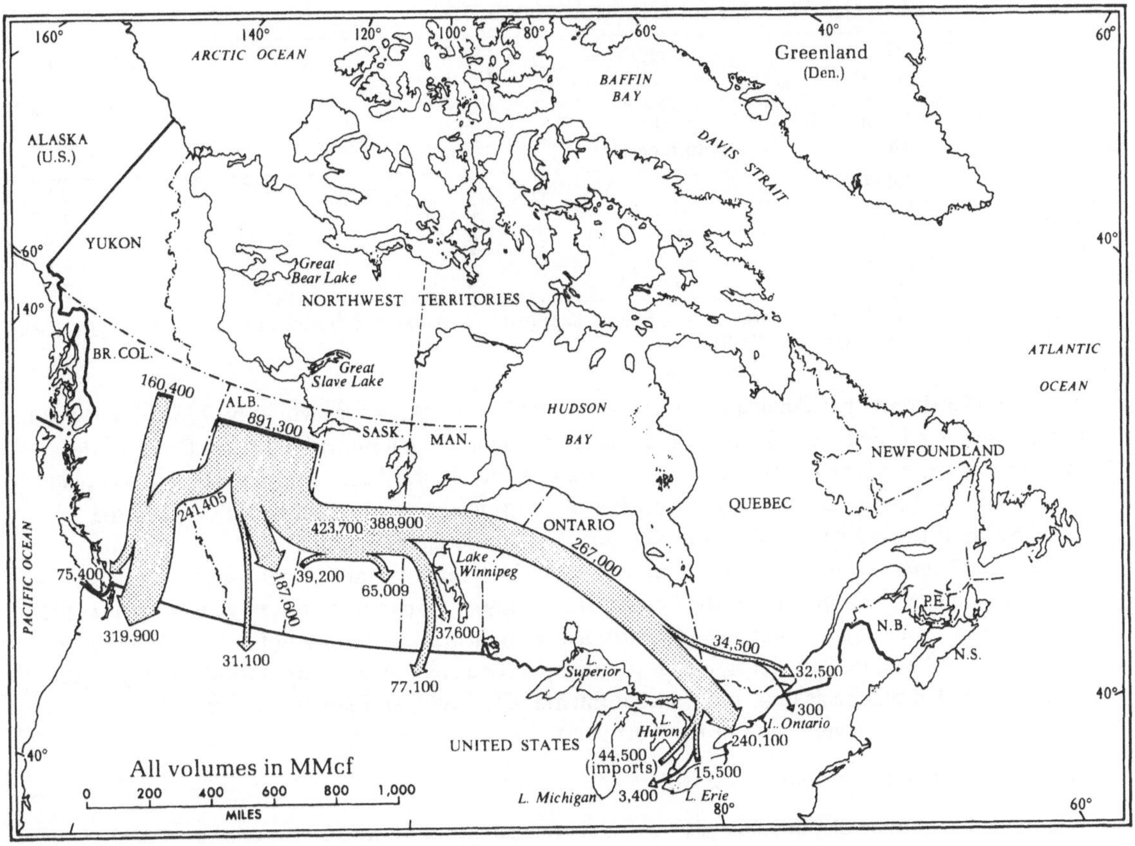

FIGURE 3-1. Canadian natural gas receipts and disposition, 1966. *Source*: Dominion Bureau of Statistics.

These prices pertain to gas imported from proved reserves of Canadian gas already (or soon to be) committed to existing pipeline projects. The quantities of gas that can be imported at, or near, these prices are limited to the quantities presented on line 1 of table 3-2 and amount to less than 4 percent of our domestic consumption.

Projected Quantities of Imported Canadian Gas

Recently a study was made by the Canadian Energy Board[1] of the quantity of natural gas that could be exported to the United States on the assumption that all the Canadian gas reserves were limited to the Western Sedimentary Basin. These quantities are presented in column 1 in table 3-2. They represent the quantities of gas we might expect to import at near current, or somewhat higher, prices as more gas is drawn from the northern areas of this sedimentary basin. These data indicate that the quantity available for import from the Western Sedimentary Basin amounts to only about 4 percent of our anticipated gas requirements as projected by the Future Requirements Committee. If we are to increase importation of Canadian gas from its present level of 3.5 percent to about 12 percent of U.S. demand by the year 1990, it will require the development of additional Canadian gas reserves in the Northwest Territories or elsewhere.

It is reasonable to expect that future growth of gas reserves will be found in the northern frontier areas as present exploration activities continue. Early in 1969 the Canadian Petroleum Association

[1]Canada, National Energy Board, *Energy Supply and Demand in Canada and Export Demand for Canadian Energy, 1966-1990* (Ottawa: 1969).

Table 3-2. Canadian Gas Available for Export

Year	(1) Western Sedimentary Basin *(tcf)*	(2) (1) as a % of U.S. requirements	(3) Northern frontier areas *(tcf)*	(4) Total (1) + (3)	(5) Total as a % of U.S. requirements
1966	0.432	2.5	–	0.432	2.5
1975	1.165	4.0	0.110	1.275	4.35
1980	1.359	4.0	0.941	2.300	6.85
1985	1.449	3.75	2.251	3.700	9.5
1990	1.626	3.6	3.774	5.400	12.0

SOURCE: Canada, National Energy Board, *Energy Supply and Demand in Canada and Export Demand for Canadian Energy, 1966-1990* (Ottawa: 1969).

published its estimate that Canada's ultimate recoverable natural reserves would be 725 bcf, of which the Western Canada Sedimentary Basin is expected to yield 270 trillion cubic feet (tcf). Thus, the northern frontier areas might be expected to yield 455 tcf. On this basis, another estimate was made by the Canadian Energy Board of the quantities of additional gas that could be exported to the United States if the reserves expected in the northern frontier areas materialize.

These additional quantities of gas (column 3, table 3-2) must be transported greater distances than Alberta gas and can be imported to the United States only at much higher prices. This is because the only known reserves of sufficient size to justify a large transmission line are not in Canada at all but are in Alaska at Prudhoe Bay. Thus, we must either justify a large pipeline from Prudhoe Bay to the United States that will be as long as a transcontinental line from San Francisco to New York or wait until additional reserves are found in the Northwest Territories of Canada.

Current Canadian Gas Transmission Costs

Some understanding of Canadian gas transmission costs to the United States can be obtained from the data presented in table 3-3. The delivered costs at the U.S. border do not necessarily agree with the average prices listed in table 3-1 because these data represent individual pipeline projects.

Table 3-3. Existing Major Canadian Export Pipeline Systems

Pipeline system	Westcoast Transmission	Trans Canada Pipeline	Alberta and Southern
Source of gas	Ft. Nelson	Alberta	Alberta
Cost of gas (*¢/Mcf*)	15	15	19
Size of main line (*inches*)	30 & 36	30, 34, & 36	36
Border crossing point	Sumas, B.C.	Emerson, Man.	Kingsgate, B.C.
Transport cost to border (*¢/Mcf*)	19.4	13.3	5.3
Delivered cost at border (*¢/Mcf*)	34.4	28.3	24.3
Final market point	Portland, Ore.	Dawn, Ont.	Antioch, Calif.
Delivered cost to market (*¢/Mcf*)	38.9	42.0	33.8
Total transmission cost (*¢/Mcf*)	23.9	27.0	14.8
Total pipeline miles	1,109	1,850	1,180
Transmission costs (*¢/Mcf/100 mi.*)	2.15	1.46	1.25

SOURCE: D. M. Coulter, D. H. Norris, and G. W. Walker, "Pipeline Designs for the Far North", *Pipeline Industry*, December 1969.

These transmission costs are all above the rate of 1.2 cents per Mcf per 100 miles that has sometimes been used as a representative transmission cost in the United States. This is to be expected because of the more difficult terrain and greater transportation costs of material and equipment.

Data reported by the Federal Power Commission (FPC) on pipeline construction costs in the United States for fiscal 1969 show values ranging from $130,000 per mile for a 24-inch pipeline to

$280,000 for a 42-inch pipeline.[2] These costs are 15 to 20 percent higher than those reported in 1968. This increase in one year is sufficient to offset the economic advantage of 6 inches of pipe diameter in the larger sizes. That is, a 48-inch pipeline built in 1969 will transport gas for about the same cost as a 42-inch pipeline built in 1968.

Some gas transmission companies are currently using annual cost escalation factors of as much as 4 percent. This suggests that very large pipeline diameters will be needed simply to offset increased construction costs and will provide little or no reduction in transmission costs. Thus, the transmission costs shown in table 3-2 for 30- to 36-inch pipelines will soon be applicable to 40- and 48-inch pipelines if construction costs continue to inflate in Canada as well as in the United States.

Projected Canadian Gas Transmission Pipelines

Three projects for pipelining natural gas from Prudhoe Bay to the United States have been announced. These are shown in figure 3-2 along with the existing pipelines running through southern Canada.

Westcoast-Bechtel Project. Westcoast Transmission Company of Vancouver, British Columbia, and Bechtel Corporation of San Francisco have announced a two-phase plan for a pipeline project extending from Kingsgate, British Columbia, to Prudhoe Bay. The first phase will include a 40-inch diameter

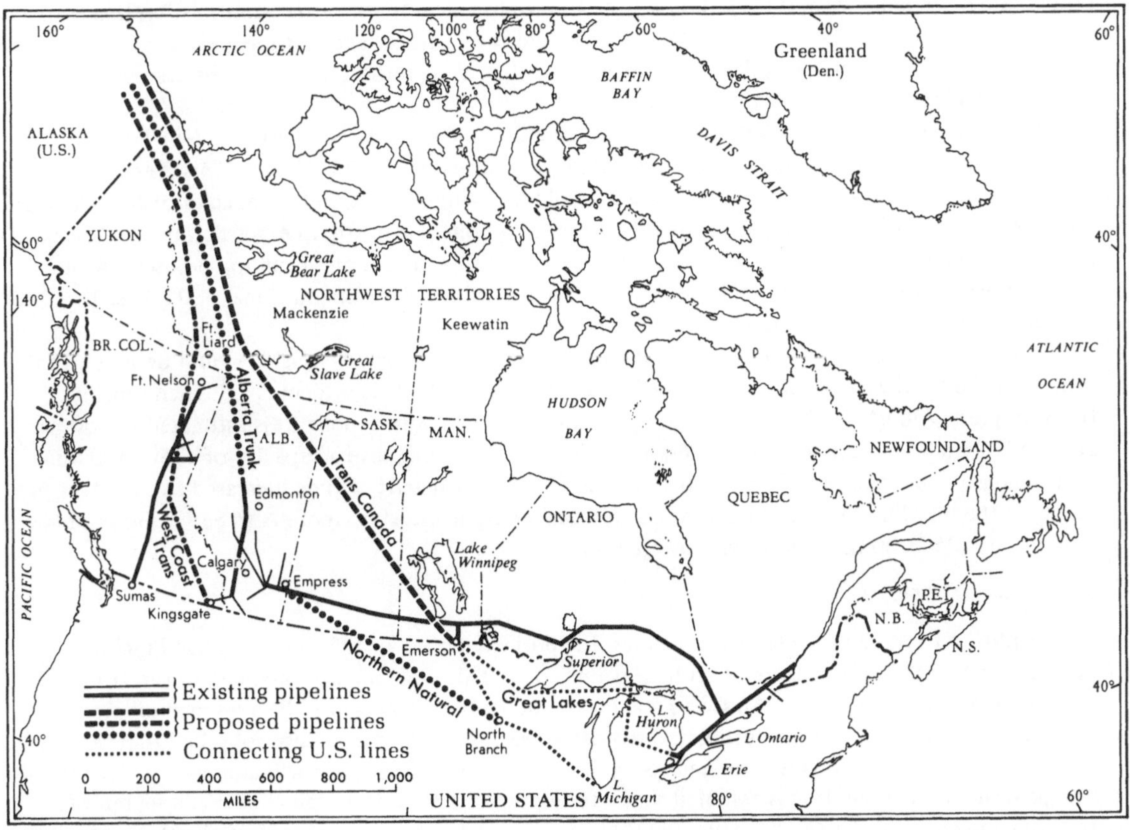

FIGURE 3-2. Proposed Canadian gas transmission pipelines.

[2] FPC, Bureau of Natural Gas, *Cost of Pipeline and Compressor Station Construction under Non-Budget Type Certificate Authorizations as Reported by Pipeline Companies in Fiscal Year 1969* (1970).

line running from Kingsgate northward through British Columbia to Fort Liard in the Northwest Territory, a distance of 995 miles. It is projected to cost $400 million and will have an initial capacity of 800 million cubic feet per day.

The second section would reach from Fort Liard to the Prudhoe Bay area via the Mackenzie River Delta, a distance of 1,085 miles. This line would have a 48-inch diameter and have a capacity of 1,650 million cubic feet per day initially, which could later be expanded to 2,500 million cubic feet per day with additional compressor facilities. This segment has an estimated cost of about $736,000 per mile, while the first section is estimated at about $400,000 per mile.

Trans Canada Pipeline Project. Trans Canada Pipelines Ltd., Natural Gas Pipeline Company of America, and Michigan Wisconsin Pipeline Company are studying a pipeline to run from Emerson, Manitoba, south of Winnipeg, diagonally across Canada to Fort Liard, then down the Mackenzie River to the delta region and across to Prudhoe Bay in Alaska. It will be 2,350 miles long and 48 inches in diameter. The line will have fifty compressor stations spaced about fifty miles apart and will be powered by gas turbines in the 12,000–16,000 horsepower range. At a total projected cost of $2.5 billion, it will cost about $1,060,000 per mile.

Alberta Gas Trunk Line Project. The Alberta Gas Trunk Line Company has exclusive rights to the gathering and transmission of natural gas in Alberta. This privilege combined with the strategic position of Alberta enables Alberta Gas Trunk Line to propose a common carrier transmission line from its present northern limits to Prudhoe Bay. This would then permit it to transport gas through its present network in Alberta to other companies for further distribution in the United States and Eastern Canada.

The first stage of the proposed project would be a 350-mile extension of Alberta Gas Trunk's Foothills Division from a point near Grand Prairie to the northwest corner of Alberta. This is estimated to cost $100 million. An additional $200 million would be spent to increase the capacity of the Alberta Gas Trunk system. A new company would be formed to build the northern extension to the Alaskan border, a distance of 900 miles, at an estimated cost of $900 million, or $1 million per mile. A third company would be formed to build the Alaskan leg to Prudhoe Bay, an additional 300 miles. This segment would cost $400 million, or $1.33 million per mile.

A project proposed by Northern Natural Gas Company of Omaha, Nebraska, would tie in with this project at the southeastern end of Alberta Gas Trunk's network. This project is a 36-inch pipeline from Empress, Alberta, to North Branch, Minnesota, north of Minneapolis. It is estimated to cost about $172 million initially for the 876-mile main line and would have a capacity of about 500 million standard cubic feet (scf) per day in 1971. By 1980 this capacity will be increased to 1,000 scf per day with the installation of $53 million in additional equipment. This amounts to $197,000 per mile initially, or $257,000 per mile when powered up.

Trans Alaskan Pipeline

In addition to these projected gas pipelines, the projected costs of the Trans Alaskan Pipeline (TAPS) can be used as a measure of arctic pipeline costs. This pipeline, now planned to run from Prudhoe Bay southward across Alaska to Valdez on the Gulf of Alaska, will be a 48-inch oil pipeline, 753 miles long. It was originally estimated to cost $900 million. It is generally conceded that before completion, this cost will have escalated to $1.3 billion or more. If so, this will amount to $1.75 million per mile of pipeline. If it is assumed that a fully powered 48-inch gas pipeline could be built for this total cost per mile, it can be estimated that the gas transmission cost would be about 3 cents per Mcf per 100 miles, and the total transmission cost from Prudhoe Bay to Valdez, Alaska, would amount to 22 cents per Mcf. This assumes annual capital costs at 15 percent of total capital costs and operating costs at 20 percent of the total.

However, some of the high cost of the TAPS line is caused by the pioneering effort to build the first pipeline through difficult and inaccessible country. It can be expected that later pipelines will be somewhat lower in cost because of the experience gained on the TAPS line and the greater availability of supporting equipment and facilities. The simultaneous construction of both an oil line and a gas line offers attractive prospects for reducing the cost of construction under arctic and subarctic conditions.

Projected Gas Transmission Costs

The data reported on these projects provide a basis for estimating the transmission cost of gas in various regions of northern Canada. Using an annual capital cost factor of 15 percent and assuming that capital costs are 80 percent of the total cost, we have estimated transmission costs ranging from 1.3-2.5 cents per Mcf per 100 miles for various segments at 100 percent load factors.

These costs, used as a basis for drawing the charts shown in figure 3-3 and figure 3-4, can be used to estimate transmission costs of gas from various points in Canada to importation points and market areas in the United States. The solid lines represent the performance of the existing pipelines presented in table 3-3. The dotted lines present transmission costs based on an analysis of the various pipeline projects proposed for northern Canada. The charts can be used to determine, for example, the transportation cost to Chicago from various points in Canada and Alaska. It can be seen that gas can be transported to Chicago from Edmonton for approximately 21 cents to 24 cents per Mcf; from Fort Liard for approximately 30 cents per Mcf; from the Mackenzie River delta area for approximately 46 cents per Mcf; and from Prudhoe Bay for approximately 56 cents per Mcf.

The production and gathering cost of the gas must be added to these transmission costs to obtain delivered costs. In the Edmonton-Calgary area production prices are currently about 18 cents per Mcf—up from about 16 cents per Mcf two years ago. Production prices farther north are unknown. In the Prudhoe Bay area, production prices may well be lower because of the large quantities to be disposed of in competition with Canadian gas. We have used 15 cents per Mcf in making estimates of delivered costs.

At Emerson, Manitoba, the delivered price of gas from Alaska would be 62 cents per Mcf. At Kingsgate, it would be 60 cents per Mcf, and at Sumas, Washington, about 58 cents per Mcf. The average cost of gas transmission over the whole route from Prudhoe Bay to Emerson, Manitoba, is estimated

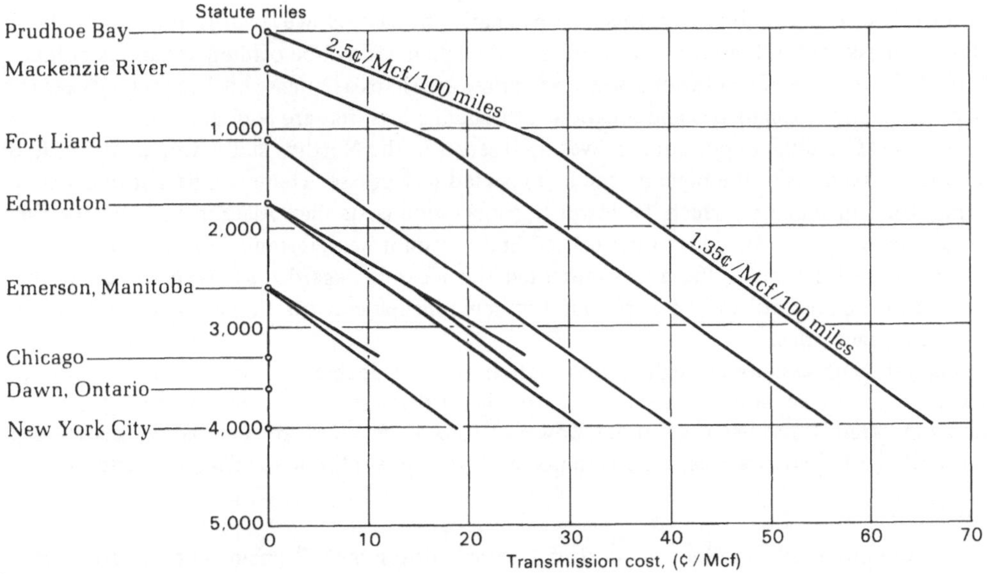

FIGURE 3-3. Estimated Canadian gas transmission costs to central United States.

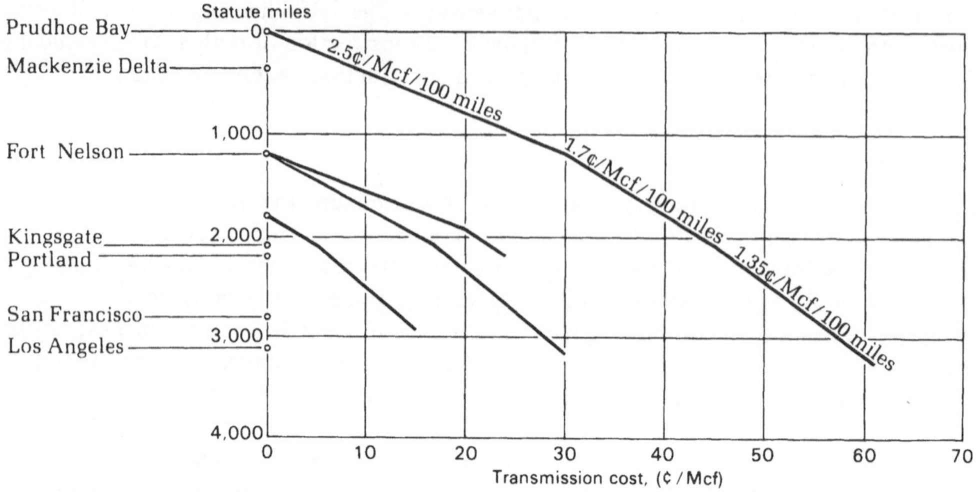

FIGURE 3-4. Estimated Canadian gas transmission costs to the U.S. West Coast.

to be about 1.9 cents per Mcf per 100 miles. This is well above the 1.5 cents per Mcf per 100 miles that was projected only a few months ago and is attributed largely to recent increases in pipeline construction costs. In the region north of the lower tier of Canadian provinces, pipeline costs are almost independent of line size, and the use of very large lines can easily be justified.

On the basis of these estimates, the delivered cost of gas in Chicago would be about 71 cents per Mcf. The Trans Canada line proposes to use a shorter route than that used in figure 3 and might bring gas into Chicago for about 68 cents per Mcf. The average transmission cost to Kingsgate, British Columbia, would be about 2.05 cents per Mcf because of the more difficult terrain. Thus, even though the distance is less, the price of gas in San Francisco would be about 70 cents per Mcf. In Los Angeles, it would be about 74 cents per Mcf.

These prices are based on the assumption that there is little or no commingling of Alaskan gas with Canadian gas. If the line is built quickly in order to dispose of Alaskan gas in quantities sufficient to fill a 48-inch line from Prudhoe Bay, this may be nearly the case. However, it is more likely that once a pipeline to Alaska is built, additional supplies of Canadian gas will be offered at various points along the route. This would result in lower average gas prices—perhaps as much as 10 cents to 15 cents per Mcf, depending on the quantities and locations of Canadian gas that are added.

As additional Canadian exploration proves up reserves in the Northwest Territories, the ratio of Canadian to Alaskan gas in the pipeline can be expected to increase. There will be a strong incentive to contract for Canadian gas, which has lower transportation costs than Alaskan gas. The net effect on the average cost of gas imported into the United States cannot be determined until these Canadian reserves are defined. However, the prices presented above can be regarded as maximum prices that can be expected to decline until some future round of new pipeline construction at higher costs serves to increase the average price.

This analysis is necessarily a rough approximation of the probable price of pipeline gas from Canada and Alaska. A precise economic analysis can hardly be justified in view of the uncertainty of many of the factors involved. It has been presented, however, to provide a basis for comparing this source of gas supply with other alternatives, such as imported LNG and synthetic gas from coal and oil.

LNG Pipelines

One method proposed to reduce the cost of gas transmission is the transmission of natural gas in liquid form. The principal incentive for using a liquefied natural gas pipeline is the fact that about 2.5 to 2.7 times the volume can be moved as a liquid in a pipe of the same diameter used in conventional gas transmission. Thus, an LNG pipeline equivalent to a 48-inch gas transmission line could be about 30 inches in diameter. These pipelines would require refrigeration and pumping stations in place of the compressor stations used on conventional gas-phase transmission lines, and the lines themselves would have to be heavily insulated. In addition, if the gas delivered is to be used in conventional gaseous form, then the cost of liquefaction and of revaporization must be included as part of the cost of the LNG pipeline. As a rough approximation, we might say that because an LNG pipeline can transmit 2.5 times the volume of an equal-size gas line, it would be attractive if it cost only twice as much as the gas transmission line but not if it cost three times as much.

While a number of design studies have been made of various configurations of LNG pipelines, experimental verification will be required of any of them before committing the huge sums of money that are required for the construction of a long-distance pipeline. Because of the time required to develop and demonstrate both the technical and economic feasibility of LNG pipelines, it is not expected that long-distance LNG transmission lines will be in operation before 1980. However, short lines may be in operation much sooner, where they can facilitate the movement of imported LNG from tankers to storage tanks and from one LNG facility to another.

Even though long-distance LNG pipelines are demonstrated to be technically feasible, there is no assurance that they will be economically attractive particularly for arctic operation. A liquefaction plant with a capacity of 2.5 billion scf per day built under arctic conditions might cost as much as 1 billion dollars. This alone would be 50 percent of the cost of a conventional pipeline. The additional costs of refrigeration equipment and more complex pipeline configuration can be expected to further reduce the savings promised by the greater capacity of a liquid-phase pipeline. Since the cost of such a pipeline is indeterminate at the present time and probably will not become a factor during the next decade, we have not estimated gas transmission costs to the United States on this basis.

Importation of Pipeline Gas from Mexico

Mexico has estimated reserves of approximately 12 tcf. Its annual production is about 575 bcf, including gas exported to the United States. The domestic consumption in Mexico is about 440 bcf. Thus, the reserve-to-production ratio, including export gas, is about 21 and, even excluding exported gas, the reserve-to-production ratio is about 27. Thus, Mexico has a supply of gas that is barely more than adequate for its own domestic needs, and unless substantial new reserves are uncovered, little relief can be expected from below the Rio Grande River for the U.S. gas shortage. For this reason, future forecasts of the price of imported Mexican gas have not been attempted, since it will have significance only in local areas in the southwest part of the country.

Nuclear Stimulation of Natural Gas Reserves

A large portion of U.S. gas reserves is locked in tight sands. These reserves are not economical to recover by conventional drilling and fracturing methods. As a result, considerable efforts are underway to explore the feasibility of releasing the reserves by the use of nuclear explosives. Two experiments have been carried out under Project Gas Buggy and Project Rulison. Two other projects, Wasp and Dragon Trail, have been planned.

Project Gas Buggy cost about $5.2 million and produced a fractured area approximately 850 feet in diameter and about 430 feet high, at a depth of about 4,200 feet in the pictured-cliff formation of the San Juan Basin in northern New Mexico. As of March 1969 total production from the well was 167 bcf, most of which was produced during a series of three 30-day production tests. By comparison, an existing conventionally completed well a little more than 400 feet away produced 85 million cubic feet of natural gas from the same formation during a nine-year period.

Project Rulison is expected to cost about $6.5 million and will create a fractured area about 740 feet in diameter, at a point about 8,000 feet below Battlement Mesa near Grand Valley, Colorado.

It has been estimated that nuclear stimulation could result in increased recovery of about 317 tcf of natural gas in the Rocky Mountain area. However, there are many questions to resolve before nuclear stimulation can be considered either a technically or economically successful means of stimulating gas production. The acceptable levels of radioactivity must not only be determined but also must gain public acceptance before such natural gas supplies can be commingled with other gas. Second, there is the matter of public objection to underground nuclear explosions. If one such explosion were to result in a minor earthquake or landslide, it could have a very strong deterring effect on future well-stimulation activities. Third, there is the question of reducing actual cost of the operations to a point where it becomes economically attractive. It has been estimated that the cost of Project Gas Buggy may be reduced to less than $1 million by improvement in the use and design of nuclear devices. If they could be further reduced to less than about $500,000, the concept might become widely attractive.

The effect of well stimulation has been investigated by the sponsors of Dragon Trail using a computerized mathematical model to determine the probable increase in gas recoverability. The results of this analysis indicate that the effect of nuclear stimulation is to reduce the number of conventional wells that would be required to obtain the same gas production. The cost per 1,000 cubic feet of gas produced would be about the same as for conventional production methods. For the long-range outlook, assuming that the problems of radioactivity are solved, it is still doubtful that the atomic stimulation technique could appreciably affect the reserve situation over the next ten-year period and will probably have only a minor effect over a twenty-year time span.

LNG FROM ALASKA AND FOREIGN SOURCES: GENERAL CONSIDERATIONS

In this section, the economics of transporting LNG from Alaska to the United States and the importation of LNG from foreign sources is discussed. Several underdeveloped countries of Africa and South America possess large reserves of natural gas. Of major importance in considering these reserves, as well as those of Alaska, as potential sources are the transportation costs and availability of shipping and port facilities.

Tanker Construction Costs

Probably the most critical item in the importation of LNG into the United States from foreign sources is the cost and availability of LNG tankers. Not every shipyard is equipped to build these specialized and expensive tankers. A single berth can produce one ship per year, although total construction time is about two years. The total lead time from the placing of an order to the delivery of a ship is running about four or five years.

The price of LNG tankers has increased 30-40 percent during the past two years. LNG importation studies made as recently as one year ago used tanker costs of about $48 to $50 per barrel for a 600,000-barrel tanker. It is reported that the *Polar Alaska* and the *Arctic Tokyo*, which were placed in service last winter, cost approximately $65 per barrel also, even though they are considerably larger.

The shortage of tankers and the lengthy construction time are the limiting factors in importing LNG into the United States. The increasing cost of these LNG tankers is forcing upward revisions of the projected price of LNG landed in the United States.

Availability of Overseas Natural Gas Reserves

Worldwide reserves of natural gas are plentiful, as indicated in table 3-4. However, since shipping distance is a cost factor, the United States is interested in those available reserves that are closest to it. In this respect, Venezuela, Trinidad, Algeria, Libya, and Nigeria have the most favorable positions. A recent discovery of natural gas in Ecuador may indicate an additional potential source in South America. However, the extent of the field that has been discovered is unknown. The known gas reserves in Algeria alone are sufficient to provide all the natural gas likely to be imported into the East Coast of the United States for many years to come.

Table 3-4. Potential Worldwide Sources for LNG Supply

Country	Estimated reserves (*tcf*)
Africa	
Algeria	130.0
Libya	24.0
Nigeria	5.0
Middle East and Far East	
Saudi Arabia	56.4
Abu Dhabi and Neutral Zone	21.0
Kuwait	39.0
Qatar	7.5
Iran	197.0
Iraq	19.0
Pakistan	35.0
Indonesia and Brunei	6.7
Western Hemisphere	
Canada	54.2
Chile	3.0
Colombia	4.5
Trinidad	2.5
Venezuela	27.8
Mexico	12.0
Oceania	
Australia	16.0

SOURCE: A. R. Khan, "A World Supply and Demand Appraisal for Base Load Usage of LNG" (Institute of Gas Technology, March 1970).

Port Facilities

There are at least a dozen ports along the East Coast that can accommodate LNG tankers of the present size. They range from Portland, Maine, to Jacksonville, Florida. The largest LNG tankers being planned, however, will draw so much water that they must come in on the tide even in our deepest harbors, such as New York or Philadelphia. Unloading facilities are currently being planned in at least a half-dozen ports and can be made ready in time to accommodate LNG tankers now under construction. It is difficult to estimate the cost of receiving terminals because of the fluctuation in the cost of land and the construction cost of port facilities in different locations. In general, the cost of unloading, storage, and vaporization of imported LNG can be expected to add about 1.5 cents per Mcf to the imported-gas price. In addition, it can be expected that port facilities will add 1-3 cents per Mcf to this figure. In this analysis, 5 cents have been added to the cost, insurance, and freight (CIF) cost of imported LNG to cover all these terminal costs.

Shipping Distances

Shipping distance is the major variable in the cost of LNG from various ports of the world. The distance in nautical miles from various foreign countries to particular East Coast ports is presented in table 3-5. The shipping distance from Prudhoe Bay through the Northwest Passage would make LNG transportation from Alaska much less competitive than LNG from North Africa or West Africa. This is a different situation from that existing in the case of oil where the cost of transportation is a smaller portion of the total cost; high royalty payments paid to foreign governments for oil make it feasible to consider oil transportation from Alaska to the East Coast.

Relationship to Base Load Markets

The high cost of liquefying and transporting LNG by ocean transport suggests that imported LNG will compete most favorably in those markets where large quantities of gas are consumed at relatively

Table 3-5. World Shipping Distances for LNG Supply Routes

Supply source	Market	Distance (*nautical miles*)	Round-trip time (*days*)
Alaska (U.S.)	Japan	3,280	18.6
	U.S. (West Coast)	2,160	13.1
Algeria (Arzew)	France (Le Havre)	1,410	9.4
	United Kingdom (Canvey)	1,540	10.0
Algeria (Skikda)	France (Fos)	400	4.5
	U.S. (East Coast)	3,470	19.5
Brunei	Japan	2,400	14.2
Indonesia	Japan	3,250	18.5
Libya	Italy (La Spezia)	990	7.7
	Spain (Barcelona)	1,060	7.7
	U.S. (East Coast)	4,540	24.7
Nigeria	Argentina	4,390	24.0
	U.S. (East Coast)	5,080	27.0
Persian Gulf	Japan	6,770	35.7
Trinidad	U.S. (East Coast)	1,930	12.0
Venezuela	Argentina	4,860	26.3
	U.S. (East Coast)	2,000	12.3

SOURCE: Conch Methane Services, Ltd.

high prices. This requirement is best met in the northeastern portion of the country along the Atlantic seaboard. The average prices paid by gas utilities for gas for resale in states along the eastern seaboard are presented in table 3-6. These prices range from about 40 cents in New Jersey and eastern Pennsylvania to about 60 cents in Massachusetts and New Hampshire. Currently planned importation programs cannot meet these wholesale prices, and it is only because of the growing shortage of domestic natural gas that users are willing to pay a premium price to bring in imported LNG. In order to keep this premium as small as possible, it is necessary to organize large-scale projects such as that being set up by El Paso Natural Gas Company.

Relationship to Peakshaving Markets

In the northeastern part of the country, between 5 and 10 percent of the total gas sent out by a utility can be classed as peakshaving gas. This is gas that is drawn from some other source than current pipeline deliveries in order to avoid the very high cost of pipeline gas used for such peak demand uses. All of the LNG brought into the United States thus far has been imported for peakshaving purposes

Table 3-6. Average Price Paid by Utilities for Natural Gas, by States, 1967

State	Price (*$ thousand*)	Amount purchased (*thousand therms*)	Cents/mil. Btu
Connecticut	41,269	714,076	57.79
Massachusetts	78,610	1,399,619	56.16
New Hampshire	3,527	57,993	60.82
Rhode Island	12,070	219,266	55.05
New Jersey	172,445	4,231,947	40.75
New York	282,580	6,433,524	43.92
Pennsylvania	469,182	12,161,344	38.58
Delaware	713	11,602	61.45
Florida	25,120	622,383	40.36
Georgia	91,649	2,670,129	34.32
Maryland	70,199	1,665,250	42.15
North Carolina	27,751[a]	1,200,163	–
South Carolina	54,488[a]	1,326,631	–
Virginia	80,240	1,862,325	43.09

[a]Dollar revenues for North Carolina are not correct but are partly combined with South Carolina. Total for both states combined is correct but not for either state alone.

and sold in competition with domestic liquefied natural gas and peakshaving gas made from liquefied petroleum gas. The market for imported LNG for peakshaving purposes has already been established in this country and is growing at a rate that appears to be limited by the availability of LNG tankers and receiving facilities to supply the demand for this gas. Since a distribution utility can afford to pay considerably more for peakshaving gas than for base load gas, it becomes feasible to import LNG at base load prices and transport it considerable distances for storage by a gas utility for peakshaving purposes. For example, LNG can be transported by truck for about 15 cents per Mcf per 100 miles. In addition, it might cost perhaps 50 cents per Mcf to store and revaporize this LNG at its point of delivery. Thus, if LNG is imported into the East Coast for 70 cents, it could be transported a distance of 400 miles inland, stored, and revaporized for a total cost of about $1.80 per Mcf. This is a price that many utilities can afford to pay for peakshaving gas. If all the peakshaving gas in an area within 400 miles of the East Coast were to be supplied by imported LNG, it would require a baseload supply of approximately 500 million cubic feet per day.

TRANSPORTATION OF LNG FROM ALASKA

The movement of natural gas from Alaska to the lower forty-eight states is a difficult and expensive task. This problem was discussed in the previous section in connection with overland pipeline transmissions; here, the alternatives of moving gas from Alaska in liquid form by ocean transportation will be considered.

Very serious consideration is currently being given to the movement of LNG to the West Coast of the United States, particularly the Los Angeles area, from the South Coast of Alaska. This does not necessarily involve gas from the Prudhoe Bay area. There are estimated reserves of about 6 tcf available in southern Alaska that are now used to supply the Phillips Marathon LNG project transporting LNG to Japan. Additional gas could be supplied to the Los Angeles area by pipelining gas from northern British Columbia to the Pacific Coast and liquefying it for transport by sea to Los Angeles. Both of these alternate sources have been examined. The Canadian alternative, of course, has the advantage of avoiding the complications of the Jones Act. Studies made by Arthur D. Little of comparative costs of shipping LNG from Alaska to Los Angeles indicate that the Jones Act results in an increase of about 8 cents per Mcf.

A comparison of the cost of importing LNG to Los Angeles on three different bases is presented in table 3-7. It can be seen that LNG transported from Cook Inlet in Alaska to Los Angeles could be vaporized at a Los Angeles terminal for about 71 cents per Mcf, and LNG transported from British Columbia to the seacoast at Stewart, British Columbia, and liquefied there for transportation to Los Angeles would also cost about the same price. If the Jones Act could be set aside, LNG could be brought into Los Angeles for about 63 cents per Mcf and would gain a competitive advantage over importation from British Columbia.

Transportation of gas from Prudhoe Bay south across Alaska, a distance of about 733 miles would, even at 2.5 cents per Mcf per 100 miles, add approximately 18.5 cents per Mcf to the cost of gas

Table 3-7. LNG Costs at Los Angeles, at 500 Million Standard Cubic Feet per Day

(cents per thousand cubic feet)

	LNG from Cook Inlet (under Jones act)	LNG from Cook Inlet (not under Jones Act)	LNG from British Columbia
Gas cost	15.0	15.0	18.0
Pipeline transport to coast	–	–	6.9
Liquefaction cost	28.0	28.0	28.0
Sea transport	22.9	14.8	12.6
Terminal costs	5.0	5.0	5.0
Total cost of LNG (vaporized)	70.9	62.8	70.5

vaporized in Los Angeles for a total of about 90 cents per Mcf under the Jones Act. This would not be competitive with overland pipeline through Canada, which might serve Los Angeles for about 70 cents per Mcf or less.

The question has also been raised of liquefying natural gas at Prudhoe Bay and shipping it by ocean transport through the Northwest Passage to the East Coast of the United States or possibly through the Bering Strait to the West Coast. LNG importation studies made by the Institute of Gas Technology have shown that LNG transported through the Northwest Passage to the East Coast would cost as much as 60 to 75 percent more than LNG imported from Algeria. The shipment of LNG through the Bering Strait would add an additional 1,400 nautical miles to the 2,000-mile ocean distance from Cook Inlet to Los Angeles. While overall costs by this route would be comparable to the cost of pipeline transportation across Alaska to a liquefaction plant at Cook Inlet and sea transportation from there to Los Angeles, either of these alternatives is more expensive than conventional overland pipeline transportation from Prudhoe Bay to Los Angeles.

LNG IMPORTATION PROJECTS

Thus far four projects for the importation of LNG into the United States have been announced—all of them to the East Coast. Several additional projects are known to be in the planning stage. It is possible that by the late 1970s, LNG importation could exceed 3 billion scf per day if sufficient LNG tankers are built.

Of the thirty-three LNG tankers that have been built or are known to be on order, only eleven will be available for East Coast shipments. These ships will be able to import LNG at a rate of about 1 billion scf per day by 1975. To increase the importation rate to 3 billion scf per day, will require about twenty additional large LNG tankers. These must be ordered during the next year or two if all the project importation plans are to be in full operation by the late 1970s. There are about twelve shipyards with LNG tanker experience (all of them European); hence the problem is one of fitting new LNG tanker orders into busy shipyard schedules.

El Paso Project

The largest LNG-import project announced thus far is that of the El Paso Natural Gas Company. In July 1969, El Paso announced a 25-year agreement with Sonatrach, a national Algerian gas and oil producing company, to import 1 billion scf per day of Algerian natural gas into the East Coast of the United States beginning in 1973. The size of the project has recently been increased to 1.5 billion scf per day. The price of this gas delivered to the United States will be about 58.5 cents per million Btu. This represents a sizable increase over the price of about 50 cents per million Btu reported earlier.

The project will require from twelve to fourteen large LNG tankers. Contracts for two of them have been announced. These will be ships with a 750,000-barrel capacity scheduled for delivery in late 1974 and late 1975. The estimated cost of the original project was $900 million, but it has probably increased somewhat because of increased tanker costs.

Philadelphia Gas Works Project

Philadelphia Gas Works (PGW) and Corporación Venezolana del Petróleo (CVP) have signed a twenty-year contract to allow PGW to import 500 million cubic feet per day of Venezuelan LNG. Total investment for the project, which should begin in the fall of 1974, is expected to be $400 million: $200 million for the Venezuelan liquefaction plant and port facilities, $102 million to $160 million for three or four tankers of 680,000-barrel capacity, and about $40 million for the Philadelphia port and vaporization facilities. The investment is thus $0.80 per million cubic feet of daily capacity.

Distrigas Project

Distrigas Corporation plans to bring Algerian LNG into the New England area, primarily for peak-shaving purposes. Recently the company asked the FPC for permission to import up to approximately

6 bcf per year of natural gas in liquefied form over a twenty-year period starting November 1, 1971. The LNG will be bought from Alocean, Ltd., a joint organization of Sonatrach and GasOcean International, and will come from the liquefaction plant at Skikda, Algeria.

To transport the LNG, Distrigas chartered the *Descartes*, which has a capacity of 1 bcf of natural gas (300,000 barrels) and was scheduled to go into service in the spring of 1971. Contracts call for the delivery of two shiploads between August 1, 1971, and March 31, 1972; four shiploads in the twelve-month period starting April 1, 1972, and six shiploads each year thereafter. Shipyard space has also been reserved for a 440,000-barrel (1.47 bcf of natural gas) tanker for 1973 and a 600,000-barrel (2 bcf of natural gas) tanker for 1974.

Columbia Gas System–
Standard Oil Company Project

Recently Columbia Gas System contracted with subsidiaries of the Standard Oil Company to import LNG at a rate equivalent to 425 million cubic feet of gas per day from Venezuela. The gas will be liquefied in a plant to be built by Creole Petroleum Corporation in a joint venture with the Venezuelan government. The plant is estimated to cost $154 million.

The LNG will be delivered to Columbia Gas at its planned facility at Cove Point, Maryland, on the Chesapeake Bay for a price of about 53.5 cents per million Btu, which is subject to future escalation. Columbia will buy the supply from Esso LNG under a twenty-year contract for 40.5 cents per million Btu, and Esso Oceanic will deliver to Cove Point for an additional 13 cents per million Btu. It is estimated that this gas can then be delivered in the Washington, D.C., area at around 60 cents per million Btu.

ESTIMATED COST OF IMPORTED LNG

Approximately two years ago, the Institute of Gas Technology made a computer analysis of the cost of importing liquefied natural gas to the East Coast of the United States from various countries in Africa and South America. The purpose of this study was to assess the feasibility of using imported LNG as an alternate fuel for power generation to relieve air pollution problems along the eastern seaboard. Since that study was made, the costs of importing LNG have increased sharply—particularly the cost of LNG tankers—between 30 and 40 percent during the past two years. It has been necessary, therefore, to revise the LNG importation prices that were determined in the course of that study. For this analysis, importation costs have been increased by amounts necessary to reflect the increased tanker costs. The revised levelized costs are presented in table 3-8 for New York City and for Savannah, Georgia. These two ports were selected as representing the range of costs that may be experienced along the East Coast.

The lowest costs are for importation from Venezuela, which on this revised basis range from 56.5 cents per Mcf to 63 cents per Mcf on the scale of operations that is being contemplated. This is about 5 cents less than the price of 53.8 cents per million Btu announced by Columbia Gas when corrected for heating value and shipping distance. This may easily be accounted for by differences in the cost of gas supply to the liquefaction plant. LNG from Nigeria and Libya are at a competitive disadvantage with Algerian gas, but could be justified as an alternative source to Algeria, at least for peakshaving applications. The estimated price of LNG from Alaska by way of the Northwest Passage is obviously much too high to be competitive with other sources of LNG or even with pipeline gas brought through Canada to the East Coast.

Thus, pipeline gas can be made available in the New York City area at prices ranging from 60 cents to perhaps 70 cents, depending on the size of the project involved. Gas brought in farther south along the coast will cost a few cents more unless shipped from South America.

These represent premium prices compared with present-day natural gas wholesale prices along the East Coast. However, by the time these LNG projects are in full-scale operation, the wholesale price of domestic gas may be as much as 10 cents higher than current prices. Since it is unlikely that the

Table 3-8. Estimated Costs of Imported LNG

	Shipping distances (*nautical miles*)	Levelized cost in ¢/Mcf vaporized at port of entry		
		Million scf/day		
		350	500	1,000
New York City				
From: Algeria	3,483	71.1	68.0	63.6
Venezuela	1,890	63.0	60.0	53.8
Nigeria	5,230	82.7	79.7	74.6
Libya	4,540	81.0	77.8	72.2
Alaska	4,500	114.0	108.0	103.0
Savannah, Georgia				
From: Algeria	3,946	74.3	71.3	66.6
Venezuela	1,466	59.7	56.5	50.7
Nigeria	4,806	79.7	76.9	71.8
Libya	5,027	84.8	81.0	75.5
Alaska	4,925	119.0	113.0	107.0

amount of LNG imported into the East Coast will ever amount to more than 10 or 15 percent of the total supply, its effect on the overall average wholesale price of gas will be relatively small. For example, if by the mid-1970s the wholesale price of domestic gas in the New York area were to increase from 45 cents to 55 cents, an additional 15 percent supplied by LNG at 70 cents per Mcf would increase the overall price by only 2 cents per Mcf.

SYNTHETIC GAS FROM COAL

During the past decade, considerable research has been done on the problem of making pipeline-quality gas from coal. The technology required to make a synthetic gas that is completely interchangeable with natural gas is quite different from the technology used to make the low-Btu manufactured gas used extensively by the gas industry before the advent of natural gas. To make a high-Btu gas from coal, hydrogen must be added to the carbon in the coal. A typical bituminous coal contains 75 percent carbon, 5 percent hydrogen, and about 20 percent waste materials, which must be removed. To produce a high-Btu gas comparable to methane, the waste material must be removed and enough hydrogen added to increase the hydrogen content of the product gas to about 25 percent.

Thus far, four processes have been developed to make synthetic pipeline gas. These include the Bituminous Coal Research Two-Stage Superpressure Process, the Kellogg Molten-Salt Process, the Consolidation Coal Company CO_2 Acceptor Process, and the Institute of Gas Technology (IGT) Hydrogasification Process. All of the processes are generally similar, the major difference being in the way in which heat is supplied for the reaction. It is entirely possible that other processes exist or could be devised that would be better than those now under development. However, the large cost of new process R&D, combined with the advanced state of development of the present programs seems to preclude any possibility of finding any radically different method of gas manufacture during the next decade. Therefore, commercial coal gasification plants will probably use one of the four processes now under development.

Bituminous Coal Research Two-Stage Superpressure Process

This process is in the earliest stage of development of the four processes. Research results to date are based on the operation of bench scale laboratory units. Coal and steam are introduced directly into a rising column of hot synthesis gas produced in the lower section of a gasifier by the slagging gasification of recycled char with oxygen and steam at temperatures of 2,500°F. or higher. In the upper section of the gasifier, the volatile portion of the fresh coals are converted directly to methane at temperatures of 1,700-1,800°F. The unconverted coal is carried out of the gasifier in the product gas stream and fed to a cyclone where it is separated for recycling in the lower section. The raw gases leaving the cyclone are subsequently purified and subjected to methanation for the ultimate production of high-Btu pipeline gas.

The early stage of development of this process would place commercialization about five years further off than at least two of the other four processes being developed. It is not possible at the present time to estimate very confidently the commercial production costs of gas from this process.

Kellogg Molten-Salt Process

In this process steam and fine coal are injected continuously into a molten salt bath where they react to form synthesis gas. One of the advantages claimed for the Molten-Salt Process is the suspension of the coal particles in a salt melt. The physical properties of the coal particles should not be as critical as the fixed-bed, moving-bed, or fluidized-bed gasifier where agglomeration and gas channeling must be avoided. Also, it should be possible to obtain better fuel utilization by using larger particles of coal, which would require less grinding equipment and power consumption than a suspension gasifier.

Although this process has several desirable features, it is subject to one serious problem that has precluded any further development work. The molten-salt mixture is extremely corrosive, particularly at the high temperatures used. To date, no satisfactory material that is both sufficiently corrosion resistant and economically feasible has been found to contain the molten-salt mixture. Process development has been carried through the bench scale stage, but the Office of Coal Research has ceased funding this project.

Consolidation Coal Company's
CO_2 Acceptor Process

The third process under development by the Office of Coal Research is the Consolidation Coal Company's CO_2 Acceptor Process. A basic feature of the process is the use of lime or calcinated dolomite to react exothermally with the carbon dioxide liberated in gasification to supply a portion of the heat needed for the carbon-steam reaction. The removal of carbon dioxide by this method enhances the water-gas shift reaction and methane formation, both of which are exothermic. The combination of these effects allows operation of the gasification system without an oxygen supply. The spent acceptor (lime or dolomite) is withdrawn from the gasification vessel and calcine separately with residual char and air supplying the necessary heat. The product is a hydrogen-rich methane water/gas that is methanated to yield high-Btu gas. A feasibility study of a commercial plant with a capacity of 250 million cubic feet per day has been completed. In July 1969 construction was initiated on a large pilot plant at Rapid City, South Dakota. The plant site in South Dakota was chosen because of the availability of large deposits of lignite in that area. Development work on this process has been directed specifically toward the use of lignite as the carbonaceous feed. To gasify bituminous coal by this process at a reasonable reaction rate, the temperature in the reactor must be so high that the coal becomes sticky and agglomerates, creating development problems.

IGT Hydrogasification Process

This process was designed to accept coals of widely varying physical and chemical characteristics and operates at a relatively high pressure. The important features of the process are shown in figure 3-5. In this specific configuration of the process, the coal is pretreated with air at mild conditions to prevent agglomeration (which is typical of most bituminous coals). The coal reacts with hydrogen at a temperature ranging from 1,300°F. to 1,700°F. and at 1,000-pound pressure. Unreacted char from the reactor is combined with steam in the gasifier and heated by electricity. The hydrogen-rich gas produced is recycled to the hydrogasifier. The product gas is methanated to produce a high-methane-content pipeline gas.

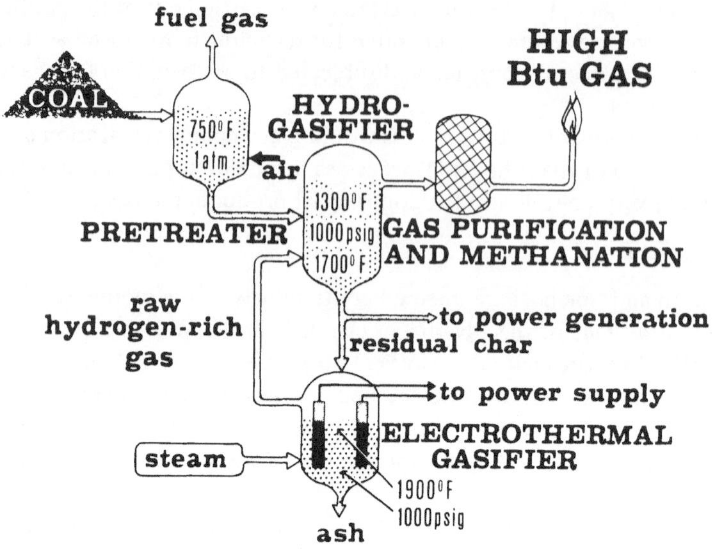

FIGURE 3-5. IGT hydrogasification with electrothermal gasifier. *Source*: Institute of Gas Technology.

The primary process reactions of carbon with hydrogen may be expressed by two chemical equations. The volatile matter in the coal reacts very quickly (ten seconds or less).

$$C^*_{vm} + xH_2 \rightarrow CH_4 \text{ @ } 1{,}300° - 1{,}500°\text{F},$$

where C^*_{vm} is the carbon in the volatile fraction of coal, and x is 1-2 hydrogen atoms.

The second reaction is the conversion to methane of the less reactive fixed carbon with hydrogen, and is relatively slow (15-30 minutes).

$$C_{fc} + 2H_2 \rightarrow CH_4 \text{ @ } 1{,}700° - 1{,}800°\text{F},$$

where C_{fc} is the fixed carbon in the coal.

An important feature of the HYGAS Process is the use of a third reaction for temperature control in the reaction zone. This reaction is the conversion of steam and carbon to carbon monoxide.

$$H_2O + C \rightarrow CO + H_2.$$

The first two reactions are exothermic and take place at two different temperature levels. The steam-carbon reaction, however, absorbs heat in the 1,700°-1,800°F. range and does not proceed at temperatures below 1,600°F. The reaction of the volatile (or active) carbon tends to raise the temperature of the bed until the fixed-carbon reaction starts. However, as the temperature increases, the steam-carbon reaction soaks up more and more of the heat released and prevents the reaction zone from becoming too hot. Conversely, if the temperature falls the steam-carbon reaction ceases, and the volatile-carbon reaction keeps the temperature of the reaction zone high enough for the conversion step. Thus, the steam-carbon reaction acts as an internal temperature regulator, and supplemental methods of heat removal are not necessary.

Several variations of this process are possible. The major difference is the way in which the heat is supplied to convert the residual char into hydrogen. One of these methods is the addition of oxygen to the dilute-phase char-steam mixture (figure 3-6). This is commonly referred to as the Texaco Process, although several other similar versions have reached various stages of development.

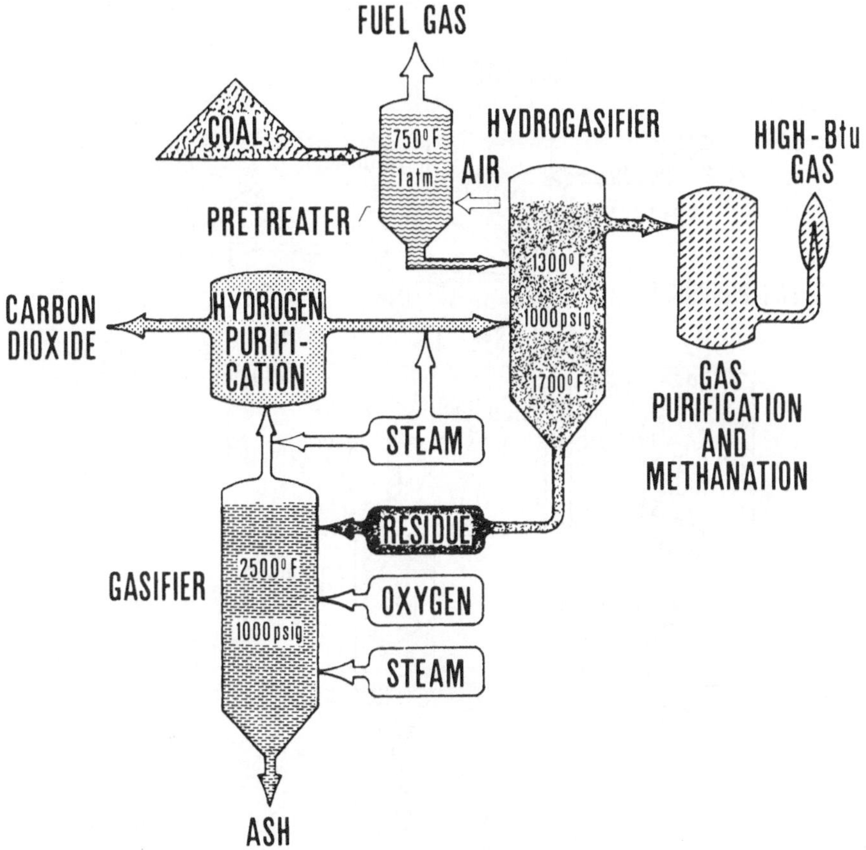

FIGURE 3-6. Variation in the IGT Hydrogasification Process –Texaco version. *Source*: Institute of Gas Technology.

Another variation of the Hydrogasification Process is the Continuous Steam-Iron Process under development at IGT by Fuel Gas Associates (figure 3-7). In this process, hydrogen is produced by having steam react with partially reduced iron oxide. The iron oxide produced is then reduced to iron and recycled. This method of hydrogen production could potentially simplify the overall process and considerably lower plant investment.

IGT is currently in the process of completing the detailed design of a 3-ton-per-hour pilot plant that will use the electrothermal process for hydrogen production. The plant will be capable of producing 1.5 million cubic feet of pipeline-quality gas per day with most coals and will be completely integrated to facilitate optimization of all process steps. The pilot plant is located at IGT facilities in Chicago. Detailed design and construction are being carried out by Procon, a subsidiary of Universal Oil Products. The approximate state of development of these four coal gasification processes is presented in figure 3-8.

Char-Oil Energy Development Process

The Char-Oil Energy Development (COED) Process is under development by the Food Machinery and Chemical Corporation and is presently in the pilot plant stage at Princeton, New Jersey. The process is based on coal pyrolysis which produces char, oil, and product gas. While the process produces some gas, most of the energy in the coal appears as char and oil, and only about 12 percent of the product mix is high-Btu gas. Therefore, the COED Process should probably not be viewed as an alternate process for supplying pipeline gas from coal. While high-Btu gas from a COED plant might

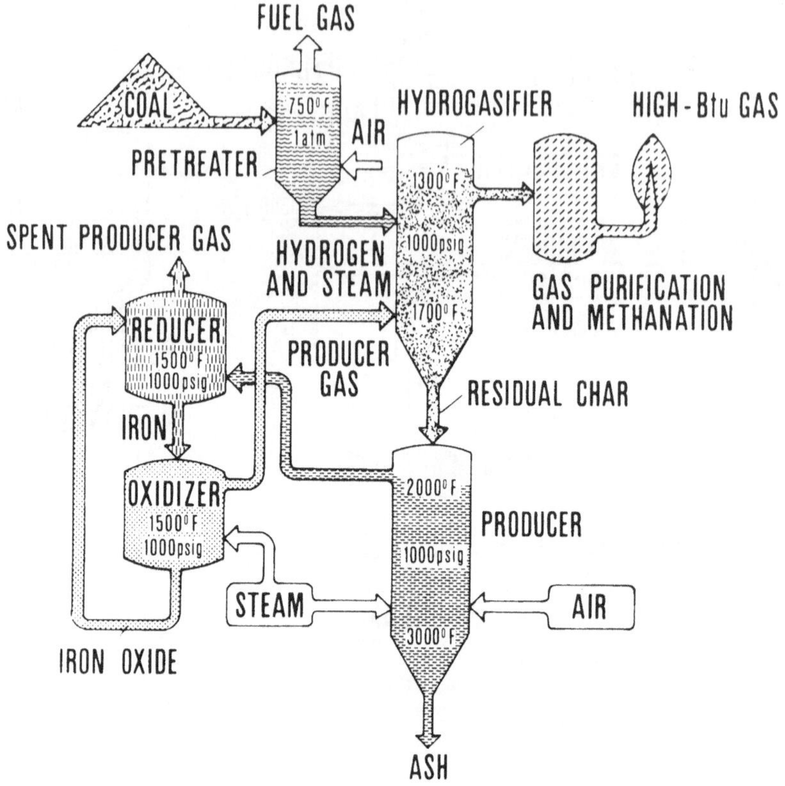

FIGURE 3-7. Variation in the IGT Hydrogasification Process –Steam-Iron Process of Fuel Gas Associates. *Source*: Institute of Gas Technology.

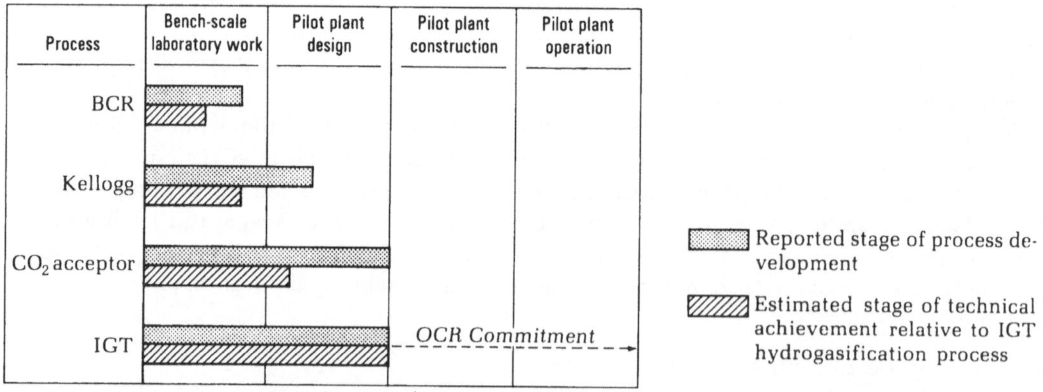

FIGURE 3-8. Relative development and technical maturity of gasification processes. *Source*: C. L. Tsaros and T. J. Joyce, "Comparative Economics of Pipeline Gas from Coal Processes" (paper presented at AGA Synthetic Pipeline Gas Symposium, Pittsburgh, Pa., November 22, 1968).

sell for less than that from a single product coal gasification plant, its price advantage hinges on the marketing of the oil and coal products. In order to produce gas on a scale required to relieve the U.S. supply shortage by this process, it would be necessary to dispose of prodigious quantities of char and oil. While this may be feasible in the early phases of development, a point must be reached eventually where it becomes necessary to alter the product mix increasing the yield of gas. In doing so, the COED Process will become more nearly like the other gasification processes.

COMPARATIVE ECONOMICS OF SYNTHETIC GAS PROCESSES

In November 1968, the IGT made a detailed analysis of the comparative economics of the various processes for making pipeline gas from coal. This analysis was made using the American Gas Association's (AGA) accounting practices and raw material costs that were as closely comparable as possible.

The following gas production processes were evaluated:

1. Two-Stage Superpressure Gasification (coal)
2. Kellogg Molten-Salt Gasification (coal)
3. CO_2 Acceptor (lignite)
4. IGT Hydrogen-Technical Steam-Oxygen (coal)
5. IGT Hydrogen-Continuous Steam-Iron (coal)
6. IGT Synthesis Gas via Electrothermo Gasification (coal)
7. Synthesis Gas via Electrothermo Gasification (lignite)

A summary of the economics of these seven pipeline gas processes is presented in table 3-9. The total investment required for gas production of 250 billion Btu per day ranges from $82 million to $147 million. All gas prices are twenty-year average values calculated by the AGA's accounting procedure. They range from 36 cents to 58 cents per million Btu, and at 16 cents per million Btu, fossil fuel costs range from 46.5 cents to 58.2 cents per million Btu. Figure 3-9 shows the percentage of gas price represented by fossil fuel investment and operating costs for each of these processes.

The effects on gas price of variation in the cost of coal investment and operating costs were calculated during the analysis. The variation in gas prices for a range of coal costs from 12 cents to 20 cents

Table 3-9. Summary of Economics of Pipeline Gas Processes

Process	Total investment[a] (*dollars*)	Capacity (*bil. Btu/day*)	20-year average price of gas at stated fuel-cost (*¢/mil. Btu*)	
			Fossil fuel	Gas
BCR two-stage gasification	142,348,000	231	14.8 (4.00) 16.1 (4.35)	56.4 58.2
Kellogg molten-salt with by-product power	135,295,000	230	14.65 (4.00) 16.1 (4.39)	54.2 56.6
Hydrogasification				
H_2 by steam-oxygen	147,221,900	251	16.1 (4.00)	57.8
H_2 by steam-iron	105,289,000	250	16.1 (4.00)	50.2
Synthetic gas, electrothermal	93,208,000	258	16.1 (4.00)	51.1
Lignite, synthetic gas, electrothermal	236,171,000	500	10.0 (1.47)	36.3
Consolidation CO_2 acceptor	82,444,000	237	10.0 (1.33) 11.3 (1.50)	39.3 41.3

NOTE: Figures in parentheses are dollars per ton of raw coal. The apparent lack of correlation with costs given on the basis of cents per million Btu is due to variations in the Btu content of coal from different sources considered in the analyses.

[a]Installed equipment or "bare" cost is 83-85 percent of total investment. Fixed investment (total less working capital) is 94-96 percent of total investment.

SOURCE: C. L. Tsaros and T. J. Joyce, "Comparative Economics of Pipeline Gas from Coal Processes" (paper presented at AGA Synthetic Pipeline Gas Symposium, Pittsburgh, Pa., November 22, 1968).

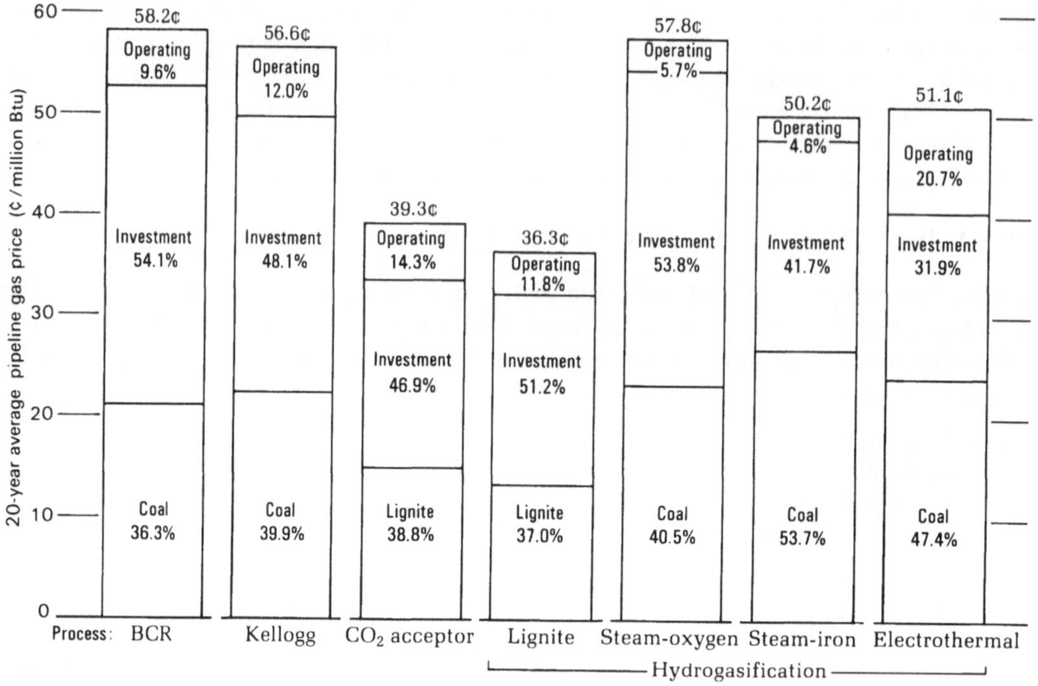

FIGURE 3-9. Components of pipeline gas prices.
Source: See figure 3-8.

per million Btu are presented in figure 3-10. Similarly, price changes caused by variation in the cost of lignite are presented in figure 3-11. Sensitivity of gas price to installed-equipment cost is shown in figure 3-12. Since the number of operators in a modern highly automated plant is quite small, the sensitivity of gas price to operating labor costs is relatively small. A 50 percent increase in labor costs would result in an increase in the price of gas of only 1-1.25 cents per million Btu.

AVAILABILITY OF COAL RESERVES

Bituminous coal beds more than 3.5 feet thick and lignite beds more than 10 feet thick account for 12 percent of the measured and indicated reserves in the United States. This amounts to 333 billion tons of coal in deposits thick enough to be mined, based on the latest U.S. Geological Survey (USGS) data. If it is assumed that 60 percent of these reserves can be recovered, the United States has approximately 200 billion tons of coal reserves that would be available at or near present prices. If it is further assumed that 50 percent of this is controlled by independent producers or public utilities or is otherwise committed, then 100 billion tons should be available for gasification. This amount of coal is equivalent to 1,500 tcf of synthetic pipeline gas—a relatively conservative estimate. Perhaps a more realistic approach would be to take the USGS figure of 1,560 billion tons of coal known to be present as determined by mapping and exploration and assume that 50 percent is recoverable. Since most authorities agree that half of this recoverable coal is minable at 1-1.5 times current costs, this leaves a potential supply of 390 billion tons that would produce about 6,000 tcf of gas. In either case, it appears that the United States has adequate supplies of coal from which to make pipeline gas.

The major problem is finding adequate supplies of minable coal in locations close to gas markets that also have adequate supplies of water available. A coal gasification plant of 250 million scf per day will require in the range of 6,000-12,000 gallons per minute of water for process and cooling needs.

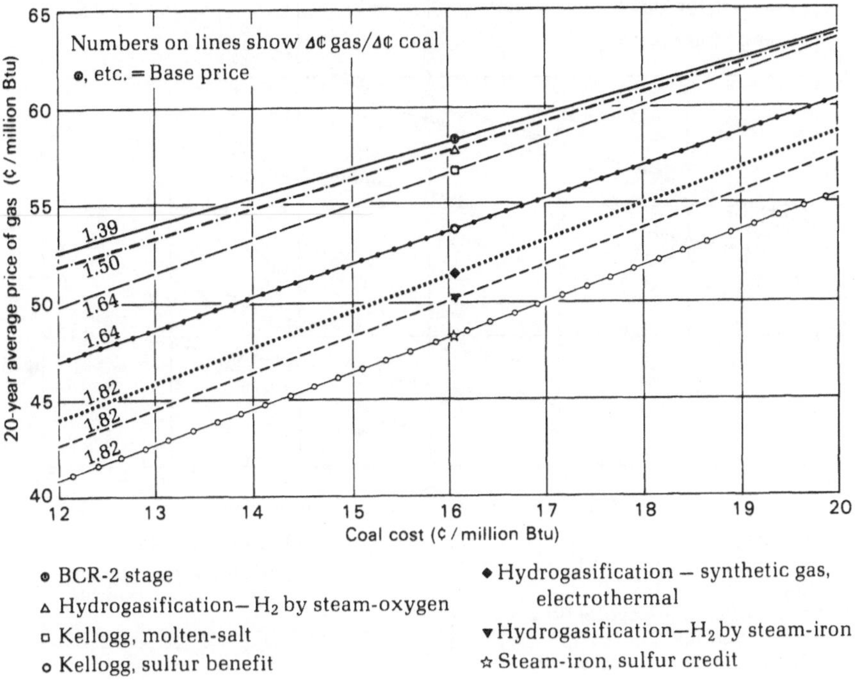

FIGURE 3-10. Sensitivity of gas price to coal cost.
Source: See figure 3-8.

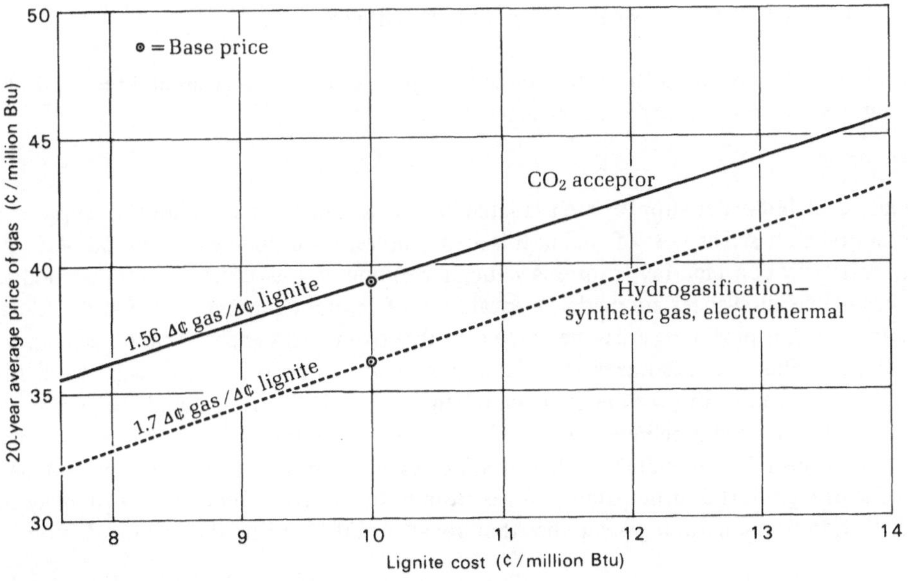

FIGURE 3-11. Sensitivity of gas price to lignite cost.
Source: See figure 3-8.

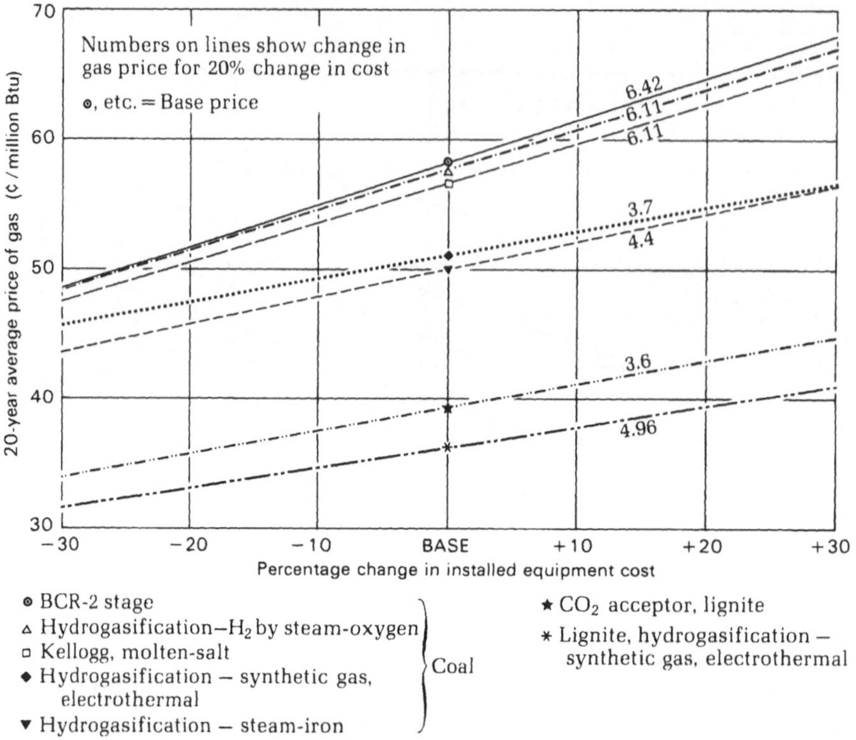

FIGURE 3-12. Sensitivity of gas price to installed equipment cost.
Source: See figure 3-8.

PROJECTED GAS COSTS IN VARIOUS COAL REGIONS

Costs for gas will depend on the nature of coal supplies in various regions and required methods for mining them, as well as on transportation costs.

Southern Illinois

Adequate coal and water supplies are available in central and southern Illinois to support many coal gasification plants. The cost of coal in this area is presently in the neighborhood of $5 per ton. However, costs vary considerably from this value, depending on the particular mining conditions that exist. If reasonable allowances are made for escalation of capital equipment costs and coal costs, it can be estimated that pipeline gas can be made available from coal for about 60-62 cents per Mcf in southern Illinois. This price is somewhat less than the projected costs of pipeline gas from Alaska and is comparable to the price of pipeline gas from northern Canada. Thus, we might expect synthetic gas to compete with imported pipeline gas in the Chicago area, with imported gas having a competitive advantage in northern border states, while synthetic gas would have a competitive advantage south of the Great Lakes region. It is anticipated that gas from both sources will be needed to make up the deficiencies of domestic natural gas supplies that are anticipated during the coming decade.

Appalachia

The supplies of coal in the Appalachian region have been heavily committed to industrial and utility customers. Those supplies as yet uncommitted are fragmented and somewhat more expensive to mine. Thus, it is more difficult to assemble the large blocks of cheap coal that are necessary to establish a coal gasification plant, and it is likely that there will be fewer coal gasification plants built in this region than in the Midwest. The availability of heavy oil along the eastern seaboard may permit the construction of gasification plants using heavy oil as a raw material rather than bituminous coal.

North Dakota

While Appalachian and Illinois coal are generally found in seams ranging from 2 to 8 feet in thickness, buried deeply enough to require underground mining, coal in the western areas of the United States is often found in much thicker beds that permit strip-mining operations. In North Dakota there are very large deposits of lignite that can be easily mined at low costs and converted to pipeline gas at the lowest prices of any process currently under development. If we apply escalation factors to the cost of making gas from lignite similar to those used in estimating southern Illinois gas prices, it should be possible to produce gas for about 45 cents per Mcf in North Dakota. This gas could be transported to the Chicago area for approximately 12 cents per Mcf and would be competitive with synthetic gas from southern Illinois in the Chicago area. It would also be competitive with gas pipelined from Alaska and from northern Canada in the regions northwest of Chicago.

Wyoming

There are very large reserves of subbituminous coal in northeastern Wyoming. This coal lies in very thick beds with thin overburdens. It can easily be stripped and mined for prices ranging from 8 cents to 14 cents per million Btu. On this basis, synthetic gas can be manufactured for about 46 cents per million Btu, which is comparable to the cost of gas from lignite in North Dakota. The distance to large gas markets is only slightly greater than the distance from North Dakota, and we can expect that synthetic gas produced in Wyoming would be competitive with other alternative sources of gas as far east as the Chicago area.

Colorado

There are large deposits of both bituminous and subbituminous coal in western Colorado, southwestern Wyoming, and northeastern Utah. There are currently no markets for these coal reserves and they could be mined and converted to synthetic gas at prices in the range of 58-60 cents per million Btu. The region is quite remote from any large gas markets, however. It could not easily compete with Wyoming coal or North Dakota lignite, but it could conceivably be pipelined to the West Coast, a distance of some 900 miles. The transportation costs would be in the neighborhood of 14 cents, which would make gas available in the San Francisco area for about 74–75 cents per million Btu.

Four Corners Area

Large coal deposits are also located in the Four Corners area and in southern Utah. Coal costs here are about 16 cents per million Btu and could be converted to synthetic gas for about 60 cents per million Btu. Transportation costs to the Los Angeles area might amount to 8 cents, which would bring synthetic gas into the Los Angeles area for about 68 cents per million Btu. This price is in the same range as pipeline gas from Alaska or northern Canada and also with imported LNG from Alaska.

PROJECTED GAS COSTS FROM HEAVY OIL

The growing concern about gas supply combined with the lack of low-cost coal near the eastern seaboard has created interest in oil gasification as an alternative supply source in this region. This interest has been further stimulated by the recent increases in the projected cost of imported LNG. Much of this interest has been spurred too by the hope that commercially available oil gasification processes using European technology could provide supplemental gas supplies at an earlier date than that projected from the development of coal gasification processes.

Residual fuel oil and heavy topped crude are the only low-cost feedstocks available for oil gasification in the United States. Even these are relatively expensive—35-45 cents per million Btu at current market prices compared with coal and lignite prices of 10-20 cents per million Btu. These oils have a high ratio of carbon to hydrogen and are difficult to gasify. At present there are no commercially available processes suitable for large-scale pipeline-quality gas production from relatively low-cost residual fuel oil and heavy crudes. Considerable development work will have to be done. This subject is being investigated at IGT and elsewhere. It is quite possible that oil gasification may prove feasible in special circumstances along the East Coast of the United States. The economics of such ventures will depend heavily on feedstock costs and credits from by-product sales. The development of this technology is in an early stage; therefore, estimates of the cost of gas from residual oil must be regarded as preliminary. However, it may be possible to process residual oil for about 35-40 cents per million Btu exclusive of the cost of the oil itself. Thus, if residual oil at $2.00 per barrel (32 cents per million Btu) is used, the resulting gas would cost about 67-72 cents per million Btu. At $2.50 per barrel (40 cents per million Btu) gas would cost about 75-80 cents per barrel. While residual oil prices are currently increasing, it should be recognized that high-sulfur oils can be used for gasification, and these oils do not command the premium currently being paid for low-sulfur oils.

Light hydrocarbons, such as naphthas, can be used to produce synthetic gas, and processes have been developed in England that could be used in the United States if adequate supplies of these light hydrocarbons could be made available. The cost of processing is much less—about 15-20 cents per million Btu. However, the market value of these light hydrocarbons on the East Coast of the United States is high enough to discourage their use for this purpose.

Even the price of residual oil poses some problems. There are three factors to be concerned about in planning an oil gasification venture:

1. If the price of oil increases, the cost of synthetic gas may increase to the point where it is more economical to gasify coal as far away as Illinois and transport it to the East Coast in conventional pipelines.
2. If the price of oil should fall, then the price of desulfurized oil on the East Coast may fall substantially below the price of natural gas from conventional sources, and some of the markets for energy will tend to convert to oil fuel.
3. Should the price of oil remain stable, an oil gasification project that would provide the same amount of gas as the present import venture of El Paso Natural Gas would require 20 percent of the total imports of residual oil into the United States. Present U.S. production of residual oils is about 6 percent of the total crude production and is forecast to decline by 3 percent by 1980.

Thus, while it appears that oil gasification may be desirable, particularly in the near future, it also appears to be faced with some long-range hazards.

CONCLUSIONS

Perhaps the most striking conclusion to be drawn from this examination of the costs of alternative supplies of gas is that all of them promise to provide gas in major market areas at prices very close to 70 cents per million Btu. Along the eastern seaboard, imported LNG will cost about 65-70 cents per million Btu, while gas from oil will cost this much or more. In addition, some quantities of synthetic gas may be made for slightly under 70 cents in the Appalachian region, but larger quantities may have to come from farther west at higher prices.

In the Chicago area, gas from Alaska will also cost in the neighborhood of 70 cents per million Btu. Synthetic gas from southern Illinois, from North Dakota, and from Wyoming may all compete with the Alaskan gas at prices ranging from 65-70 cents in the North Central region. Canadian gas imported from the lower provinces of Canada will have a price advantage but if commingled with Alaskan gas will serve to reduce the overall average price of imported gas to something less than 70 cents per million Btu.

On the West Coast, imported LNG and Alaskan pipeline gas are estimated to arrive in the Los Angeles area at about the same price, again 70 cents per million Btu or slightly higher. Synthetic gas from the Four Corners area may be piped into the Los Angeles area for a few cents less. Farther north along the West Coast, Canadian pipeline gas gains a decided advantage over both imported LNG and synthetic pipeline gas from Colorado or Utah.

A second significant conclusion is that nowhere in the United States can any alternate source of gas compete with domestic supplies of natural gas on a price basis. Thus, all of our alternate gas supply sources can augment but cannot reduce the demand for conventional domestic gas supplies. This situation will remain true even if the price of domestic gas at the wellhead is increased substantially. The only exception to this is the North Central region, including the states of North and South Dakota, Minnesota, and Wisconsin, where Canadian gas imported from Alberta and possibly synthetic gas made from lignite can begin to displace conventional domestic gas supplies. This means that natural gas producers need not fear the development of alternative gas sources as a threat to the market for existing or future supplies of domestic natural gas.

Both imported LNG and pipeline gas from Canada and Alaska are limited in their ability to fill the projected deficiencies of gas supplies. While synthetic gas from coal has the potential of filling the gap, it will be many years before this potential can be realized. Thus, we will be heavily dependent on our conventional natural gas sources for many years to come.

Each of the alternative gas supply methods involves large capital investments. A comparison of the capital requirements necessary to provide 1 tcf of gas annually from each of these alternative sources follows:

Imported LNG–$2.7 billion
Alaskan pipeline–$2.3 billion
Coal gasification plant–$2.2 billion (including mine)

It should also be recognized that the cost of finding 20 tcf of natural gas and setting up to produce 1 tcf annually would require comparable amounts of capital.

4 PRODUCER REGULATION: A COMMISSIONER'S VIEWPOINT

*Charles R. Ross**

To be asked to Monday quarterback oneself is to be asked to perform a most difficult task. But the assignment becomes a challenging one when viewed as an opportunity to perhaps help others in the Federal Power Commission (FPC) to avoid some of the errors made in the past.

As one who served on the commission from September 1961 to September 1968, I will try to describe the scene as the new commission under the Kennedy administration took over from that under the Eisenhower administration. Deteriorating public confidence in the FPC demanded action. In this connection, it is useful to summarize the goals of the commission and the remedies it sought to put into effect in order to reduce to manageable proportions the problem of establishing just and reasonable rates for the production of natural gas and at the same time to provide the nation with an adequate supply of natural gas for interstate commerce. I will then analyze whatever successes and failures resulted from FPC actions and will offer recommendations for the future. To put it more bluntly: Is the interstate market threatened with a shortage of gas; if so, what caused it, and will higher prices solve the problem?

My statements obviously are based on my own personal experience. Although the Kennedy and Johnson commissions were spoken of collectively, FPC policy during those years was usually the result of an adversary hearing involving the applicant, the staff, and other interested parties, followed by an initial decision supervised by an examiner for review by the commissioners.

From the last half of 1961 through 1965, the commission was chaired by Joseph C. Swidler of Tennessee, who is now the chairman of the New York Public Service Commission. Chairman Swidler was succeeded by Lee C. White of Nebraska in the spring of 1966. White served until the summer of 1969, when John N. Nassikas of New Hampshire was named chairman by President Nixon.

Howard Morgan of Oregon served until the summer of 1963, and he was replaced by David Black of Washington, who served until the summer of 1967 when he resigned to head the Bonneville Power Administration. He was replaced by John A. Carver, Jr., of Idaho, who is still on the commission. Lawrence O'Connor of Texas was appointed in the summer of 1961 and continues to serve. Jerome Kuykendall, the chairman during the Eisenhower years, served through 1961, when he was succeeded by Harold Woodward of Illinois. Woodward served until his death in the summer of 1964. He was succeeded by Carl E. Bagge, also of Illinois, who was appointed in the spring of 1965 and who also continues to serve. My seat on the FPC was taken over by Albert B. Brooke, Jr., of Kentucky, who was appointed by President Johnson and reappointed by President Nixon.

The names and numbers of the players are important, for policy and responsibility ultimately rest on the members of the commission. Each member is an individual, and the final policy is determined in many cases as a result of the interactions of these individuals. Each FPC member naturally has his own idea of what is in the public interest and is sensitive about his contribution, but above all he is concerned that the policy pursued by the FPC is in the nation's interest.

*Former commissioner, Federal Power Commission.

It is obvious that my account of these important years will be colored to some extent, even though I have tried to avoid bias. To minimize the danger of coloration, I have relied on documented sources as much as possible.

REGULATION OF NATURAL GAS PRODUCERS–BACKGROUND

For those who may have forgotten, the regulatory scene in Washington in 1960 and 1961 was not one that generated a great deal of confidence in the administrative process. To those of us who were being introduced to federal regulation, it seemed that confidence could not sink much lower. In any event, as a former chairman of the Vermont Public Service Board, I was generally aware of the depths to which the FPC had allegedly sunk. Soon after the election of John F. Kennedy, it was obvious that changes would be made in the regulatory agencies. It was no secret by that time that the FPC was in trouble. The Supreme Court had reversed the commission quite forcefully, and the regulation of producers was a festering sore.

The scope of the problem was not confined to the industry nor to the followers of regulation. In fact, efforts to provide relief through the legislative route had failed twice in Congress. President Truman vetoed one attempt to exempt producers from regulation, and President Eisenhower vetoed the second attempt when it was disclosed that the industry had sought to influence the vote of Senator Francis Case of South Dakota.

That the regulation of producers was a controversial issue is borne out by a reference to the Natural Gas Investigation of 1948 by the FPC (Docket no. G-580). This investigation resulted in two reports, one by Commissioners Leland Olds and Claude Draper and another by Commissioners Nelson Smith and Harrington Wimberly. Two of the recommendations in the Olds-Draper report are significant even today:

IV. The Federal Power Commission should be provided with the necessary funds and should promptly undertake a thorough and continuing survey of the country's natural gas reserves.
V. The Federal Power Commission should watch the field price situation to determine when the trend toward concentration of control of gas reserves reaches a point where effective competition is destroyed and eventually monopoly prices are substituted for competitive prices.–Should such a situation develop there would appear to be no recourse but Federal regulation of all sales of gas in interstate commerce.[1]

The Smith-Wimberly report, on the other hand, suggested an approach that is currently receiving some attention:

If the Federal Power Commission were to be authorized to fix the prices of arm's length sales of natural gas by producers and gatherers to interstate pipeline companies, the result would be to establish Federal authority over a substantial part of all field and well prices for natural gas. Control of oil as well as gas would become involved, since their production is to a large extent interrelated. Fixing the prices on sales of natural gas by producers and gatherers would also impinge upon the functions of the States relative to oil and gas conservation and the protection of correlative property rights of producers.

The Commission's rate-making authority, by which any improper payments for gas by an interstate pipe line can be denied, does not require price-fixing powers or jurisdiction over sellers of gas in the producing fields. There are, moreover, established Federal statutes and agencies to deal with any possible dominance or monopoly of gas supplies in the fields, just as there are with respect to oil, coal, steel, or any other product. It would, of course, continue to be the duty of the Commission to recommend antitrust proceedings against any monopolistic price-fixing or similar practice detrimental to the public interest.[2]

[1] Leland Olds and Claude L. Draper, *Natural Gas Investigation Report*, FPC, Docket no. G-580 (1948), p. 13.

[2] Nelson Lee Smith and Harrington Wimberly, *Natural Gas Investigation Report*, FPC, Docket no. G-580 (1948), p. 8. Compare the language in the statement about the commission's rate-making authority with that of recently proposed Rule R-389 instituting investigation in the Permian Basin.

Smith and Wimberly further discussed the controversial matter of direct sales and end-use control. In fact, they suggested that pipeline companies ought to forgo the opportunity to sell dump gas and should concentrate on selective development. For example, they stated:

A general shift of emphasis from interruptible to firm service is necessary if the full usefulness and premium value of gas for industrial purposes is to be realized. With this upgrading and repricing of industrial markets for gas, high system load factors will be less important to the successful operation of pipelines. The opportunities for developing year-round industrial loads appear to be almost unlimited.

Upgrading industrial markets for gas also is required in order to compensate for rapidly mounting demands for seasonal space heating and to assure that the expanding requirements of domestic customers can be supplied at reasonable rates.[3]

In light of this background, it is really not surprising that the issue of regulation of producers reached the Supreme Court.[4] While it may be surprising that the Court came to the conclusion that the Natural Gas Act required the FPC to assume jurisdiction over the regulation of wellhead prices, it was not unexpected that the FPC should end up devoting a considerable amount of time to finding a solution.

A significant statement by the Court should not go unnoticed, since it dominated the briefs and the oral arguments in succeeding years: "Protection of consumers against exploitation at the hands of natural gas companies was the primary aim of the Natural Gas Act. . . . Attempts to weaken the protection by amendatory legislation exempting independent natural gas producers from federal regulation have repeatedly failed and we refuse to achieve the same result by a strained interpretation of the existing statutory language."[5]

During the period following the Phillips Petroleum Company case, the Eisenhower commission attempted to resolve the problem, although the commissioners continued to advise Congress to exempt producers. It was inevitable that cases would pile up and that dockets would become clogged. Some members of the staff suspected that industry was not particularly anxious to help resolve the procedural morass.

In the meantime, the commission was being reversed in the courts as a result of appeals by the New York Public Service Commission and the Eastern Intervenors, a group of eastern distribution companies that were disturbed over the rising prices and attempts to evade regulation.

On September 28, 1960, the commission finally issued its decision in the Phillips case.[6] Originally started on September 28, 1948, the case really is not concluded yet. Nonetheless, the FPC decision, even though it was not definitive, had a profound effect on regulation, for it highlighted the procedural morass and gave headline hunters some wonderful copy. It rejected the concept of individual company cost-of-service and provided a useful summary of the producers' arguments—why regulation was unworkable generally and why cost-of-service, with its problems of allocation, specifically would haunt the commission.

In an attempt to change the course of regulation, the commission issued the now famous policy statement of September 28, 1960, which played an important role during the interval between that date and the final issuance of the first area rate case on August 5, 1965. The policy statement provided a new concept that was to be developed in later years, namely the two-price system, though at that time it was believed that eventually the two-price system could be abandoned. The policy statement and this unilateral approach had several drawbacks that bothered me and apparently bothered the distribution companies as well. The commission was sensitive to the issue, for it made clear in its policy statement that the price levels "are for the purpose of guidance and initial action by the Com-

[3] Ibid., p. 13.
[4] *Phillips Petroleum Co. v. Wisconsin*, 374 U.S. 672 (1954).
[5] Ibid.
[6] FPC, *Reports*, vol. 24 (1964), p. 537.

mission and their use will not deprive any party of substantive rights or fix the ultimate justness and reasonableness of any rate level."

THE KENNEDY COMMISSION

Thus, the stage was set for another attempt by a new lineup. Under Kennedy, the FPC had its job cut out for it. All members were anxious to make producer regulation work. Unlike the members of the Eisenhower commission, we thought we could do it. It is true that, with the exception of Commissioner O'Connor, none of us had special expertise in the oil and gas industry, although Commissioner Woodward had had some general regulatory experience with the Illinois Commission, and Commissioner Morgan and I had both served on state commissions.

The challenge was ours, and I can still remember talking with Chairman Swidler, after my appointment had been announced, about the difficulties we could see ahead. Actually, some of us thought we could not do much worse, and we were determined that some action was better than the indecisiveness we had perceived in previous years.

I suppose that Chairman Swidler, because of his long and distinguished service with TVA, and Morgan and I, because of our belief in the necessity of virgorous regulation in order to protect the interests of the public, were influenced by the often repeated quote from the CATCO case that "*the Act was so framed to afford consumers a complete, permanent and effective bond of protection from excessive rates and charges*."[7] Commissioner O'Connor, a former accountant for an oil and gas firm and thus from a far different background than the three of us, nevertheless was motivated by a strong desire to make the commission work. His knowledge of the industry proved to be extremely helpful during the years I was on the commission and, while he approached the job somewhat differently than the three of us, he was a good balance wheel. In fact, at times I felt that he may have leaned too far away from identifying with the producers because of his background. In any event, during the commission meetings, when policy was debated and controversial issues were resolved (which was done in front of all key staff personnel with the exception of those involved in the trial of a case, unlike the procedure of our predecessors), Commissioner O'Connor was an effective force and attempted to keep us on the right track.

Commissioner Woodward, on his part, seemed more often to reflect the standard views advanced by the members of the Eisenhower commission. Generally, he was likely to be on the other side of Swidler, Morgan (to be replaced by Black), and Ross. While Woodward believed in regulation, I did not have the impression that he was quite as concerned with the interests of the consumer as we were. He may have felt that the three of us were moving too fast and too far.

PROBLEMS FACED BY THE KENNEDY AND JOHNSON COMMISSIONS

There is no question the commission as a whole felt that the future of regulation itself was at stake. The Eisenhower commission's famous statement, "Thus, if our present staff were immediately tripled and if all new employees would be as competent as those we now have, we would not reach a current status in our independent producer rate work until 2043 A.D.—eighty-two and one-half years from now,"[8] acted as a stimulant to the new commission, particularly since the statement was quoted with great delight by James Landis in his well-known study of regulation prepared for President Kennedy. Even more widely quoted, the Landis report placed the commission in a "put up or shut up" position. Landis hit the nail on the head in many respects; more important, he put the spotlight of publicity on

[7] *Atlantic Refining Co. v. Public Service Commission*, 360 U.S. 378 (1959).

[8] 24 FPC 546 (1960).

the deterioration of the administrative process. To have someone of his experience say, "The Federal Power Commission without question represents the outstanding example in the federal government of the breakdown of the administrative process,"[9] was to parallel the effects of some of the reports by Nader's Raiders.

Chairman Swidler's prompt response may have frozen the commission to some extent, since shortly after taking office Swidler indicated in a speech to the gas industry that he intended to give area pricing a whirl. Area pricing was not his only concern, nor the commission's. All of us knew that in order to make regulation work we had to deal with the problem of indefinite escalation of prices through the device of spiral escalation or favored nations, or what have you. We also had to face up to attempts by the industry to circumvent regulation itself. The most blatant were the attempts to dispose of holdings by "sales in place"–the Bastian Bay, Ship Shoal, and Rayne Field cases.

To some of us, it seemed that the constant repetitive statements about the producer backlog indicated that the industry felt it could swamp us by numbers. The staff was realigned in an attempt to deal with the procedural problem. An Office of Economics was established, a new head of the Bureau of Natural Gas was appointed, and the commission girded itself for action. The numbers game was on.

In addition to the problems already mentioned, the commission was faced with old and controversial problems of pipeline production. An indication of the complexity can be discerned from the Southern Natural case (28 FPC 323 [1962]), in which I wrote the opinion for the majority and filed a separate supplemental statement. Commissioner Morgan concurred in the result and filed a separate statement, as did Commissioner O'Connor, and Commissioner Woodward concurred in part and dissented in part. This was the commission's second attempt to deal with the recurring issue of incentive to the industry. It had surfaced in the El Paso Natural Gas Company case (28 FPC 688 [1962]), which Chairman Swidler authored, and all of us were aware that these cases would reflect our ideas on producer regulation generally.

Among themselves and in trade journals, members of the industry were arguing that a decision was needed and that without it they would abandon any further efforts to stay in the production business. The industry, though, was badly split, since El Paso, for example, wanted to be regulated strictly on a cost-of-service basis, and most of the other pipelines wanted fair market prices or at least an area price. Personally, I believe that before any regulated firms receive an incentive with the blessings of the regulator, it should be established that the benefit of such a contribution would be dedicated to the consumer. For some reason, why, I am not sure, we were never able to resolve the issue by the time I left. The principal reason, probably, was that almost everyone thought that if producer regulation could be solved, pipeline production would fall into place. It did serve, however, as an introduction to the Permian case.

Another skirmish, which built up during the early 1960s and turned into a battle royal, was the matter of accelerated depreciation and compulsory flow-through. There is no point in discussing those decisions in detail, but it was obvious that the new commission was taking a far different tack from that of its predecessor. The Cities Service case (30 FPC 158 [1963]) was another involving taxes and was viewed as a critical one to the industry. This case (in which the commission was subsequently affirmed), like the case involving the United Gas Pipe Line Company (32 FPC 687 [1964]), dealt with the problem of consolidated returns and the allocation of income tax savings. It was obvious to all of us that we had incurred the wrath of the industry, since a rash of bills was introduced in Congress to rectify such "stupidity." The cases are important, not only because to the majority of the commissioners they represented fair treatment for the consuming public, but also because they served to draw together the members of the majority against the attack of the industry and may have increased our concern for the consumer.

[9] James M. Landis, *Report on Regulatory Agencies to the President-Elect*, submitted by the chairman of the Subcommittee on Administrative Practice and Procedure to the Senate Committee on the Judiciary, 86 Cong. 2 sess. (1960), p. 54.

In my opinion, throughout this period the industry suffered badly because of its misguided attempts to force changes at the commission. The following actions all served to cause the commission to question the industry's credibility: constant blatant attempts to pull power plays in Congress; repetitive instances of crying wolf; unwillingness to disclose complete and accurate industry data; failure to make the Natural Gas Advisory Council as effective an advisory voice as the Electric Advisory Council; the holding back of the pipeline companies from seeking to contest the higher wellhead rates and their threat to pull out of pipeline production if they did not get what they wanted; the delight of the industry at the rough handling that Chairman Swidler received at the hand of Albert Thomas, chairman of the House Appropriations Subcommittee on regulatory agencies; obvious efforts to influence President Johnson to dump me, as well as attempts to name the industry's own people to the commission; and the opposition to Canadian imports as a supplement to our reserves.

It may very well be true, as Professor James McKie has said, that a regulatory commission tends to extend its regulatory boundary externally as well as internally—what he describes as the "tar baby effect."[10] Actually, with new aggressive and activist-oriented regulators, this is almost inevitable if a commission is thought to have passed through the mature stage of its life cycle, or if the public has lost confidence in the agency. The so-called tar baby effect, however, has a counterpart in the industry. The American tradition of free enterprise has been that the less government interference the better. Many Americans have grown up to become important businessmen under the philosophy that the government is fair game; that business is simply a contest between the businessmen and the government, that one should get away with all one can, and that it is only a crime if one gets caught. Under these rules, the rewards go to the clever ones who can avoid regulation. Businessmen hire expensive counsel to warn them of the danger areas and to run interference for them. It is not a question of making regulation work but of how to circumvent it openly, if possible. If it cannot be done that way, then industry plays the trump card by collecting on its campaign contributions.

In a developing struggle of this kind, in which the ultimate stake was the type of regulation for producers and/or the necessity for regulation, it is not surprising that sides were formed. After all, FPC administrative procedure is essentially adversary. Throughout the period and even before, the Eastern Seaboard Intervenors, the New York Public Service Commission (NYPSC), the California Commission, and the Wisconsin Commission constituted a formidable combination, since together they represented not only the voice of regulation (and federal regulators do listen to and are sensitive to the views of their state compatriots) but also the industry. The NYPSC and the distribution industry, both represented by competent counsel, were a good match for the producers. The conflict had become institutionalized. I frankly enjoyed it, since, if there was to be a confrontation, I wanted to be sure the sides were as equal as possible. It was helpful to realize that in addition to staff there were well-financed parties seeking to keep each other honest. Unfortunately, I am afraid the lawyers were the only ones to win. I frequently sensed the record was developed more for the purpose of achieving some small victory over opposing counsel than for assisting the commission.

GUIDELINES FOR IN-LINE PRICES

The battle line was clearly drawn on the matter of what to do until the commission could get the Permian case decided. For several years, the "Battle of Interim or In-Line Prices" occupied most of our attention. Here I departed from my colleagues, sometimes alone, often with Morgan or Black. During this interim period, one of my concerns, which also troubled other members of the commission and some of the intervenors, was the impact such interim prices might have in subsequent area rate proceedings. Despite allegations to the contrary by the industry and its trade periodicals, all of us

[10] James W. McKie, "Regulation and the Free Market: The Problem of Boundaries," *Bell Journal of Economics and Management Science*, vol. 1, no. 1 (Spring 1970).

were acutely aware that any final price could not substantially depart from a recognized interim level that might be established. The commission obviously could not set interim prices at levels substantially below those the industry had been relying on.

In this context, the original policy statement issued on September 28, 1960, became critical, because the prices that were established unilaterally by the commission were generally the highest contract prices then on file. Since, during the early 1960s, there was no severe gas shortage scare, the idea of allowing market price to work itself into the regulatory scheme without more detailed comment or debate was subject to criticism by some of us. Talks with the staff about the method of arriving at the "guideline prices" were not very reassuring, since no serious effort was undertaken to determine the appropriateness of such guideline prices.

AREA PRICING

During this period when we were attempting to hold the line, the commission gradually began to wrestle with the actual problem of area rates. Following the examiners' initial decision in the Permian case, the fat was in the fire. On the one hand, a final decision was needed in the Permian case that would include specific guidelines on area rates. On the other hand, other area rate cases were calling for decision, too. The commissioners discussed the possibility of proceeding with the settlement of the Southern Louisiana case while Permian was still under appeal, but finally decided not to because of the danger of complicating the considerations involved in the Permian case.

Then a third proceeding, the Anadarko-Hugoton case (Docket no. 64-1), was instituted. A majority of the commissioners involved in this case decided to proceed with it—over my objection. I was opposed to their decision because I had become convinced that area pricing had to be handled on a nationwide basis.

I was concerned, too, because we failed to insist on tax and reserve data, despite the decision to use a nationwide questionnaire. I said then and I repeat today, "Adequate reserve data are prerequisite in determining price. . . . How is it possible to know if price is properly performing its function, namely, assuring an adequate supply of gas, if the volume of gas found in response to known prices remains a mystery?" (Docket no. 64-1, November 27, 1963). The majority thought that my proposal would unreasonably delay the proceeding and would be administratively unfeasible, and that there would be no difficulty in developing adequate records in the area proceedings.

Finally, in August 1965 the commission announced its decision regarding the area rates for natural gas in the Permian Basin. In this opinion, we adopted a two-price system, basing both prices essentially on costs. This approach had two prime purposes. First, we were impressed by the evidence that the industry could reasonably predict the results of its drilling, that is, whether gas or oil would be found. We hoped that by specifying a higher price for new gas we would be providing an incentive to the industry to continue to expand its supply. In addition to establishing a two-price system, we also stipulated a subsequent review for determining the adequacy of the price.[11]

The Permian case once again occupies center stage. The commission instituted a further proceeding to review the Permian rate (Order 389, June 17, 1970). Subsequently, the commission gave notice of an enlarged investigation and proposed rulemaking, expanding the scope nationwide. In the latter proceeding, the purpose was "to determine the terms and conditions which will result in an adequate supply of natural gas for consumers at the lowest rate consistent with maintaining an industry structure capable of providing and motivated to provide service with its attendant risks.[12]

The commission further stated that it desired to have the parties: (1) submit nationwide cost data, based on a 12 percent rate of return; (2) respond to questions of rate of return as discussed in prior

[11] For a more detailed description of this case and of the subsequent Southern Louisiana case, see the excellent description by Ralph Spritzer in chap. 5.

[12] FPC, Docket no. R-389A, July 21, 1970.

court decisions; (3) respond to the question of what weight should be given to gas contract prices, terms, and inducements or commodity value; and (4) respond to the question of whether the market mechanism, *in the light of pipeline regulation*, will adequately protect consumer interests, by providing data on relevant intrastate sales.

It is a little disheartening to realize that the same old story seems about to be retold. Again we are faced with a piecemeal approach. The commission now has so many balls in the air, it is hard to tell what is going to happen.

The public is confronted with speech after speech by the commissioners. Producer regulation is described as outmoded by Commissioner Bagge,[13] Commissioner Carver said that is not quite so,[14] and Chairman Nassikas said, "It has been suggested that the answer to regulatory problems is to deregulate producer pricing. I do not agree."[15]

The commission has proposed a course of action in Illinois and Appalachia,[16] but until recently it has been unable to agree on the proper procedure following remand of the Southern Louisiana case. The court was not very helpful, either, and it has now decided to try to settle the Anadarko-Hugoton case over the objections of the intervenors. On November 6, 1970, the FPC staff and most of the parties agreed to another settlement in the Southern Louisiana case. In the meantime, according to the trade press and even according to those well-known oracles of industry, the *Wall Street Journal* and *Barron's*, the nation is about to run out of gas because of men like Swidler, Morgan, Black, Ross, and others who threw a monkey wrench into the system.

Finally, best of all, the academic world and the foundations are speculating about what went wrong and why. Even sister bureaucratic agencies, such as the Office of Science and Technology (witness this series of papers for RFF), and the secretary of the interior are jumping into the scene. A national energy policy, which has been outlawed by the petroleum industry because of a fear that the policy might be dominated by the coal industry, is now the "in thing." Whether it is because coal and oil are in the same bed is unknown, but in any event the push is on. The Federal Power Commission in the meantime, after finally agreeing to push for a study, is not sure how to handle it now that Congress is willing. Should the FPC proceed with the study and hold up the decisions, or should it issue its decisions and make sure the study substantiates them?

THE PROBLEM OF INCOMPLETE AND INSUFFICIENT DATA

It is hoped that out of all this something constructive will develop. In my opinion, the commission first of all needs accurate information about reserves on a field-by-field basis so that the response to price can be determined. For too many years the commission's plans to broaden its informational base have been prohibited by statute.[17] Also, too often the commission has failed to act affirmatively by using its investigatory powers to seek additional information. How can the nation have confidence in any decision when the industry has all the information? It is like playing cards when your opponent has the deck stacked. Even the figures obtained from the new, improved reporting set-up under the auspices of the Gas Industry Committee leaves a lot to be desired. Today, as in the past, there are some who express concern about whether the alleged shortage of gas is as serious as the public is led to believe.[18] Chairman Nassikas has mentioned the problem of the unreliability of data.

[13] Carl E. Bagge, "Broadening the Supply Base: A Proposal to Eliminate Producer Price Regulation" (address before the Midwest Gas Association, Colorado Springs, February 24, 1970).

[14] John A. Carver, Jr. (remarks before the Tennessee Gas Association, Nashville, May 15, 1970).

[15] John N. Nassikas, "An Outline of National Energy Policy: Some Personal Reflections" (address for the FPC 50th Anniversary Program, Washington, D.C., June 3, 1970).

[16] FPC, Order no. 435, Docket nos. R-389 and R-389A, July 15, 1971, p. 8.

[17] See FPC annual reports, 1962 to present.

[18] Bruce C. Netschert, "The Alaskan North Slope and East Coast Energy Markets" (remarks at the National Convention of National Oil Jobbers Council, Boston, May 12, 1970, mimeo.).

It is difficult to understand why the federal government failed to insist on full disclosure of all available information about U.S. offshore reserves. For a number of years the Outer Continental Shelf Lands Act of August 7, 1953, was almost an act in name only. All of us on the commission were well aware of the tremendous potential of these lands. In fact, I can still remember the excitement that pervaded the bid opening in the spring sale of 1962, which Commissioner O'Connor and I attended. We listened as the prospects of these lands were explored. We read the glowing accounts of the prospects in the *Oil and Gas Journal*. We were hounded in those early years because there were insufficient pipelines to tap the expanding discoveries. Repeatedly we were told of large volumes of gas.

In fact, in 1963 the commission went so far as to drastically modify its procedures. Rather than get involved in worrying about whether there were specific reserves available to comply with the contractual requirements, the commission magnanimously agreed to accept the seller's warranty.[19] At that time, it was considered unthinkable for anyone to doubt a large oil company's ability to find and deliver the gas. (This is interesting, because it hardly reflects the viewpoint of the industry about five years later.) No one, including members of the industry, seemed particularly worried about a shortage then. The problem was too much gas, not too little. We were worried about take or pays (when the buyer contracts or pays for an annual volume of gas whether it is required or not, usually reserving the right to take the paid-for gas in succeeding years). Transcontinental Gas Pipe Line Corporation was a good example, and the new pipeline, Transwestern, which our predecessors set up as competition to El Paso, was trying to get out from under some contracts in the Permian area.

Although we tried hard, our staff was never successful in getting the industry to cooperate with us when we wanted more information. In addition to our attempts to convince Congress of our need for legislation,[20] we suggested that the pipeline operators divulge their long-range plans, but here again we met a stone wall. Staff members were castigated for even thinking about it. Regulators who wanted such information were obviously tar babying, since they would use it to blueprint an industry for the nation. The commissioners could not convince the industry that, at least for the pipelines, it made good sense to think that far ahead. We had already experienced problems of underestimating demand by the power industry and its problems in providing sufficient reserve capacity.

The commission, partly because of its decision to rely on warranties, decided to institute an annual reporting system of gas supply of interstate pipelines, effective July 1, 1964. At the same time, the commission, again showing its reliance on industry's statement that supply was no particular problem, issued "a statement of policy indicating that the previous requirement of a twelve-year deliverability life would not necessarily be applied for established pipelines which have active gas procurement problems and with pipeline extending into production areas in which exploration is actively continuing."[21]

The problem of adequate data haunted us in other areas. We were unable to come up with a decision in the so-called Red Snapper (Tennessee Gas Pipe Line Co.) case,[22] and I, as the author of the opinion, pleaded with industry to try again. We wanted a new opportunity to propose a rational, logical, and orderly scheme for developing the most economical offshore pipeline system. It was clearly impossible for us to certify a pipeline when we did not know what it intended to tap. We desperately wanted the hit-and-miss approach of plastering the bottom of the Gulf of Mexico with a

[19] See the *Annual Report* of the FPC for the fiscal year ending June 30, 1963, in which the commission stated (p. 113): "A number of unusual new developments occurred during 1963 in connection with purchase arrangements for several large 'blocks' of gas reserves. A major producer, in agreeing to sell more than 4 trillion c.f. of natural gas to an interstate pipeline buyer, made no dedication of specific reserves or acreage to the sale as usually occurs in a gas purchase agreement. Instead, the seller extended its corporate warranty to provide the total volumes specified in the contract, deliverable to the buyer, at the same point of delivery and price regardless of sources of production. This 'utility-type' approach by a major producer represents a guarantee which extends beyond the usual arrangement whereby a producer dedicates only such gas reserves as may be produced from specific fields or acreage over the life of the productive properties."

[20] See the most recent legislative attempt, S. 3900, introduced by Senator Lee Metcalfe, June 2, 1970.

[21] See FPC, *Annual Report* (1964), p. 110.

[22] Opinion no. 528, 38 FPC 691 (1967).

variety of small individual lines to be terminated. Again, despite an innovative computer approach, industry won out, and we are left with a pipe dream unless the Interior Department shows a strong heart.

COMPETITIVENESS OF THE INDUSTRY

The problem of securing facts, like many of the other problems confronting the commission, bears on one of the most fundamental issues involved in using market price as a norm for pricing natural gas. Underlying the whole concept of regulation of producers is the issue of whether producers are typically competitive—whether resulting market prices would be the product of bargaining by hotly competitive sellers and equally hotly competitive buyers. Is it possible to have arm's-length bargaining?

In the Permian case, we said:

> The pipelines are far from being in a bargaining position which would validate the theory that a market price will necessarily be a just and reasonable price. On the contrary, the entire history of pipeline purchasing activity, since the end of the El Paso monopoly in the Permian Basin, has been characterized by the overriding needs of the pipelines to contract for the large blocks of uncommitted reserves essential to maintain their competitive position in developing markets (see *United States v. El Paso Natural Gas Co.*, 376 U.S. 651 [1964]) and their inability to accomplish this objective except at ever increasing prices. The abstention of the pipelines from these proceedings does not indicate that any significant change can be expected in the future. The fact that many pipelines are themselves producers with an interest in generally higher producer prices also cannot be overlooked.[23]

Several years later we reaffirmed this conclusion in a different area, the Southern Louisiana case, and stated:

> We considered this problem in Permian, and concluded that the pipelines purchasing gas in the Permian Basin were so disadvantaged in their bargaining position as to invalidate the concept of regulation by contract price. We reach the same conclusion here. This record discloses that eight independent producers in Southern Louisiana accounted for over 50 percent of 1960 sales under FPC rate schedules (interstate sales). The record here does not disclose any data showing the amount of uncommitted reserves controlled by any producer or group of producers, but the pattern of sales commitments and annual deliveries indicate that a small number of sellers control a large percentage of the gas reserves in Southern Louisiana. We point out that while the necessary information concerning uncommitted reserves is exclusively in the possession of the producers themselves, they have made no showing to rebut a conclusion that there is a concentration of control over the uncommitted reserves in Southern Louisiana.
>
> The large interstate pipeline purchasers in Southern Louisiana periodically require the long-term commitment of large blocks of reserves, as we have pointed out. At any one time the latitude available to the pipelines as to reserve sources or purchasing timing is very circumscribed. Only a few producers may have the necessary volumes available for initial commitment at the moment. Moreover, pipelines pass their gas purchase outlays on to distributors, and ultimately to consumers, through the pipelines' regulated, cost-based, resale rates, which is a factor tending to dilute their bargaining motivations. This entire situation enhances the bargaining power of the sellers and often tends to make price a second level consideration to the purchaser.[24]

I am well aware that the conclusions set forth in these two opinions are hotly debated by most economists. Traditional statistics of the measure of concentration of control in manufacturing industries lead one to believe that the producing industry is highly competitive; the large number of companies competing for leases and the large numbers of producers who have natural gas for sale are reassuring. However, I cannot help but be influenced by the industry's pattern of behavior and by the results I observed while on the commission.

I strongly feel that the oil and gas producing industry is in a class by itself and should not be compared with any other industry in determining whether or not there is competition. Further, it is my

[23]Opinion 468, 34 FPC 159 (1965), p. 22.

[24]Opinion 546, FPC Docket no. AR 61-2, September 25, 1968, p. 18.

opinion that this represents an excellent case of the producers' parallel price behavior, which reflects their power. Why else has the producing industry been able to dictate such unusual concessions for the industry as a whole, such as favored nation clauses, spiral escalation clauses, and indefinite escalation clauses? Also, when one stops to realize that most of the independents must rely on the generosity of the majors in the distribution of farmouts and that many of the majors are themselves economically intertwined as joint operators in different areas, it then becomes harder to accept the conclusion that this is a typically competitive industry.

I confess also that I was impressed by the monolithic behavior of the industry in its fight for higher prices, the unwillingness of the pipeline industry to fight back, and the willingness of the bulk of the industry to let a few companies act as its spokesmen in legislative, regulatory, and industry councils. Were the purchasers in this environment better able and more willing to bargain effectively for natural gas, I might be more inclined to rely on the results of such bargaining to protect the public. That is not the case, however, and I am not anxious to substitute the commission's judgment for what it thinks a fair price to the producers would be unless the price granted has a cost-oriented base. Otherwise, market price will simply mean what the traffic will bear, and the traffic has no alternative. On balance, then, considering the nature and the practice of the producing industry and the pipeline industry operation (the latter is on a cost-of-service basis) and the parallel types of behavior within each branch, I conclude there is not yet sufficient evidence of competitive behavior by those companies producing substantial amounts of natural gas to warrant a conclusion that the nation should rely on such prices without further regulation.

EFFECT OF AIR POLLUTION ON DEMAND

The demand for gas is said to be growing by leaps and bounds. The latest report by the Future Requirements Committee[25] projects an annual compound growth rate of 6.3 percent for the three-year period 1968-71. Over a twenty-year period, 1970-90, the committee projects an annual compound growth rate of 3.6 percent with commercial and industrial uses increasing more rapidly than residential.

Behind these cold statistics is a grim story of a nation squandering and plundering one of its most vital and precious resources–its atmosphere. Uncontrolled avariciousness, which Adam Smith wanted to rely on in order to protect consumers, has worked to the consumers' disadvantage in this instance. What better dumping ground for pollutants than the very atmosphere that surrounds us? It has the unique capability of not being subject to the law of capture, and since the atmosphere has been hard to reduce to a property right, it is available for free use. As soon as the public became aware of the seriousness of the deterioration of the atmosphere, the natural gas industry was called on for relief.

The famous Gulf Pacific case[26] is an example. Here, the Gulf Pacific Pipe Line Company, a subsidiary of Tennessee Gas Transmission Company, sought permission to transmit natural gas from a group of producers in Texas who were making a direct field sale of a large block of gas for use as boiler fuel by Southern California Edison and the City of Los Angeles, who proposed to use this gas to fuel their electric generating plants. El Paso Natural Gas and Transwestern Pipe Line, on the other hand, proposed to transport a smaller amount of natural gas and sell it to the local distributors in southern California for resale. In the opinion of some of us on the commission, the Tennessee parties obviously tried to play on our sensitiveness to the air pollution problem, hoping that we would grant the Gulf Pacific group a certificate in order to eliminate the smog problem in Los Angeles.

The commissioners, however sensitive to the air pollution issue, recognized that the smog problem in Los Angeles was due primarily to the automobile and that the FPC should not indiscriminately

[25] *Future Natural Gas Requirements of the United States*, vol. 3 (Denver Research Institute, University of Denver, September 1969), p. 5.

[26] Opinion 500, 36 FPC 176 (1966).

grant each and every application as soon as someone raised the question of air pollution. We were very much aware that we were going to have problems with sulfur dioxide in New York and other industrial cities in the East and Midwest. It was obvious, even at that time, that if natural gas were to be substituted for other forms of fossil fuel because of air pollution, the nation would run short in no time at all. It was for that reason that I specifically concurred in the Consolidated Edison case (38 FPC 906 [1967]), in which we granted increased quantities of natural gas to help relieve the sulfur dioxide problem in New York. This was an attempt on my part to convince the majority to fully admit what they were doing, namely, considering the end use. This was done with the hope that in future cases the attention of the nation could be directed to the fact that U.S. reserves of gas could be dramatically reduced if an attempt is made to satisfy the exploding demand for gas resulting from a growing consciousness of environmental problems.

I was afraid that there was just not enough gas to go around in the short run. Therefore, I was concerned about the ability of the industry to supply the committed domestic demand as well as the demand where gas obviously was needed to minimize a serious air pollution situation.

The alternative to an attempt to promote discussion about the fair and equitable method of allocating a limited supply of gas would have been to allow the price of gas to increase in response to this sudden increase in demand. As a matter of fact, this is exactly what many people now conclude is the proper solution to the problem. Such an increase in price would have a rationing effect and would ultimately dampen demand until a point of equilibrium would be reached. A price increase would have the further advantage of promoting research and development efforts to find alternative environmental solutions.

I felt at that time and still feel that a proposal to allow a free market price for gas would be a much too simplistic remedy. There is no question that the present demand for gas for environmental reasons is almost insatiable. It is my firm belief that for some time higher prices will not significantly slow up demand. At the moment, for example, there is not an adequate supply of an alternative fuel that can match the pollution-free performance of natural gas. To allow the producing industry, the pipelines industry, or the distribution industry to fully capitalize on the situation of scarcity would be inequitable to say the least. In this situation, the short-term solution, although undesirable in theory, is to ration supply so as to assure the advancement of the nation's interest in a pollution-free environment.

This new and growing use of natural gas precipitated by public legislation throughout the country has radically changed the character of the industry. Even under a perfect market, there would have been a shortage and also a demand for interim price control, had there been none. Certainly, knowing what is known now, the price of natural gas ought to reflect this increase in demand so as to assist in the rationing effort. However, any such reflection should be under controlled conditions. An overshooting of the equilibrium price might well discourage the public efforts in the war on air pollution. Were the price to be too excessive, the country could balk at the cost of a pollution-free environment. At the moment, I am convinced that, if we are to err, we should err on the low side in order to maintain momentum in reducing air pollution. I would much rather do it this way and require society to subsidize the industry for errors on the low side than run the risk that such public benefits would be forgone. During the interim period of scarcity, while exploration and development are geared up to respond to either the increase in price or subsidy, or both, and until the alternative sources of fuel under freely competitive conditions become available, I would continue to insist that the FPC exert end-use controls with the advice and assistance of the Environmental Protection Agency and the Council on Environmental Quality.

PRICING OF INTERRUPTIBLE GAS

In retrospect it is clear that we should have been profiting from our attempts to price interruptible gas. In the early days when I was on the commission there was one factor that tended to keep the demand for natural gas in the ballpark—the intervention of the coal industry in certificate and rate proceedings. The coal industry frustrated our attempts to expedite proceedings and forced us to seek answers to difficult pricing questions, because it thought the only competitive advantage it had was a lower price. Thus, it sought to force the commission to set higher and higher prices for gas, particularly industrial, constantly trying to get us to examine the whole problem of dump, or valley, gas and whether we had properly allocated the costs associated with it. The coal industry, however, never seemed to win a case. For that reason, plus roadblocks thrown in the coal industry's way by the commission—and quite likely because the petroleum industry was starting to take over the coal industry—interventions became less and less frequent. Soon we were reading in the press how the coal industry was supporting the gas industry's efforts to get higher prices. Of course, one could say this was being done with the hope that the coal industry's competition would be less, but I do not think so.[27]

In my opinion, the commission never realistically attempted to solve the problem of establishing fair rates for the sales of pipeline gas. The Atlantic Seaboard formula, the basic method for establishing rates for interuptible gas, was always accepted in principle. Surely there were cases in which the rates were tilted one way or another, but the underlying economic rationale was never discussed, and the consequences of our rate decisions were never analyzed. Our failure to know what we were doing undoubtedly was partly responsible for the increased quantities of gas sold for boiler fuel and other industrial uses at very low interruptible rates. In the eyes of many, the present shortage is partially a shortage of cheap industrial gas, which industry has gotten in the habit of treating as firm gas since it was practically never interrupted.

JURISDICTION OVER DIRECT SALES

There also was an associated legislative oversight that the commission sought to correct for several years. The Natural Gas Act clearly spelled out the commission's jurisdiction over sales of natural gas for resale. It was also clearly apparent that the commission had no jurisdiction over direct sale by pipeline. The commission staff was aware of the fact that it probably could influence the price of these direct sales in its allocation of costs between jurisdictional and nonjurisdictional sales, but this was a cumbersome method and furthermore did not prevent a pipeline company from selling directly if someone were willing to pay the price. I suppose the commission's desire to control direct sales is another example of the tar baby effect. Whether it is or not, and particularly if pipeline regulation is going to be the only method of protecting the consumer, I feel strongly that direct sales should be regulated. It is interesting to note the following comment in the 1964 *Annual Report*:

> In the Commission's view, the placing under its regulatory jurisdiction of such direct industrial sales by both pipeline and producers would help provide protection to industrial consumers from excessive rates and from discriminatory rates, would strengthen the position of the local distributing utilities in obtaining adequate service and gas supplies for their industrial and domestic customers, and would prevent inflationary pressures on producer prices which may be caused by the larger industrial customers competing with pipelines in the field for producers' supplies. It had been thought when the Natural Gas Act was adopted in 1938, that competition with other fuels, such as coal, would exert a self-regulatory effect and obviate the need for public regulation of the direct industrial sales of natural gas. Generally, this has not proved to be the case.[28]

Commissioner O'Connor, on the other hand, opposed such jurisdiction on the grounds that "regulation of industrial sales is not needed due to the equal bargaining strength of the parties concerned."[29]

As the complexion of the commission changed, its desire for such remedial legislation became less and less. In the 1967 *Annual Report*, the commission (Chairman White and Commissioner Carver were

[27] Richard J. Gonzalez, gas consultant, former Humble Oil executive, and leading defender of the industry, in a recent comment in the *Oil and Gas Journal*, June 29, 1970, stated that interfuel competition between oil, natural gas, and coal provides "effective control over domestic fuel prices." With this emphasis on interfuel competition, it is too bad that the coal industry no longer wants to fight for its fair share of the market.

[28] FPC, *Annual Report* (1964), pp. 8–9.

[29] Ibid., p. 8.

new members) merely noted its previous legislative recommendations made in the 1966 report. By the time of the 1968 *Annual Report*, the request for jurisdiction over direct sales had become a matter of history.

While there is a tendency for a regulatory commission to overreach, it should not back off for fear of criticism from recommending legislative change when experience indicates that the present act is ineffective. The interstate resale market made our vast network of interstate pipelines possible and provides a large part of the funds to pay for the exploration and development of new gas provinces. Hence, it is ironic that this market should be at the mercy of the direct industrial customer and that the gas, which is discovered because of what the interstate market made possible, can be diverted to the intrastate market. This same interstate market, which has converted whole citywide low-pressure manufactured gas systems into high-pressure systems, now is faced with a prospective gas shortage, partly because substantial amounts of gas—discovered with funds from the interstate market—are going to the intrastate market. Parenthetically, this is why the offshore lands are so important, because the commission, if it so desires, can restrict these sources for use in the interstate resale market.

In other words, the situation is more serious today than ever. The producers, and there are comparatively few, have the pipelines at their mercy. The demand for increased quantities to satisfy the air pollution crisis, the desire of pipeline companies to make direct sales, and the commission's inability to price interruptible gas realistically have led to a substantial rise in the demand for gas. The pipeliners are falling all over themselves in promoting higher prices for the very product they are supposed to be selling in competition with the power and coal industry.

Such actions, of course, raise all sorts of interesting questions. Are they trying to create a climate of fear? I was pleased to note that the New Jersey Public Service Commission failed to buckle when the large electric utility holding company system announced it did not intend to go ahead with the purchase of additional generating units unless it got certain increases, even though utility service would be in serious danger. In announcing restrictions on sales to large industrial consumers, the pipelines seem to be seeking support from prospective purchasers in order to justify the purchase of more gas that would be sold at dump prices and yet not be interrupted frequently. The extent to which residential customers would be hurt is seldom spelled out. I am surprised that a major pipeline like Peoples Gas Light & Coal, which has a producing subsidiary, could be caught in such a bind so fast. Maybe the FPC was too lenient in assuming that Peoples' management would have the foresight to maintain a twenty-year supply. Possibly a pipeline is more interested in high producer prices because of its interest in the production activities of its own system, knowing that every increase can be passed along as increased rates.

I do not intend to single out Peoples particularly. Pipelines generally are so involved with the producers and the positions they take are so identical that one wonders if this is the typical dog-eat-dog free enterprise system.

In fact, it also makes one wonder why the Justice Department does not scrutinize the activities of the producing and pipeline industry more closely. As a result of the many joint bidding and drilling ventures, sometimes it is hard to know which company is the dominant one. Generally, the industry is monolithic, and when some of the small independents talk about the majors, it is not hard to believe stories that industry policy is dominated by the few. Many independents also complain about the unfairness of trying to compete against diversified pipeline companies that can allocate costs to the pipeline business that might more truly belong to oil and drilling.

GAS SHORTAGE

It is still hard for me to believe, despite the increased use of natural gas for industrial purposes to alleviate our environmental problems, that the nation is faced with a serious short- or long-term gas supply deficiency. A review of FPC annual reports during my tenure on the commission is interesting and may explain why I am so surprised.

In the Kennedy commission's first annual report, covering the fiscal year ending June 30, 1961 (the last year of the Kuykendall regime), we noted that the reserve life index declined from 21.1 years to 20.1 years and that proved recoverable reserves increased from 262.6 trillion cubic feet (tcf) to 263.8 tcf. In the 1962 report, the commission noted a further decline in the reserve/production (R/P) ratio and an increase in proved recoverable reserves. However, at the request of the Natural Gas Advisory Council, we devoted a good deal of space in the report to a discussion of the desirability of deliverability requirements.

The report for fiscal 1963 revealed that there had been a slight improvement in the R/P ratio, and the commission proposed new reporting requirements to provide continuing information on natural gas reserves of pipelines. The 1964 *Annual Report* showed an increase in proved recoverable reserves, though not as substantial—only 3.9 tcf. A chart even illustrated the sporadic relationship between net production and new discoveries, revisions, and extensions.

In the 1965 *Annual Report* the section on reserves was replaced with one on exploration and development. The industry's alleged ability to hunt separately for gas fields was noted, and the giant fields discovered since 1945 were listed. Under "Gas Supply," the commission stated that proved recoverable reserves rose 6 tcf. All seemed well. The commission also observed that some pipelines were in a position of oversupply.

According to the 1966 *Annual Report*, exploration during 1965 continued at a satisfactory level. Discoveries exceeded those of the previous year by 1 tcf and exceeded net production by some 5.2 tcf. The decline in the R/P ratio was only 0.4 years. Interstate pipeline companies had under contract 194 tcf, 67 percent of the American Gas Association's (AGA) estimate of national proven reserves.

The following year there was a 6 percent reduction in the R/P ratio to 16.5 years. Discoveries and additions, however, still exceeded net production. Interstate pipeline reserves increased to 207 tcf.

The report for fiscal 1968 listed fewer discoveries and fewer wildcat wells drilled. We noted, however, that the industry did put up some $1.7 billion in lease bonuses for offshore acreage from May 1967 to May 1968 as an indication of willingness to meet the challenge of finding new gas and oil provinces. While net production of 18.4 tcf set a new high, additions still amounted to 21.8 tcf. Thus, things still looked more or less standard as I prepared to leave the FPC in September 1968.

For the fiscal year ending June 30, 1969, the FPC reported that reserve additions dropped a dramatic 37 percent. Therein lies the problem. For the same period, the AGA reported that there had been less drilling and that the association had revised downward some earlier reserve estimates. In the 1969 report the commission also observed that pipeline reserves declined for the first time since the report had been instituted forty-nine years earlier. There was a further decline in the R/P ratio, which the commission sought to explain by stating that in part the decline reflected "the economic tendency to seek an optimum reserve inventory representing the lowest investment cost consistent with the ability to supply market contract needs."

Is it any wonder that I wonder? All of a sudden, considering the time lag in exploration, the nation is confronted with a tremendous crisis—no natural gas. The industry kept telling us it needed higher prices. Nevertheless, there was no specific evidence during that period to indicate that the amount of exploration effort was falling off.

There are a lot of unresolved questions about the alleged shortage of gas, including that of the credibility of the industry itself. First of all, I wonder if the industry is telling this same story to the investment counselors? It is not unusual for the industry to cry shortage in order to influence the regulatory commissions and then turn around and reassure the investing community as to the financial well-being of the industry and "pooh-pooh" any shortage that might imperil the capability of a pipeline—to stay around during the next twenty years to pay off its debentures, for example.

Second, I wonder if the industry could conceivably be on a slowdown. In industrial circles such actions are not impossible. What better way to get an increase in price than to scare the commission with a short-term reduction in drilling? Bruce Netschert of National Economic Research Associates believes the reasonableness of the alleged shortage may be questionable. In a speech to the national convention of the National Oil Jobbers Council on May 12, 1970, he stated:

What is not understandable is the great trumpeting by the industry and its trade press over the fact that the year-end reserve total for the United States declined.

He goes on to say:

It is the height of daring for the industry to add to already-mounting public skepticism about its unverified reserves data by drawing conclusions shakier even than the data on which they rest. George Shultz, the Secretary of Labor and Chairman of the Cabinet Task Force on Oil Import Control has already advised a Senate Subcommittee of the danger of "downright manipulation in the short run," and of the inability of the Administration to "distinguish between a fake disaster and a real one. This does put a premium on improving the quality and reliability of statistical information as recommended in the report, so as at least to free the program managers from conclusory [sic] industry assertions."

I have often wondered if it is possible for the industry to hold back on some discoveries for a short time, depending on its prediction of future prices. This certainly would work to industry's advantage. It is well known that in order to improve bidding on adjoining tracts, companies will sit tight on a discovery, and it is also well known that tremendous quantities of gas have been held from the market waiting for better prices. Witness the King Ranch and KATY fields, which were held off the market for many years.

Of course, without being dishonest, the industry can always take the most conservative position possible in estimating reserves, and hearing after hearing has demonstrated in the past that this is the case. AGA figures have been proved exceedingly conservative time after time. Even in this present gas scare, I think the industry is failing to level with the public. In all the talk about North Slope oil, there has been little public mention of the tremendous quantities of gas that have been discovered, though Canadian interests are already talking about a pipeline to the Mackenzie River area and the Prudhoe Bay field.[30] As far as I can determine, these Alaskan reserves, which are owned by the United States and Alaska, have not been considered in estimates of the tremendous decline in additions. Furthermore, I often wonder about the readjustment downward of estimates. It would be helpful if this kind of information could be widely publicized so that it could be disputed if found to be overly conservative.

It is obvious, however, that in some years exploration for natural gas may fall off irrespective of the price of natural gas. In fact, I believe that much of the present shortage is caused more by the financial and economic conditions of the country than by the artificially low FPC established prices. Even before I left the commission, the natural gas pipeline industry had hired Leon Keyserling in order to convince "us liberals" that the industry was desperate because of the tightening economic squeeze. Things have gone from bad to worse since. Trade papers indicate the difficulty of wildcatters in finding drilling funds whether for oil or gas. With money costing 10 percent and more, I would expect the industry to take it easy.[31] Cynic that I am, and having observed the industry for a number of years, I am convinced that it is entirely possible that the industry has decided to forgo short-term profits in the hope that if a shortage scare can induce Congress or the FPC to grant the industry's demands for regulatory relief such short-term deficiencies will be more than compensated for with long-term profits.

Tied in with the question of a possible slowdown in natural gas drilling is the whole interrelationship of oil and gas, foreign and domestic. Despite our finding that there seemed to be an ability to drill specifically for natural gas, it is obvious that if oil drilling falls off, gas additions will be affected.

[30]See *Oil Daily*, June 20, 1970.

[31]It is also not beyond the realm of possibility that the recent tax changes may have had an impact. If one accepts the industry's thesis on the effect of changes in depletion, it would be reasonable to expect some falloff in domestic activity. Furthermore, the recent concern over oil spills and the possible liability of the industry obviously has made life more complicated for the industry.

Since the Permian case, I have been struck by the fact that there seems to be less reliance on the directionality theory, at least in rank wildcats. The emphasis seems to be more on whether there is a chance for a "big strike" in either oil or gas, and usually one strike follows another. Witness the Arctic and the North Sea. In any event, as very promising areas open up, everyone seeks to get into the act. Funds are diverted from offshore or from the continent to these more promising areas. I still remember the hollering by the small independents during the early 1960s, because the majors had abandoned the continental United States for offshore. They complained bitterly that drilling prospects and farm-outs had dried up. Of course, this inability to predict the use of natural gas revenues complicates the problem of determining whether an increased price will provide additional natural gas or whether the revenues will be diverted to searching for oil—or to exploration in Canada, the North Sea, Libya, Egypt, Nigeria, or elsewhere. It also accounts for the speech on a "use it or lose it" policy by the chairman of the board of Northern Natural, W. H. Strauss, at the commission's fiftieth anniversary.

Whatever the reason, there seems to be a short-term decline in additions to our natural gas supply. Frankly, I do not think it is a serious long-term problem, as some people do, unless the nation decides to use natural gas as the sole means of solving the air pollution problem. I feel that there may well be reason to use it for spot relief in some critical areas. It should not be indiscriminately used as the principal means of solving the problem, however, and forecasts for the future should not project substantial amounts for boiler fuel purposes, whether for electric generation or similar purposes. Our fossil fuels are far too important to be used for such purposes.

In my opinion, natural gas consumption cannot be left to free market forces when the supply of alternative products, such as oil, are being artificially controlled. It distresses me to read of suggestions that the price of natural gas should be determined by the price of alternative fuels, such as oil, when the domestic price of oil is so far above the world price and when the supply is limited by import control and state regulatory practices.[32] Furthermore, it is no longer as reassuring to look to coal prices. Now that the petroleum industry seeks to take over the coal industry, one wonders if the price of coal will ultimately be artificially inflated by one means or another. The use of such alternative fuels as a measuring rod would provide an illusory yardstick.

INCREASED PRICE AS AN INCENTIVE TO INCREASED PRODUCTION

One theme characterizes all discussions by industry about what will solve the problem. One spokesman after another has said that an increase of 5 cents per thousand cubic feet (Mcf), or better yet, 10 cents per Mcf would solve all their problems. I would feel a lot better if they would go on to say that such an increase would necessarily lead to an increased supply.

The professional economist has sought to intervene in the discussion. This type of question appears to be right up their alley. In recent years, economists have been far less bashful about second-guessing the commission. On the other hand, I can remember pleading with our staff economist to get someone in the field of economics to write about the subject and to encourage him to find outside economists who could testify on the matter. At that time, the literature was conspicuous by its absence, and I believe it is still hard to find competent witnesses.

Now the commission is accused because it has failed to show how it arrived at the price. It is convicted for failing to prove that the FPC price would have elicited the supply needed for future consumers. Its attempt at econometrics has been soundly and rightfully criticized by some and laughed at by others. The commission is told that allocations make the problems of setting prices on a cost

[32] In "Crude Mischief," *Barrons*, June 29, 1970, p. 1, the following statement appeared: "Last week, when the Texas Commission, in a deliberate effort to prop-up petroleum prices, decreed a sharp cut in output for the third straight month, the Administration went a good deal further. Such restriction policies, a spokesman said, may lead to a challenge by the newly created anti-inflationary Regulation and Purchasing Review Board, or a reappraisal of the federal statutes which grant them legal sanction."

basis absolutely impossible. Now that there is an alleged shortage, there is no reluctance to jump on the bandwagon in berating regulation.

At the moment, I think all of us would do well to confess that we are not sure *what* price will do the job. The commissioners with whom I served were well aware of the need to have an adequate supply of gas for existing customers and, within reason, for future customers. We considered making consumer-oriented supply our principal concern. We were not naive enough, however, to think that our crystal ball was necessarily perfect. We were sure of two things. One, we were confident that the price was not too far out of line as far as average costs were concerned, even given the complex problems of allocation. These problems did not bother us as much as they did some of the bystanders and some of the lawyers. We were reasonably sure that our allocation decisions could not be too far off. There had been much discussion of the pros and cons of each issue, and we had the benefit of a mass of argument. We were aware that in using averages, unquestionably someone was going to get hurt. We knew that the name of the game was the giant field and that every producer worth his salt was willing to risk a number of small inefficient discoveries for the one big one. These "giants" and their impact on costs had to be measured against the H. L. Hunt interests and others who had had some high-cost discoveries by which they hoped to jack up the ceiling prices.

Second, we were also aware that the prices, which admittedly were cost oriented, were not too far off from the prevailing market price in 1960. This market price itself was the end product of a series of artificially escalated rounds of price increases; in fact, it caused a short-term surplus of gas in the early sixties. In other words, the industry in the middle sixties was willing to sell large quantities of gas on a long-term contract basis for a price that seemed to approximate the cost-based price we arrived at. Then, because we were aware that such prices had served to induce industry to go offshore and to drill the deep wells of the Delaware Basin of western Texas and southeastern New Mexico, we felt we could not be too far off.

We could not guarantee the result, nor could anyone. Even today, members of the industry make no claim that they have the exact incentive that will produce certain additional reserves. As men of common sense, they say simply that the industry cannot help but do better. Whether the price is increased 5 cents, however, or 10 cents, or even 1 cent will make a lot of difference to the revenues paid by the consumer. Even J. Rhoads Foster, president of the consulting firm Foster Associates, said that regulators should not assume that it is possible to predict the precise quantity of gas that will become available at any price.

> One should not lose sight of the complexities of the real world in a search for certitude. The influence of price on supply should be viewed as a matter for consideration by exercise of informed judgment in the light of reason and practical experience, not as a subject easily susceptible to precise statistical measurement.[33]

In his testimony in the reopened Southern Louisiana proceeding, Dr. Foster stated that many things other than price will influence drilling—monetary inflation, changes in real cost of import resources, availability of import resources, changes in alternative investment opportunities, development of new producing areas, size of prospective reservoirs, changes in business confidence, degree of directionality of exploratory effort, and change in industry technology. I probably should add changes in commission personnel, staff as well as commissioners.

All of us have an idea of what the correct level might be. With the hope that there might be a method to check the general level of prices, I encouraged Harold Wein of Michigan State University and Haskell P. Wald, chief of the FPC Office of Economics, to pursue econometrics.[34] The saying

[33] J. Rhoads Foster, *Foster Associates Report*, no. 743 (June 11, 1970), app. p. 14.

[34] In my separate statement in the Permian decision, I said: "I am therefore looking forward to the continuing use and refinement of the application of econometrics and other advanced analytical techniques to the difficult task of producer regulation. Surely we should strive for more objective evaluation of economic matters in contrast to relying upon our 'feel' in such situations."

"Rome wasn't built in a day" applies here. The use of econometrics in assisting in price regulation has benefited from some of the constructive comments received. It is getting better all the time, and I am convinced, more than I was in the Permian case, that ultimately it will be an extremely valuable tool. If it is good enough now to prove we were wrong during the sixties, perhaps econometrics will be accurate enough to help us find a solution for the future. It would have been helpful if the commission had chosen the Appalachian Basin for experimentation, that is, to remove all ceilings and try to project the equilibrium price.

I have already mentioned a matter that bears directly on incentive. It is difficult for me to endorse an incentive price when I have no assurance that the revenues it will produce will be used for exploring and drilling for natural gas. I am afraid it could go for other purposes previously mentioned, some of which are far removed from the business of supplying energy to the nation, or for construction of homes or other diversification attempts.

Furthermore, it seems to me that the nation is already providing certain special tax allowances that should serve as incentives to the industry to engage in oil and gas drilling, and I refer to the expensing of intangibles as well as to the depletion allowance. When is enough too much? What controls are there if the price is too high?[35]

Some people support the idea that interfuel competition will serve as a restraining influence on artificial rises. In such discussions, one must constantly remember that oil is one of the principal competitors of natural gas. Since there is the problem of allocation of costs between oil and gas, it is not unreasonable to believe that the industry would operate so as to maximize its overall return, irrespective of the burden on the gas customer. The shifting of costs from oil to gas would be logical under some circumstances, or vice versa.

In addition, the effectiveness of interfuel competition will depend on the area of the United States in which the competition occurs. In some areas there would be none. Furthermore, while the commercial and industrial customers might shift, the residential customer is caught.

To say we should rely on interfuel competition is to say we ought to deregulate the industry.[36] To do so at the moment would be disastrous, which Chairman Nassikas recognizes even though Commissioner Bagge did not. Commissioner Carver also agrees that deregulation is not practical. To force pipeline companies to go looking for gas in an uncontrolled market when, according to them, there is such a shortage, would cause prices to skyrocket. It would be like announcing that sugar is going to be rationed within a week. Obviously industry will overshoot. By how much? I do not know and I doubt anyone does.

I guess I frankly do not believe that an industry, essentially dominated by a few giants, that is marketing the most basic product the world knows–energy–should be allowed to charge what the traffic will bear, particularly when the ultimate recipient of the bad news is almost helpless to countermand such a decision. The consumer would bear the price but most unwillingly, since without regulation there is little he can do about it. While the average regulator has a tendency to worry about the ultimate residential customer, even some commercial and industrial customers are helpless at times when confronted with only one pipeline.

THE FUTURE

If regulation is not to be abandoned, what can be done to alleviate the situation, particularly the alleged shortage of natural gas? I hope that my discussion of the problem and of FPC successes and

[35] See also discussion in the *Imperial Oil Review*, June 1970, about Canada's white paper on taxation, in which Imperial Oil suggests a form of depletion allowance that "would promote exploration by rewarding successful effort rather than subsidizing spending as the white paper proposes."

[36] See Richard A. Posner, "Natural Monopoly and Its Regulation," 21 *Stanford Law Review* 548 (1969), who seeks to establish that the benefits of regulation are dubious, and Joseph C. Swidler's reply in 22 *Stanford Law Review* 519 (1970).

failures may prove illuminating if nothing else. As far as I am concerned, I confess that the falling off in the supply of natural gas is a disappointment and that it does demand corrective action—first for the short term and then for the long term.

It is still hard for me to believe that the FPC erred in either the Permian Basin or Southern Louisiana cases in the level of prices it established at that time. I have seen no evidence that the industry would have been in financial distress if the exploratory effort for gas had continued. I am still confident that the industry would have made a reasonable profit, possibly not as much as in earlier days, but then no industry is doing well now. Even though energy is extremely important to our economy, I can see no reason why the industry should be guaranteed against loss, which seems to be its attitude. Nor do I see any reason why the industry should be entitled to be released from determinations of the reasonableness of the prices for gas in old contracts. In essence, the industry is still shouting for a guaranteed escalator. In part, it would be a windfall, pure and simple, and would simply reflect the fact that the natural gas industry is so important that we must grant it all kinds of concessions.

Since time seems to be of the essence, since the industry seems united in its determination not to increase its level of exploration and development for natural gas, and since whatever gas is discovered will be dedicated to the intrastate market because state authority will not establish any limitations on what the traffic will bear, I suggest the following short-term solutions:

1. The Department of the Interior should actively pursue a policy to develop the reserves of natural gas that are known to exist on the outer continental shelf. While he was representing Consolidated Edison, former Chairman Swidler suggested in a letter to Assistant Secretary of the Interior Harrison Loesch that: (a) additional offshore sales be scheduled as soon as possible, (b) that producers holding leases be required to accelerate exploration, (c) that the present basis of competitive bidding be revised, and (d) that no producing relief be granted unless approved by the Federal Power Commission.

In this connection, I am pleased to note that press accounts of the recently released report of the Public Land Law Review Commission likewise urge certain important changes, including that of having the federal government cease to rely on state conservation agencies for regulations governing the development of the adjoining outer continental shelf.

The Department of the Interior, in its excellent report on the U.S. continental shelf seems to indicate that the federal government must seize the initiative and not wait for informal expressions of interest on the part of the industry on such factors as considerations of the state of technologic development, availability of equipment, and industry's financial capability.[37] When one pauses long enough to consider the immensity of this resource in years past—nearly 4.5 *billion* has been paid to the federal government as a result of oil and gas operations on the outer continental shelf (OCS)—it is apparent that we are talking big money.

Any discussion about the OCS must include the need for the broader use of data that the Interior Department has in its possession so that a better energy policy can be determined. If exploration data cannot be made available, then possibly the U.S. Geological Survey should be given the job of providing basic preliminary geological data that would be freely available, rather than providing it on the basis of contract as it is done now.

Furthermore, the Interior Department and the FPC together with the Office of Emergency Preparedness should immediately reopen studies on a government-owned transmission system that could tap these new reserves, which more and more often will be coming from deeper and deeper waters. Because of environmental considerations, we no longer can afford the luxury of having each producer owning his own pipeline. Neither should rate base considerations dictate the transmission system on the OCS. It should be done either through a government corporation or a common carrier system designed and planned with direct government supervision.

[37] U.S. Department of the Interior, Bureau of Mines, *Impact of Petroleum Development in the Gulf of Mexico*, information circular 8408 (1969).

The FPC should immediately announce a policy requiring that natural gas reserves discovered on federal lands be used in the interstate market alone. Also the reserves should not be used for the generation of electrical power except if the user stipulates that such use is for a limited period only, that is, until he can secure pollution control equipment for other fossil fuels or until he intends to retire such stations.

2. I suggest that the commission immediately conduct an intensive survey of the uses to which natural gas is now being put. In this survey, I would concentrate on the direct sale and interruptible markets, noting particularly whether the interruptible sales in fact had been interrupted and for how long. In such a survey particular attention should be given to competing fuel prices and to who controls the competing fuels. There are some people, including myself, who believe that some of the gas scare language emanates from industrial customers who are complaining because for the first time interruptible gas is just exactly that—interruptible.

3. I think the commission should resurrect its request to control direct sales.

4. The FPC should institute a proposed rule suggesting certain priority uses. This would be valuable in that it would concentrate the nation's attention on its lack of wisdom in consuming a scarce and valuable resource too rapidly. It might provide guidance to pipelines that are besieged with requests for new supplies. Who has the first right? Do existing customers come first or not? Actually, in the public utility field no one really has a specific right to new service, even though he is willing to pay whatever price is demanded. Existing customers have usually been granted special relief. That the establishment of some sort of emergency priority would be worthwhile is borne out by the plans of the electrical industry to ration power if need be or to reduce voltage, as it already has.

5. Until the FPC is freed from the task of regulating producer prices, it should not capitulate to the industry. It should not put the burden on the consumers by going the rule-making route in its eagerness to increase the price. The consumer now is faced with a rule-making decision that is essentially a predetermined result or with a settlement arranged between the interests, the consumer excepted. Due process to *all* is not an unreasonable request of any regulatory commission. In its haste to make waste, the FPC could at least pay passing heed to one segment of the public—the consumer.

6. The commission should give consideration to recommending legislation that would prevent any producing corporation that sells most of its gas in the interstate market from selling substantial amounts of gas in intrastate commerce.

7. As long as advance payments are recognized by the commission, I would recommend expanding the practice by increasing such payments to small independent producers so as to loosen up the right of entry into the production business. Somehow, the nation must break up the lack of effective competition in the sale of natural gas at the wellhead. Granting incentives to small producers is one possibility.

8. I applaud the decision of the Federal Trade Commission to investigate the natural gas industry and the increasing combination of petroleum companies with companies engaged in the coal and uranium industries. The FPC should assist its sister agency in every way possible.

9. The commission, *on its own initiative*, should institute an investigation of the practice of diversification by existing companies and, in fact, could well propose the breakup of existing conglomerates. If the natural gas pipeline business is as bad as it is said to be, maybe such companies would be glad to give up such a low-return business.

10. The commission should also institute a public hearing in conjunction with the office of Russell Train, the chairman of the Council on Environmental Quality, to consider possible recommendations for lessening the environmental impact of natural gas production and transmission activity. Since our future lies seaward, and since the industry is already under the gun from the environmentalists, let us bring the issue out into the open. Such a hearing could review the problems of drilling accidents; oil spills, whether from ship or transmission lines; the desirability of nuclear stimulation or storage; improved transmission practices; and how natural gas should be used in order to alleviate dangerous pol-

lution practices. Like the electric power industry, the natural gas industry is quickly finding the environmental route a hard one.

11. The commission should enlarge its proposed natural gas survey so as to make it a joint government effort. Unless something is done immediately, there will be a race between the Interior Department and the FPC to see who will do the job and get credit for it. Nowadays, though, we have two new entrants, the Office of Science and Technology and the Office of Emergency Preparedness. They, too, should be brought in. The credibility of the study, however, will be directly proportional to the quality and type of the participants. Industry spokesmen should not dominate the show as they wanted to do with the American Petroleum Institute or the National Petroleum Council. It is a job for government after having received advice from industry. Of course, there is a good possibility that the legislative branch may well seize the initiative, and if they do, it serves the bureaucrats right. In fact, a congressional investigation might be more public oriented. Any such study, regardless of who makes it, should encompass international problems that I shall mention later.

12. The commission should call industry's hand right now and raise the price of natural gas after a national hearing. It should grant an incentive, so much per Mcf of gas discovered, on the condition—and only on the condition—that the applicant file with the commission a proposal that he intends to drill a wildcat for natural gas in such and such an area and that he proposes to dedicate the gas to the *interstate* market. The incentive should not apply to any development wells but should be limited to rank wildcats. I would also consider granting incentives to small independents. Such incentives would serve until the time when Congress could more realistically determine what is a fair price, if any, for the nation to pay for encouraging exploration.

13. Our interstate pipeline system is one of the world's finest, but that does not mean it could not be improved. The finality of the location of such a pipeline, the tremendous sums of money accounted for by transportation, and the inability to respond to new demands quickly, all point to a need to constantly review whether such a system is still the world's finest. As a result, I would recommend holding a series of public hearings to inquire into the need for an authority to compel interconnections under certain conditions, as well as common carriage. It is also necessary to determine whether there is an undue amount of crosshaul and duplication.

Economies of scale should also be examined as well as technological developments. Without an outlet, finding natural gas is meaningless. Furthermore, the existing location of a pipeline may well have a direct bearing on where further exploration is conducted.

CONCLUSION

Since March 1959 I have either been engaged in regulation as a member of a regulatory commission, both state and federal, or I have been attempting to teach a course in regulation at the University of Vermont. This experience leads me to conclude that regulation of the producer should not be abandoned, although the past record of regulation leaves sumething to be desired. Any time government, for whatever reason, intervenes in our economic process, it becomes exposed as the one agent responsible for all our economic ills. The responsibility for mistakes cannot be charged to millions of consumers who have collectively expressed their desires for goods and services. Any large economic unit, whether the U.S. government or General Motors or American Telephone and Telegraph can misjudge the needs and requirements of the nation, and when this happens corrective action must be taken in light of experience gained.

Regulation of the natural gas industry (and this is how the problem must be viewed and not solely as a problem of regulating the price of natural gas at the wellhead) remains a desirable and necessary means of protecting the man in the street from an economic environment in which the producers and the sellers hold most of the trump cards. In establishing regulation, we must continue to seek means by which the consumer is afforded the benefit he would have had under competition. Regulators must not hesitate to seek legislative assistance in making industry more competitive or in achieving results that would occur if there were competition.

Accepting the premise that natural gas is in short supply, that the problems in allocating costs make it difficult to determine the proper price for producing natural gas, that there is a growing and almost inexhaustible demand for natural gas because of environmental reasons, that additional incentives are needed to find this gas, that regulation has not in and of itself been able to produce a satisfactory supply, I propose that the United States Government establish a public corporation open to private participation to engage in the exploration and development of oil and gas. The operations of the public corporation should be limited to the public lands of the United States, including the outer continental shelf, together with all lands lying under the seas that are not under the jurisdiction of the United States.[38]

This proposal to establish the government in the business of finding oil and gas stems from the failure of the industry to find sufficient gas today; the necessity to proceed with greater and more rapid development of some of our *publicly owned* resources, such as oil shale in the Far West and oil and gas on the OCS; and the fact that the future of the United States and of the world may well be determined by our ability to develop in an orderly and rational fashion the resources under the sea.

I am convinced that more accurate information is needed about what is going on. If it is necessary to form a corporation in order to break the confidentiality barrier, then let it be so. I also am convinced that, as our environmental concerns deepen and we become more conscious of the utter necessity of developing our resources cautiously with concern not only for the stockholder but also for the general public, we need a yardstick corporation that can engage in areas under guidelines. It would be unfair to ask a private corporation to do this.

I am further convinced that a publicly owned corporation could make a material contribution to our efforts to secure for the nation an ample, secure supply of natural resources. The time has long since passed for conducting nationalistic wars to insure private industry's right to exploit the world's resources.[39] The United States must accept the fact that it has no primary right to develop the resources of the developing nations or of the sea. These resources and the rate of exhaustion are properly the concern of all, not just the concern of the affluent who desire to maintain their standard of living. The children of tomorrow, whether in South America, Africa, or the Far East, have a stake not only in how the resources of the world are developed but also in how they are used, for they are the ones who will have to live with our mistakes.

Our efforts to provide sufficient quantities of natural gas for the nation during the sixties in order to meet all conceivable demands made on this resource are a good example, though literally on a very small scale, of the problems the world will face in the decades ahead. May we profit from our experience.

[38]For an illuminating discussion of the proposed means and methods for the development of the resources that lie under the international seas, see Center for the Study of Democratic Institutions, *Ocean Enterprises: A Summary of the Prospects, and Hazards, of Man's Impending Commercial Exploitation of the Underseas*, Center Occasional Paper, vol. 2, no. 4 (June 1970). Since the nation's ultimate fate may well depend on these resources, particularly oil and gas, this paper and the proceedings of the Pacem in Maribus conference sponsored by the center in Malta, June 1970, should be required reading.

[39]George Stocking, *Middle East Oil–A Study in Political and Economic Controversy* (Vanderbilt University Press, 1970).

5 CHANGING ELEMENTS IN THE NATURAL GAS PICTURE: IMPLICATIONS FOR THE FEDERAL REGULATORY SCHEME

*Ralph S. Spritzer**

Gas regulation–in particular, the control of prices in the field–is under heavy attack. This in itself is hardly new. The regulation of producers has been a matter of bitter controversy since 1954 when the Supreme Court, at the behest of a "consumer" state (Wisconsin), ruled that the sales made by a major producer (Phillips Petroleum Company)[1] to interstate pipelines were subject to the jurisdiction of the Federal Power Commission (FPC).[2] For some years thereafter, the industry's opposition took the form of a concerted effort to reverse the decision by legislation–an effort that barely failed, and then only because the measures approved by Congress were twice vetoed.[3] The claim that control of producers' prices was bound to prove unworkable gained added credence when, in 1960, the commission concluded that it had best abandon its six-year endeavor to comply with the Phillips remand by use of traditional public utility rate-making techniques and consider resort to a radically different approach.[4]

A new set of commissioners arrived in the capital in 1961[5] and took up the task of shaping and implementing the suggested change in approach: to devise price ceilings for the industry in each of the principal producing areas instead of pursuing further the laborious task of fixing rates for individual companies by separate investigations of their respective costs. In significant measure this has been done.[6] However, the commission's more recent efforts have engendered the outcry that although the

*Professor of Law, University of Pennsylvania Law School.

[1] Phillips was one of the largest independent producers of natural gas. The term "independent" is used to describe the nonintegrated producer, i.e., the producer that is not involved in the transportation of gas by pipeline, though it may be engaged in extensive gathering of gas in the fields. Most producers fall in this category, although some enterprises are engaged in the gas business at more than one level of the three-tiered structure consisting of production, transportation by pipeline, and local distribution.

[2] *Phillips Petroleum Co. v. Wisconsin*, 347 U.S. 672 (1954).

[3] The so-called Kerr bill (S. 1498, 81 Cong. 1 sess., 1949) was vetoed by President Truman in 1950 on the grounds that the ownership of gas reserves is concentrated, that pipelines "cannot easily move from one producer to another in search of lower prices," and that the public interest requires the prevention of excessive charges. (*1950 Public Papers of the President of the United States* [Washington, D.C.: Government Printing Office, 1965], pp. 257–58.) The Harris-Fulbright bill (H.R. 6645, 84 Cong. 1 sess., 1956) was vetoed by President Eisenhower some six years later, despite the president's sympathy for the bill, when it became known that unsavory techniques had been used to obtain its passage. (*1956 Public Papers of the President of the United States* [Washington, D.C.: Government Printing Office, 1958], pp. 256–57.)

[4] *Phillips Petroleum Co.*, 24 FPC 537 (1960). The commission's termination of the original Phillips proceeding was sustained by the Supreme Court in *Wisconsin v. FPC*, 373 U.S. 294 (1963).

[5] Although the five commissioners are appointed for five-year terms on a staggered schedule, a series of circumstances enabled President Kennedy to make five new appointments during his first year in office.

[6] The first area rate case, involving the fields of the Permian Basin in southwestern Texas and southeastern New Mexico, was decided by the FPC in August 1965 (*Area Rate Proceeding*, AR 61–1, 34 FPC 159), and its order was ultimately affirmed in *Permian Basin Area Rate Cases*, 390 U.S. 747 (1968). The FPC's second decision, involving the Southern Louisiana area, was handed down in September 1968 (*Area Rate Proceeding*, AR 61–2, 40 FPC 530) and recently affirmed by the Court of Appeals (*Southern Louisiana Area Rate Cases*, C.A. 5th, no. 27492 et al.,

agency may have achieved an administrative and legal success–increased mastery of its docket and critical victories in the courts–the economic results of its handiwork are apt to prove disastrous.

It is said that the United States faces a coming crisis in supply, with deficiencies only a few years off. The FPC has indeed stabilized prices but, it is argued, at a level insufficiently high to meet the financial requirements of the industry. The proof of the pudding, we are told, is that there has been a slackening of exploration and development in the face of a burgeoning demand for natural gas that can only be satisfied by greatly accelerated efforts. The elicitation of adequate supplies of natural gas at reasonable prices is of course the central objective, however much opinions may differ as to the means best calculated to achieve it.

THE REGULATORY SCHEME

In the discussion that follows, it is assumed that economic regulation of gas producers is here to stay, that there is no immediate likelihood that Congress will jettison the regulatory scheme, and that the legal responsibility to fix just and reasonable rates for gas in the field will continue to rest with the FPC. Proceeding from that assumption, an attempt will be made to form some estimate of the character and dimensions of the supply issue that has come recently to overshadow all others in the minds of both the regulators and the regulated. This, of course, leads to further questions. To what extent is the supply problem caused or affected by deficiencies in regulation? What changes in governmental policy or in the techniques of regulation would tend to remedy or alleviate it? What significance do gas imports and changing technology have for the regulatory regime? Preliminarily, however, it is helpful to consider, albeit in summary form, the premises and the structure of the Natural Gas Act and the methods that the commission has thus far employed in seeking to execute the statutory directives.

Pipelines

When the Natural Gas Act was adopted in 1938, the focus of concern in Congress was the pipeline segment of the industry, even though there was keen awareness that problems in the producing areas were also acute. In the new and more important producing fields, those problems were quite far removed from questions of consumer price protection. Gas and oil had been found in huge quantities in Kansas, Oklahoma, and Texas.[7] Although gas in nonassociated wells could be capped and held for the future,[8] gas associated with oil was an unavoidable by-product of oil production.[9] And the pipelines had not been developed to the point where they could begin to absorb the available quantities of the by-product.[10] Gas was accordingly a glut on the market. Waste–dramatically illustrated by the flaring

---

March 19, 1970). In its opinion in the Permian case, the commission stated that the Permian Basin accounts for approximately 11 percent of all gas moving in interstate commerce. Southern Louisiana, according to the Fifth Circuit's opinion in that case, produces approximately one-third of the nation's gas supply. Other area rate cases are pending before the agency. Moreover, it should be noted that interim price controls designed to "hold the line" pending completion of area proceedings have been in effect throughout the past decade.

[7]The earlier development of natural gas had been centered in the Appalachian region, with most of the production coming from West Virginia, western Pennsylvania, and southeastern Ohio. Production in this area, however, had passed its peak by the end of World War I. See the opinion of Justice Jackson, *FPC v. Hope Natural Gas Co.*, 320 U.S. 591, 629–34 (1944).

[8]A serious complication–one that contributed heavily to gross inefficiencies of production–was the danger of drainage. Under the rule of capture, a landholder might draw off fugitive resources from his neighbor's properties by initiating production. Overdrilling and hasty methods of extraction resulted. Ultimately, the states responded by conservation measures (prorationing statutes, well-spacing requirements, laws governing the engineering aspects of drilling and extraction, etc.).

[9]Oil is inert, and the energy provided by gas or by a water drive from beneath the reservoir provides the principal means of lifting it to the surface.

[10]Although several long pipelines extending northerly to midwestern cities were built in the 1930s, the big explosion in the pipeline industry occurred after World War II. Improved technology–in particular, the feasibility of using pipe of larger diameter–played an important role.

of gas into the air and by the use of gas in carbon black plants located near the fields—was commonplace.[11] The problems associated with production and waste were purposefully left to the states; in its opening section, the federal statute declares that its provisions shall not apply "to the production and gathering of natural gas" or to the local distribution of natural gas—matters that Congress regarded as primarily of local concern and subject to regulation by the states.[12]

The interstate pipelines, on the other hand, were beyond the effective reach of the states. The Supreme Court had made plain that wholesales of gas made by an interstate carrier to a local distribution company were immune from economic regulation by the receiving state (although the subsequent in-state sales by the distribution company to the ultimate consumer were conceived to be within the state's regulatory control).[13] If, therefore, a pipeline monopolist boosted its rates to a captive customer company or discriminated in the rates charged distribution companies in different cities, the state would be without means to safeguard the interests of the consumer.[14] The federal statute, the House Report stated, was designed to occupy the area in which "the Supreme Court has held that the States may not act."[15]

The structure of the act fits the conventional mold of public utility regulation. A natural gas company is defined as one engaged in the transportation of natural gas in interstate commerce or the sale in interstate commerce of such gas for resale (sec. 2). The company's engagement in these activities—likewise its construction or extension of facilities for these purposes—requires a certificate of public convenience and necessity from the federal commission (sec. 7).[16] The commission is accorded rate-making powers and the duty of correcting unjust, unreasonable and unduly discriminatory rates, charges, practices, and classifications (secs. 4, 5).[17] It may also order extensions or improvements of facilities and require physical interconnection with distributors, subject to the proviso that the new or increased service will not impair adequate service to existing customers (sec. 7). The commission is additionally granted broad powers to supervise accounting procedures, to require the preparation and production of records and reports, to adopt rules, and to conduct investigations and hearings.

These powers, it should be noted, embrace an authority to allocate or ration supplies. Thus, if a pipeline proposes to make new industrial sales or to deliver added volumes to an existing distributor or industrial customer, further certification is required.[18] This may involve evaluation of competing demands and judgments as to the comparative importance of the end uses to which the gas would be put.[19] Similarly, in deciding whether it should approve the design of pipeline rates, the FPC must consider the effects on different classes of customers.[20] Since gas is a fuel with highly desirable charac-

[11] FPC studies indicate that in the 1920s and early 1930s the waste of natural gas exceeded total domestic consumption. (S. Doc. 12, 70 Cong. 1 sess. [1936], pt. 84–A, pp. 93 ff.)

[12] Natural Gas Act of 1938, 52 stat. 821, as amended, 15 U.S.C. pars. 717 ff.

[13] See *Missouri v. Kansas Gas Co.*, 265 U.S. 298 (1924); *East Ohio Gas Co. v. Tax Commission*, 283 U.S. 465 (1931); compare *Public Utilities Commission v. Attleboro Co.*, 273 U.S. 83 (1927).

[14] Many such complaints came from the area served by the depleted Appalachian fields where, in Justice Jackson's words, the "public came to feel that the companies were exploiting the growing scarcity of local gas." (*FPC v. Hope Natural Gas Co.*, 638.)

[15] H. Rept. 709, 75 Cong. 1 sess. (1937).

[16] Abandonment of facilities or service likewise requires FPC approval (sec. 7).

[17] Sec. 4 empowers the FPC to suspend a rate increase for a maximum period of five months pending investigation. If the investigation is not completed during that period, the rate may be put into effect, but the company is subject to the requirement of refunding amounts collected in excess of the rate ultimately determined to be just and reasonable. Under sec. 5, the FPC may institute a rate investigation at any time; its order, however, has only prospective effect.

[18] The power of certification gives the FPC substantial control of direct sales by pipelines to industrial customers, even though such sales are not subject to its rate-making authority. The FPC may deny authority or impose conditions, e.g., a condition requiring that the pipeline's rate cover the direct customer's allocable share of fixed costs. (*South Georgia Natural Gas Co.*, 23 FPC 795 [1960]).

[19] See *FPC v. Transcontinental Gas Corp.*, 365 U.S. 1 (1961).

[20] In his opinion in the Hope case, Justice Jackson strongly urged that the FPC design rates so as to discourage industrial consumption and conserve natural gas for the household consumer. (320 U.S. 634–37.)

teristics[21] (the more so in a time of increasing concern with problems of air pollution), since new increments of demand will doubtless be heavy for the foreseeable future, and since the rate at which added supplies will be forthcoming is a matter of at least some uncertainty, it is apparent that the commission's function of allocating gas at the distribution end of the line is importantly and intimately related to the problem of eliciting adequate production at reasonable prices.

Producers

The Gas Act contained a latent ambiguity. The FPC is accorded jurisdiction over "any person" engaged in the sale of natural gas in interstate commerce for resale–language obviously broad enough to comprehend a producer's sale to an interstate pipeline. Congress, however, has excepted "production and gathering." If the exception is read to cover all facets of the producer's business, the federal commission is excluded. If it is read, on the other hand, as applicable only to the physical aspects of production and gathering, the federal commission may regulate the terms of a producer's interstate sale, while the state controls such matters as well spacing, techniques of extraction, protection of correlative rights of landholders, etc.

The FPC in fact refrained from asserting a broad claim of jurisdiction over producers.[22] Nonetheless, the commission did have some occasion in its early years to grapple with the problems of costing the production of gas. Some pipelines had substantial interests in producing properties, ordinarily held through affiliated companies. In regulating such a pipeline's rates to its customers, the FPC was faced with the need of putting a figure on the cost of gas that the pipeline obtained from its affiliate.[23] One method would have been to value the gas entering the pipeline on the basis of the going field price. The other method–the one actually adopted–was to treat the pipeline and its affiliate as a single enterprise and to apply the traditional cost-of-service approach from beginning to end. In this the commission was sustained, although dissenting justices of the Supreme Court objected that the consequence was to make regulable in fact what was not regulable in law.[24]

The larger question of jurisdiction over independent (nonintegrated) producers was not squarely decided by the courts until 1954 when a divided Supreme Court upheld Wisconsin's claim that the FPC had erred in ruling that it lacked jurisdiction to determine the reasonableness of the field prices charged by the Phillips Petroleum Company.[25] The Court's majority read the Gas Act as giving "the Commission jurisdiction over the rates of all wholesales of gas in interstate commerce, whether by a pipeline company or not and whether occurring before, during or after transmission by an interstate pipeline company."[26] The words of exemption were held to refer only to the "physical activities, facilities, and properties used in the production and gathering of natural gas."[27]

The immediate result of the commission's newfound jurisdiction was a flood of certificate and rate filings. The pilot rate-making case–Phillips remanded–dragged on until finally terminated by the commission in 1960 on findings that Phillips's past charges had not been excessive and that the evidence of record, based on a 1954 test year, was too stale to serve as a basis for prescribing rates prospectively.[28] More importantly for present purposes, the FPC concluded for a variety of reasons–among

[21] Ibid.

[22] From an early date, however, there were dissents from individual commissioners. See, e.g., dissenting opinion of Commissioner Scott, *Columbian Fuel Corp.*, 2 FPC 200, 209 (1940). The Smith-Wimberly report urged clarification by Congress. (*Natural Gas Investigation*, Docket no. G–580, p. 155.)

[23] Hope case, and *Colorado Interstate Co. v. FPC*, 324 U.S. 581 (1945).

[24] *Colorado Interstate*, 618 ff.

[25] *Phillips Petroleum Co. v. Wisconsin.*

[26] 347 U.S. 682.

[27] Ibid., 678.

[28] *Phillips Petroleum Co.*, 24 FPC 537 (1960). The FPC's decision to terminate the case was ultimately sustained in *Wisconsin v. FPC*, 373 U.S. 294 (1963).

them, that it was not administratively feasible to regulate a host of producers by individual cost-of-service proceedings—that gas pricing should be instituted on an area basis rather than a company basis.[29]

Concurrently with its disposition of the Phillips rate case, the commission took the first step toward an area approach by adopting a Statement of General Policy, which it proposed to follow pending the completion of formal area proceedings.[30] For each designated geographic area, the FPC set two guideline prices. One price—the higher figure—represented the top price that the commission would regard as permissible in deciding whether to certificate a new sale.[31] The other marked the price level to which the commission would refer in deciding whether to suspend a price increase filed pursuant to an existing contract. Although the FPC's explanation of the basis for selection of the specific prices is less than precise, it appears that the price for new sales was based on the highest price at which sales in the area had been previously certificated and the "suspension" price determined by reference to the weighted average price for all sales in the area.[32] These prices, announced by the commission without any participation by interested parties, did not of course purport to determine "just and reasonable rates." They were designed only to serve as provisional guides to the exercise of the FPC's power to certificate sales and to suspend increases. They were, in short, a means of establishing a "freeze"—albeit one subject to modifications—while formal proceedings were going forward.[33] Their practical effect, however, has been marked. Prices in all of the producing areas have remained relatively stable throughout the last decade, although only two of the area proceedings (Permian Basin and Southern Louisiana) have reached the stage of final decision by the agency.

The FPC's Permian Basin decision, issued in August 1965, marked completion of the agency's first formal effort to fashion a new system of price regulation for producers.[34] At the outset of its opinion, the commission rejected a contention that it should permit contract prices negotiated by producers and pipelines "at arm's length" to govern, at least so long as the ratio of proven reserves to production did not become inordinately high. In so doing the commission suggested various reasons for concluding that price competition among producers was ineffective:

[29] Among other things, the FPC stressed the difficulties of allocating joint costs as between gas and oil and the fact that the business of producing gas, unlike that of the conventional public utility, is a highly speculative enterprise without predictable relationship between investment and yield. It also observed that under the rate base method the producer who takes the initial risk in exploring and developing a field may get the least return, since those who enter later will undoubtedly be compelled to pay more for their rights.

[30] Statement of General Policy, no. 61-1, 24 FPC 818 (1960), as amended, 18 C.F.R. sec. 2.56 (1970).

[31] In setting these guides, the FPC was strongly influenced by the Supreme Court's decision of the previous year in the "CATCO" case, in *Atlantic Refining Co. v. Public Service Commission*, 360 U.S. 378 (1959). In that case, the FPC had issued an unconditioned certificate to producers of Louisiana offshore gas at a price substantially in excess of any price previously approved in that producing area. Earlier, the FPC had refused its approval, offering a certificate at a lower price with leave to the producers to file an increase to the contract level after twenty-four hours (an approach that would have enabled the producers to collect the higher rate while preserving to consumers the possibility of a refund if the rate were ultimately found unreasonable [see fn. 17]). But when the producers balked, threatening to withdraw the supply from interstate commerce, the FPC receded. The Supreme Court reversed. Acknowledging that the FPC was not obliged to conduct a full-blown rate proceeding in advance of granting a certificate, the Court nonetheless expressed concern over the sharp rise in prices that had occurred during the 1950s and concluded that the FPC should not certify an initial sale at an out-of-line price. An initial out-of-line price, it observed, would carry no refund obligation. Moreover, rate investigations of producer's rates under sec. 5 of the Gas Act were proceeding at snail's pace.

The FPC's 1960 guidelines represent an effort to implement the mandate of CATCO by keeping price "in line" during the course of the inevitable delays inherent in full-blown rate proceedings.

[32] See *Wisconsin v. FPC*, 373 U.S. at 316 (dissenting opinion).

[33] The term "freeze" is perhaps something of an exaggeration since the FPC's power to suspend is limited to five months. However, the FPC's power to withhold certification of new sales, or to condition new certificates upon the producer's acceptance of a price moratorium (see *United Gas v. Callery Properties*, 382 U.S. 223 [1965]) has held spiraling rates near the levels authorized by 1960 contracts.

[34] *Area Rate Proceedings*, AR 61-1, 34 FPC 159 (1965). On review, the FPC's decision was modified by the court of appeals (*Shelley Oil Co. v. FPC*, 375 F. 2d 6 [10th Cir. 1967]) but reinstated in full by the Supreme Court (*Permian Basin Area Rate Cases*, 390 U.S. 747 [1968]).

1. It found indications in the record that, although producers of gas were numerous, holdings of uncommitted reserves were heavily concentrated in the hands of a few companies.[35]

2. The long-prevailing industry practice is to enter twenty-year contracts for delivery of gas from proven reserves. The pipeline industry is capital-intensive and has high exit costs. It therefore has strong incentive to secure large blocks of assured supply for the long term (a tendency reinforced by the reserve and deliverability requirements traditionally imposed by the FPC in pipeline certification proceedings).[36] Serving an expanding and relatively price-inelastic market, the pipelines also have basis for confidence that they can "pass along" increases in the price of purchased gas. State-imposed "conservation" measures, including gas allowables, create artificial restrictions on supply. The purchasing options open to a pipeline are at least somewhat restricted by its location in relation to sources of supply. Many pipelines have interests in producing properties and an interest in higher field prices. These factors, in combination, tend to create, at least periodically, a seller's market, with pipelines bidding up the price of large blocks of gas and conferring huge economic rents on the owners of producing properties.

3. The recent history of pricing patterns in the gas fields has been one of price uniformity and confirms the conclusion that producers have undue bargaining power. "Typically," the commission observed, "where a large producer has been able to command a higher price for a big block of gas needed by a pipeline entering the market or preparing for a major expansion, this has had the effect of raising the price to all other purchasers by all other sellers, big and small, and regardless of their costs or profit ratios."[37] Price increases, moreover, did not merely set the tone for the future. By virtue of price escalation provisions, including most prominently the favored nation clause,[38] the increases were transmitted to gas flowing under old contracts, thus nullifying *pro tanto* more advantageous terms previously obtained.

Central to the commission's decision was the adoption of a two-price system anchored to a determination of costs. One ceiling was established for "old" or "flowing" gas (defined to include gas-well gas sold under pre-1961 contracts and *all* oil-well gas); a second, higher ceiling was fixed for "new" gas (that is, gas-well gas sold under a contract executed after January 1, 1961).[39] The industry, the commission found, had developed a substantial capability to engage successfully in directional drilling, that is, to search for gas reservoirs separately from oil reservoirs. Thus, it was feasible to employ a price mechanism that would provide incentives to engage in the independent search for gas without according the same "premium" price to gas already found or to gas that might be found as a by-product of the search for oil.

The price for new gas, determined from composite industry data, was based chiefly on the then current national cost of finding and producing gas from gas wells.[40] It included a 12 percent rate of

[35] The FPC lacked the data to make specific findings as to shares of reserves but noted (34 FPC 182) that the producers, "in the exclusive possession of the facts," had failed to rebut the inference drawn by the FPC. The Supreme Court has observed that the relevant market for natural gas, in which price is determined, is the market for uncommitted reserves (*United States v. El Paso Natural Gas Co.*, 376 U.S. 651, 660 [1964]). See also Edward J. Neuner, *The Natural Gas Industry* (University of Oklahoma Press, 1960), p. 6.

[36] The FPC has had a twelve-year deliverability rule for pipelines, although it has permitted exceptions. *Area Rate Proceedings*, AR 61-1, 34 FPC 184.

[37] Ibid., 182.

[38] Although there are many variations of detail, the favored nation clause characteristically provides that the contract price shall be escalated to the level of any higher price paid by the purchaser to any other producer in the area. In 1962 the FPC had proscribed contractual provisions authorizing indefinite escalations in price. That action, however, applied only to contracts executed after April 2, 1962, and left existing escalation clauses in effect. See *FPC v. Texaco Inc.*, 377 U.S. 33 (1964).

[39] Although the FPC rendered its decision in 1965, it chose a January 1, 1961 dividing date on the following grounds: that the adoption of dual guidelines in the FPC's 1960 Statement of Policy had encouraged the expectations of producers; that directional drilling first became significant in about 1960; and that the FPC data were derived from a 1960 test year, and costs had remained constant from 1960 to the date of decision.

[40] This approach to the pricing of new gas enabled the FPC to finesse some of the difficult problems inherent in allocating joint costs between gas and oil.

return and was designed to provide producers with sufficient inducement to replenish reserves and meet anticipated growth of demand. The price for old gas was determined primarily on the basis of the historical cost of gas-well gas in the Permian Basin.[41] It similarly included a 12 percent allowance for return on capital. The differential between the rates for new and old gas proved to be 2 cents per thousand cubic feet (Mcf)–16.5 cents as opposed to 14.5 cents (including Texas taxes).[42] The FPC's order also established a moratorium on filing any increased rate in excess of the area rate until January 1, 1968.[43]

In large measure, the scheme and methodology developed in the Permian case were followed in the Southern Louisiana proceedings decided by the commission on September 25, 1968.[44] Southern Louisiana has huge offshore reserves and accounts for almost one-third of all gas sold in interstate commerce.[45] The case therefore has had exceptional significance for all segments of the industry. The decision and the aftermath have particular relevance for the claim that the FPC pricing policies are generating a crisis in supply.

The pricing structure adopted in the Southern Louisiana case differed from that in the Permian case in that the commission established a three-tiered rather than a dual system. So-called first vintage prices fixed ceilings for gas sold under pre-1961 contracts and for all oil-well gas. Second vintage ceilings pertain to gas-well sales consummated between January 1, 1961, and October 1, 1968 (the approximate date of the FPC order). Third vintage prices are for subsequent sales of nonassociated gas. For offshore gas in the federal domain, which is not subject to the Louisiana severance tax of 2.3 cents, the commission set rates 1.5 cents lower for each vintage.[46]

The highest price allowed–that for "new" or third vintage gas–was lower than the provisional guideline prices for initial certification that has been in effect since 1960 and substantially lower than the prices many producers were collecting subject to refund.[47] Yet, the price ceiling was substantially above cost (including a 12 percent rate of return) as determined by the commission.[48] The commission found that there was no evidence of a rising trend in unit costs ("where costs have increased, they are offset by the large amounts of gas found per well drilled and per foot drilled")[49] and explained its

[41]The ceiling for old gas-well gas was applied to all oil-well gas on the theory that oil-well or casinghead gas had no higher value (in fact, deliverability of casinghead is more restricted because of the relationship to the demands of oil production) and that its cost is doubtless lower if one assigns exploration and discovery costs to oil.

[42]For Permian gas from New Mexico, the figures are 15.5 cents and 13.5 cents *plus* taxes.

[43]The FPC order had other aspects that need not be elaborated here. Among these were provisions for refund (numerous sec. 4 cases having been consolidated with the area proceeding), provisions relieving small producers of various regulatory burdens, and provisions for adjustment of price where gas varied significantly (in pressure, purity, or heating value) from the standard established for pipeline-quality gas.

[44]*Southern Louisiana Area Rate Proceeding*, AR 61-2, Op. no. 546, 40 FPC 530.

[45]The gas is of high quality and more than 80 percent comes from nonassociated gas reserves.

[46]The prices fixed (per Mcf) were as follows:

Vintage	Onshore	Federal offshore domain
First	18.5	17
Second	19.5	18
Third	20.0	18.5

In the examiner's decision, rendered in December 1966, a ceiling was recommended for new gas one-half cent lower than that adopted by the FPC. The producers had urged prices ranging from 24 cents to 27 cents.

[47]The guideline price was 21.5 cents, later reduced to 21.25 cents including taxes (24 FPC 902; 26 FPC 661). Numerous producers have been collecting, subject to refund, rates in the range of 23.5 cents. See *United Gas v. Callery Properties*, 382 U.S. 223 (1965). The FPC estimated that its Southern Louisiana order would result in rate reductions totaling $49 million annually.

[48]The FPC found the cost of new gas to be 16.5 cents (without tax) and 18.8 cents for gas subject to Lousiana tax. As indicated in fn. 46, the ceiling is 18.5 cents on gas from the untaxed federal domain and 20 cents on taxed gas. The "premium" is larger on the offshore gas because the risks and uncertainties are regarded as greater in that area.

[49]40 FPC 633.

above-cost ceilings as a move to take care of fluctuations and uncertainties during the extended period of moratorium[50] and to provide added stimulus to necessary exploratory and developmental activity.

The aftermath of the Southern Louisiana decision has been marked by growing indications of a commission retreat. In its initial opinion, the FPC observed that the cost information in the record was more representative of onshore than offshore production and that the question whether offshore costs are higher should be considered anew when further evidence is available. In its opinion on rehearing, some six months later, it decided to institute a new proceeding for gas to be sold in the future from offshore.[51] It also decided to permit rate increase filings, subject to suspension and refund, up to 20 cents (1.5 cents above the ceiling) with respect to new gas in the federal offshore domain. The declared purpose was that the institution of the new proceeding should not discourage the formation of new contracts during its pendency. On December 15, 1969, the commission, noting claims that a serious deficiency in supply was developing, decided that the new proceeding should be enlarged to include the entire area, both onshore and offshore, and the prices to be applied to *all* vintages of gas.[52]

Such was the posture of matters when, shortly thereafter, the court of appeals heard argument in the review proceedings involving the Southern Louisiana decision. A threshold question was whether the court should review the first-round proceedings, rather than remand it in light of the initiation of a broad-based second-round proceeding.

The court concluded that it should go forward:

> It is true that the Commission now recognizes the possibility of a serious supply deficiency and that it did not recognize this possibility in the decision we are reviewing. This awareness, however, and the Commission's prompt action upon it, militate in favor of affirmance rather than reversal. The new proceeding is evidence of the Commission's ability to adapt the regulatory process to changing circumstances. We conclude that the indications that there is a supply deficiency should be examined in this second-round before the Commission.[53]

Examining the commission's determinations of cost (defended by the FPC and one group of gas distribution companies, while challenged by producers and certain consumer interests as too low and too high respectively), the court found them fully supported. The court also approved, over the objection of consumer interests, the commission's action in pegging its ceilings above cost. It noted that offshore reserves are of growing importance; that drilling in deep water is a relatively new undertaking and characterized by uncertain risks; that when prices are being limited for a protracted period by means of a moratorium it is reasonable to allow a margin for error and for changes in costs; and that price stability may be "more in the public interest than a rate . . . lower by a few percentage points." The court's principal concern was that the "noncost additions" might be insufficient to elicit the added supplies that would be needed in the future to accommodate growing demand. It criticized the commission's "failure to relate its noncost additions to specific probabilities concerning influences upon supply or to make any showing as to the demand they are needed to meet." The commission, it goes on to state,[54] should estimate future needs, make determinations as to the increased difficulty of finding greater volumes of gas, and estimate the supply that a given rate will elicit. Although it concluded that the FPC findings were deficient in these respects, the court affirmed on the grounds that area regulation was still in an experimental stage and that the second-round proceedings provided an opportunity to adapt the price structure to changing circumstances.

The producers promptly applied to the court of appeals for rehearing, urging that the court remand the case to the commission instead of affirming its order. The distribution companies and consumer interests vigorously opposed this, but the commission, on May 27, 1970, voted to acquiesce in the

[50]The FPC initially provided for an indefinite moratorium but, in its opinion on rehearing, fixed the terminal date for third vintage gas at January 1, 1974 (Opinion no. 546-A, 41 FPC 301, March 20, 1969).

[51]Ibid.

[52]*Order Enlarging Investigation and Proposed Rulemaking*, AR 69-1, December 15, 1969.

[53]*Southern Louisiana Area Rate Cases*, 428 F. 2d 407, 421 (5th Cir. 1970).

[54]Ibid., pp. 427, 443-44.

proposal.[55] On June 16, 1970, the court of appeals declined to remand the case but asserted that its original opinion left the agency free to "reopen any part of its orders" and to make "retrospective as well as prospective adjustments."[56]

Another recent development is an announcement by the commission that it is now prepared to review the ceilings established for the Permian Basin by its decision of 1965. This was accompanied by a statement that the staff of its Bureau of Natural Gas (BNG) had made a preliminary estimate indicating a significant increase in industry costs.[57]

THE "CRISIS" IN SUPPLY

Throughout the tortuous course of the Southern Louisiana proceedings, the producing industry, contending for a price 25 percent or more higher than thus far allowed by the commission, has argued that a failure to authorize a substantial escalation of the price level would have grave consequences—inadequate exploration and development of new sources of gas and a dwindling supply to meet growing demands. Recent evidence, it is said, confirms these predictions.

The primary source of data relating to reserves is an industry group, the American Gas Association (AGA). Its most recent report shows[58] gas reserve and production statistics for 1960-69 (see table 5-1). According to these figures, the production of gas during the last decade rose from approximately

Table 5-1. Total U.S. Gas Reserves, 1960-69

Year	Reserves added: Extensions and revisions	Reserves added: New discoveries	Net production	Estimated year-end proved reserves	Reserves/production ratio	Findings/production ratio
	(billion *Mcf*			)		
1960	7,293	6,601	13,019	262,326	20.1	1.07
1961	10,259	6,908	13,379	266,274	19.9	1.28
1962	13,185	6,299	13,638	272,279	20.0	1.43
1963	12,587	5,578	14,546	276,151	18.9	1.25
1964	13,343	6,909	15,347	281,251	18.3	1.32
1965	14,776	6,543	16,252	286,469	17.6	1.31
1966	14,163	6,058	17,491	289,333	16.5	1.16
1967	16,109	5,695	18,381	292,908	15.9	1.19
1968	10,775	2,922	19,373	287,350	14.8	0.71
1969	4,562	3,813	20,723	275,109	13.3	0.41

NOTE: Mcf given at 14.73 pounds per square inch absolute. Figures exclude Prudhoe Bay area reserves because of nonavailability of data.

13 trillion cubic feet (tcf) to more than 20 tcf. During the same period, the ratio of proven reserves to production (R/P) declined from 20.1 to 13.3. For 1968 and 1969, moreover, the figures show that reserve additions fell below production. Projections based on AGA data have led the BNG to conclude that the R/P ratio may drop to 10 by the end of 1973[59]—a level it describes as uncomfortably low and apt to lead to some regional deficiencies of supply.[60]

[55] *Wall Street Journal*, May 28, 1970, p. 10, col. 3. The FPC's response, filed with the court of appeals, noted that two commissioners dissented from this position.

[56] *Southern Louisiana Area Rate Cases*, 444 F. 2d 125, 127 (5th Cir. 1970).

[57] *Wall Street Journal*, June 18, 1970, p. 7, col. 2. On July 17, 1970, shortly after completion of this paper, the FPC announced that it was initiating a rule-making proceeding to determine whether it should promptly make a temporary upward adjustment in producer price ceilings throughout the nation. It stated that the interim increases, if authorized, would apply only to contracts executed subsequent to the announcement and would be subject to further examination in individual area proceedings. (*New York Times*, July 18, 1970, p. 10, col. 3.)

[58] *Report of the Committee on Natural Gas Reserves* (April 6, 1970).

[59] BNG, *A Staff Report on National Gas Supply and Demand* (September 1969), chap. 2.

[60] Problems of supply may emerge while proven reserves remain substantial. "In general, gas reserves are useful and available to a buyer only when they are within reach of market areas via existing or proposed transmission

Several observations are in order.

1. The industry's concept of proven reserves is a very conservative one. It is designed to provide "a definite measure of the amount of oil or gas demonstrated by past development and considered beyond reasonable doubt to be recoverable from presently known fields by methods in actual operation and under conditions currently prevailing." Thus, "the amount of oil or gas known to be recoverable may be only a small fraction of the amount that will ultimately be recovered. As drilling proceeds in subsequent years and as more information is gained, the estimates of recovery will usually increase until development is complete."[61]

2. The reports or data submitted by individual producers to the AGA are treated by the association as confidential. None of the submissions have been subjected to any kind of outside check.

3. There is room for a very wide exercise of judgment by a producer in determining whether or when reserves have been "proved."

4. In its Permian decision, the FPC questioned the wisdom of preoccupation "with a 'hypothetical' reserves to production ratio" and suggested that the focus of concern should be on finding sufficient gas to replace that consumed each year. It noted that it is an extremely costly matter to maintain a high level of proven reserves many years in advance of their probable use.[62]

5. The principal thrust of the producers' arguments, advanced in the area rate proceedings, is that a high incentive price, as distinguished from a cost-based price, is necessary in order to elicit increased exploratory efforts. Thus, producers have had strong incentive, in order to make good their contentions, to slacken their exploratory and developmental efforts, to make pessimistic appraisals of additions to reserves, and to postpone new contractual commitments to pipelines.

Thus, there are persuasive reasons for believing that the picture of impending crisis may be overdrawn. In an economic analysis recently submitted to a congressional committee, members of the National Economic Research Associates (NERA) put the point as follows:

> It is our firm opinion that this developing crisis is artificial in the sense that, as the gas producers themselves agree, there is plenty of gas left in the ground. . . . Our point in labeling the developing supply crisis as artificial is that the gas producers—which is to say the oil companies—expect that their point of view will prevail with the Federal Power Commission and so have everything to gain by holding off on gas finding and development until they get the price they ask. In making this point we do not accuse the industry of overt agreement. Clearly each producer had a parallel expectation and behaves accordingly.[63]

This is not of course to deny that the elicitation of increasing supplies is a matter of moment. Gas currently satisfies approximately one-third of the nation's energy requirements.[64] Total energy consumption grew at an annual compound rate of 5 percent between 1965 and 1969,[65] and the rate of

---

facilities, are deliverable at the desired daily rates of production, and are not already committed to a long-term sales contract" (ibid., p. 14). The BNG notes that some estimates put the amount of undiscovered gas in the lower forty-eight states and the adjacent offshore continental shelf as high as five times the present proven reserve inventory. These estimates do not include the very large potential in Alaska and its continental shelf area (ibid., pp. 5–6).

[61] National Petroleum Council, *Proved Discoveries and Productive Capacity of Crude Oil, Natural Gas and Natural Gas Liquids in the United States* (1961), pp. 8, 9.

[62] 34 FPC 184. It has also been observed that directional drilling enables the industry to "schedule" discoveries to a far greater degree. "If it is within the power of the gas industry to decide, roughly, how much gas it will discover during some period in the future, there is far less need for a large idle stockpile of reserves. The decline in the uncertainty attending the quantity and timing of future gas discovery reduces commensurately the value of a cushion between reserves and production." (Milton R. Russell, "The Adequacy of Natural Gas Reserves," *Public Utilities Fortnightly* [October 14, 1965], p. 79.)

[63] Bruce Netschert, Abraham Gerber, and Irwin Stelzer, "Competition in the Energy Markets" (submitted to the Subcommittee on Antitrust and Monopoly of the Senate Committee on the Judiciary, 1970, mimeo.), p. 6.

[64] It is estimated that gas and oil together currently provide approximately three-fourths of the country's primary energy requirements. Coal provides 21 percent and water power and atomic energy 4 percent.

[65] Netschert, Gerber, and Stelzer, "Competition in the Energy Markets," p. 3. The underlying data are derived from publications of the U.S. Bureau of Mines.

growth for natural gas was even higher. Gas consumption increased at a compound rate of 6 percent in the 1960s and rose approximately 9 percent in 1969.[66] There can be little doubt that the national demand for energy will continue to grow at a rate faster than the population and that natural gas, which is a preferred fuel for many purposes, will continue to share importantly in the growth if adequate supplies are forthcoming at prices that do not sharply alter the competitive relationship with other fuels.

THE REGULATORY RESPONSE: PRICING POLICY

It is clear that we are in a period of strong and increasing demand for natural gas and that, at least according to industry figures, additions to proven reserves in the past two years have fallen significantly below current consumption. Moreover, there is no doubt that transmission and distribution companies have been experiencing difficulty in contracting for added increments of supply in support of their plans for expansion.[67] On the other hand, adequate gas for a good many years to come is in the ground. Its location, in large part, is known, even though much of it does not fall within the industry's strict classification of "proven reserves." The industry is also equipped to utilize greatly improved techniques of search and development. The question arises: Is the apparent decline in exploratory and productive activity and in contractual commitments the result of a defect or breakdown in the regulatory regime?

It should be emphasized that the pricing scheme adopted by the commission in its area rate proceedings was one consciously designed to elicit new supplies. The FPC rejected the idea of a single area price based on historical costs in recognition of the fact that the cost of much flowing gas was less than the cost that would be incurred in adding new supplies. The separate ceilings fixed for new gas resulted from an appraisal of the current costs of finding and producing gas-well gas not yet committed to the market. The appraisals do not appear illiberal (although, admittedly, circumstances may have significantly changed since the data of record were gathered), and they have withstood all attacks in the review proceedings that have followed the Permian and Southern Louisiana decisions. Moreover, in the Southern Louisiana case, the price level for new gas included a cushion calculated to provide a margin for error and to cover the possibility that unit costs, which had not risen during the year immediately prior to the decision, might show some upward trend during an extended period of price moratorium. Also, in that proceeding, the commission allowed a somewhat higher (0.8 cents per Mcf) price for offshore gas in the federal domain on the assumption that deep-water drilling might entail greater risks and higher unit costs.[68] Whether additional adjustments should be made in these offshore areas was the question initially set for further consideration in the second-round proceedings.

The failure of the FPC's price policies to induce the commitments of new gas that the pipelines are avidly seeking does not prove that the price levels are inadequate to compensate for the costs and risks involved. It may merely indicate, as suggested by the NERA associates, that producers have good reasons for believing that they will soon do better.[69] The history of the commission's actions since its initial decision in the Southern Louisiana case has been one of growing irresolution. An outpouring of speeches by the commissioners (some of them noted by the court of appeals in its Southern Louisiana opinion) gave rise to the conviction, widely expressed by industry journals, that the commission, as

[66] Ibid., p. 5.

[67] See remarks of Commissioner Lawrence O'Connor before the Pacific Coast Gas Association, FPC release, September 10, 1969, p. 3.

[68] As noted earlier, offshore gas in the federal domain was priced 1.5 cents below onshore gas, although it is favored by a tax differential of 2.3 cents.

[69] This is not to say that no other factors have played a role in recent R/P trends. Professor Owen Davidson has argued, for example, that there is a marked correlation between rising interest rates and declining R/P ratios. See testimony of Dr. Davidson submitted in the reopened *Southern Louisiana Area Rate Proceeding* in April 1970. (Transcript, FPC Docket no. AR 69-1.)

constituted at present, is fully prepared to sweeten the price.[70] Indeed, the only serious question in the minds of those participating in the Southern Louisiana proceedings is: How much? Bargaining negotiations looking to possible settlement of issues in the case have been going on for some time. It is an open secret that the producer parties are strongly pressing for agreement on an increase of no less than 5 cents per Mcf and that they are confident of significant gains.

If the producers' estimate of the situation is correct, they would indeed be ill-advised to commit new supplies under the traditional twenty-year contract. Unless the FPC departs from the practice followed in the first phase of the Southern Louisiana case, any new price set for new gas will apply prospectively, that is, to gas dedicated to the market subsequent to the increase. Moreover, Southern Louisiana, which is the single largest area of supply and the focus of industry attention, will doubtless set the tone for other areas. In short, the commission's open manifestations of doubt as to the sufficiency of present prices have had all the elements of a self-fulfilling prophecy.

What prescription may one offer for the future? This writer would urge the following:

1. The FPC should adhere to the basic principle that has been the root of all of the producer proceedings during the past decade—the principle that price ceilings will be anchored to costs, even though it may be wise to provide some margin of flexibility.[71] The commission has developed a workable and now familiar methodology for costing gas. There is no persuasive reason for believing that this method, applied with reasonable liberality and allowing a return on capital fully commensurate with that earned by industries encountering similar risks, will fail in the long run to elicit increasing supplies. This is not to argue that supply can be expected to satisfy the full measure of the potential demand. The FPC doubtless will be obliged to allocate increments of natural gas among uses and users. It can hardly be anticipated, for example, that the supply will permit unlimited sales of gas for use as boiler fuel. However, the commission has always been required to make comparative judgments of this kind. It passed on competing demands for new or expanded pipeline service long before the advent of producer regulation, and it would be required to do so if the regulation of prices at the wellhead were abandoned.

There is, on the other hand, every reason to suppose that if the commission departs from a controlled analysis grounded on costs it will simply flounder—without moorings and wholly exposed to the winds of pressure.

Producers' witnesses in the reopened Southern Louisiana case have recently urged that ceiling prices should be substantially raised for both new and flowing gas without regard to gas costs, using the price of alternative fuels as a general guide and allowing "market forces" to balance supply and demand.[72] Essentially, however, this is a formula for de facto deregulation. Whether to deregulate, however, is a decision for Congress rather than the commission, which, under Phillips and subsequent decisions, has the duty to establish prices at the lowest reasonable level. The agency can satisfy that objective only by seeking a basis for pricing in an appraisal of costs.

[70] See, for example, remarks of FPC Chairman John Nassikas to the American Petroleum Institute, FPC release, November 11, 1969. Mr. Nassikas found "increased warnings of a diminishing gas supply" and offered assurance that the FPC would exercise its "informed judgment" on the basis of "supply-demand-price relationships" (pp. 6–7). More recently, Commissioner Carl Bagge has stated to the Midwest Gas Association his opinion that market forces should be permitted to operate "unfettered by regulations." (FPC release, February 24, 1970, p. 17.)

[71] Commissioner John A. Carver, Jr., has observed that the FPC has an "area of judgment" and that it may fix prices within a "zone of reasonableness" while remaining within the framework of "Court-approved costing methodology." ("Remarks to the Southwestern Legal Foundation," FPC release, March 5, 1970, p. 8.)

[72] *Oil and Gas Journal*, May 18, 1970, p. 42. One may note in passing that the principal alternative fuel is oil and that its price is strongly supported by state "market-demand prorationing" and federal import controls. In its report, *The Oil Import Question: A Report on the Relationship of Oil Imports to the National Security* (February 1970), the Cabinet Task Force on Oil Import Control estimated that in 1969 American consumers paid about $5 billion more for oil products than they would have paid in the absence of import controls. The task force concluded that by 1980 the annual cost will approximate $8.4 billion if the program is continued in its present form.

It should also be emphasized that while the gas industry (which is also the oil industry) contends vehemently that gas is underpriced in relation to oil, the oil industry (which is also the gas industry) tenaciously guards a system of market and import controls that maintains oil prices at a grossly inflated level. If the country is to rely on market forces to achieve equilibrium, surely the same rules of traffic should apply on both sides of the street.

2. If, as suggested above, the prospect of a significant increase has contributed to the reluctance of producers to make new commitments, it follows that the FPC should proceed with all possible speed to establish a firm "new gas" price in the second-round Southern Louisiana proceeding. That issue may be separated from other aspects of the case and decided by interim order.[73] Until it is decided, the prevailing uncertainty will not be dispelled.[74]

3. By the same token it is vital that the commission's new prices introduce a period of stability rather than a third round of expectations and gamesmanship. The ceiling should be one designed to last for a protracted, perhaps indefinite, period.

4. Stability is not inconsistent with periodic adjustments if the adjustments are made on a predetermined and more or less automatic basis. As observed earlier, the commission has developed a methodology for costing gas that dates back to the examiner's decision in the Permian case.[75] The commission should now address itself to the task of devising a formula or index that would enable it to approximate changes in the unit cost of finding and producing gas and to make periodic (perhaps semiannual or annual) adjustments in the ceilings for new gas.[76] The adjusted ceiling would apply to sales thereafter contracted, serving essentially the same purpose as a periodic cost-of-living adjustment to an agreed wage scale. Adjustments to a firm price, made at regular intervals on the basis of a formula reflecting trends in costs, would normally be of small dimension, would minimize uncertainties, and would eliminate the incentive to "sit on" gas in the expectation that a new proceeding might bring about a fundamental change. The establishment of an acceptable index would be an appropriate issue for decision in a litigated area case, but its administration, once established, would not require elaborate supplementary proceedings.

OTHER ASPECTS OF GOVERNMENTAL POLICY BEARING ON THE SUPPLY OF NATURAL GAS

Although the main focus of the gas debate has been on the FPC's pricing of domestic supply, there are other issues of governmental policy—some of them immediate, others in the offing—that are relevant to a coordinated program for eliciting adequate supplies of gas in the future. A detailed analysis of these issues is beyond the scope of the present paper. The discussion that follows is designed to identify certain of these questions and to make some tentative assessment of their scope and import.

[73]The suggested procedure was approved in *FPC v. Tennessee Gas Co.*, 371 U.S. 145 (1962). The issues involved in the pricing of old vintages would appear to differ markedly from those that relate to the pricing of new (uncommitted) supplies. As to the former, it is difficult to see how any factor other than a substantial change in current costs of production would justify a modification of the price ceiling. As to new gas, the critical question is what price will provide adequate incentive to explore and drill.

[74]The remanded Southern Louisiana case was decided by the FPC more than a year after completion of this paper. In its Opinion no. 598, issued July 16, 1971, the FPC very substantially raised the area ceilings. Thus, on gas sold under contracts dated on or after October 1, 1968, the ceiling was lifted to 26 cents, with provision for further escalation in 1974.

[75]Examiner Seymour Wenner's opinion analyzes costs under the following headings: The Cost per Mcf of Drilling and Equipping Successful Wells; Lease Acquisition Costs; Investment in Other Production Facilities; Production Operating Expenses; Dry Hole Costs; Geological and Geophysical Expenses, Lease Rentals, and Exploratory Overhead; Adjustment of Net to Marketed Production; Regulatory Expense, Royalty Payments, and Working Capital; Allowance for Growth; Liquid Credit; Production Taxes. (Initial Decision, AR 61-1, 34 FPC 306, 377-88.)

[76]By way of example, a simplified index might be built on data showing costs of drilling per foot and reserves added during a given time frame. The efficacy of any index obviously depends on an adequate sampling of reliable data. The FPC, however, is not without authority to require the necessary data if it is prepared to exercise it.

Federal Leasing Policy

Although less than 1 percent of the continental shelf area has been tested by drilling,[77] it already seems plain that this area will prove to be our largest source of natural gas for the future. Offshore tracts adjacent to Louisiana and Texas now account for approximately one-eighth of proven reserves.[78] There are favorable prospects of large discoveries elsewhere, notably off the Eastern Seaboard and in the Alaskan shelf areas, and it has been estimated that, by 1980, 40 percent of the nation's production will come from the offshore fields.[79]

Much of this potential is within the federal domain—the so-called outer continental shelf lying seaward of the marginal belt ceded to the states in 1953 by the Submerged Lands Act.[80] Under the provisions of the Outer Continental Shelf Act of the same year,[81] the secretary of the interior is charged with the responsibility of leasing tracts within the federal domain. In this connection, he may prescribe rules and regulations designed to conserve resources, to prevent waste, and to protect correlative rights. He may also grant permission to explore and survey, and confer rights-of-way on pipelines.

The statute provides that the leasing of mineral rights shall take place under a system of sealed competitive bidding.[82] The secretary may conduct the bidding on either of two bases: (1) an "open" cash bonus (that is, one to be determined by the bidding) plus a fixed royalty of not less than 12.5 percent; or (2) a fixed bonus, with bidding as to the amount of royalty (but not less than 12.5 percent). In practice, the secretary has proceeded under the first alternative and has fixed royalties at $16\frac{2}{3}$ percent.[83] The statute further provides that a lease shall cover a compact area not exceeding 5,760 acres and that it shall run for five years or as long thereafter as oil or gas may be produced in paying quantities, or drilling or well reworking operations as approved by the secretary are conducted thereon.

The act provides no guidelines by which the secretary is to select areas for leasing. Moreover, since marine exploration by the United States Geological Survey (USGS) and the Bureau of Mines is extremely limited, the Interior Department's Bureau of Land Management (BLM), which conducts the leasing program, relies almost entirely on the industry for geological and geophysical information. The industry's submissions are made pursuant to the terms of their permits and leases. A close student of the continental shelf administration has observed:

> In many cases USGS does not attempt to obtain all of the data to which it has a right of access because of the lack of staff or facilities to properly examine and analyze the same. As a result, USGS often does not have sufficient geophysical or geological data to evaluate tracts being considered for lease offer. BLM relies, therefore, to a great extent upon industry nominations in its tract selection process.[84]

It also appears that in practice the secretary has exercised little authority over drilling and production. Authorized by the Outer Continental Shelf Act "to cooperate with conservation agencies of

[77]BNG, *Staff Report on National Gas*, p. 52.

[78]Ibid.

[79]This estimate has been made by Commissioner O'Connor (*Oil and Gas Journal*, July 7, 1969, p. 95).

[80]43 U.S.C. 1301 ff.

[81]43 U.S.C. 1331 ff.

[82]As an initial matter, one might question the wisdom of developing the resources of the outer continental shelf by a system of leasing tracts to the highest bidder. One might argue that the United States should have undertaken development in its own right by creating a public corporation with authority to "contract out" work, or that a more rational development might have been achieved pursuant to a "chosen instrument" policy. There is little likelihood, however, of a basic departure from the scheme adopted in the 1953 act and the discussion here proceeds within that framework.

[83]Robert B. Krueger et al., *Study of the Outer Continental Shelf Lands of the United States*, prepared for the Public Land Law Review Commission (Washington, D.C.: Government Printing Office, 1968), vol. 1, pp. 207–08.

[84]Robert B Krueger, "The Development and Administration of the Outer Continental Shelf Lands of the United States," *Rocky Mountain Mineral Law Institute* (Fourteenth Annual, 1968), p. 692. See also Krueger et al., *Study of the Outer Continental Shelf*, vol. 1, pp. 602–06.

adjacent states," the secretary has acceded to arrangements by which the state agencies in Louisiana and Texas are effectively applying their own controls–including well-spacing requirements and determinations of the permissible rate of production ("allowables")–to oil and gas wells in the federal domain.[85]

As noted above, a lease continues after the initial five-year period if the property is producing in paying quantities, or if drilling or well reworking operations approved by the secretary are conducted. The secretary's regulations declare that drilling or reworking operations shall be deemed approved if the lessee "diligently prosecutes" them.[86] Critics of the administration state that in fact there is no meaningful enforcement of the statutory requirement and that a producer disposed to "sit on" a lease may do so on the basis of negligible developmental activity without fear of termination.

To state the matter quite summarily, it appears that the secretary's stewardship of the outer continental shelf has been directed mainly to the conduct of lease sales calculated to yield substantial revenue; that there is thus far little evidence of the development of policies and programs geared to the country's energy needs; and that the Interior Department's discharge of its administrative responsibilities, both at the planning and operational levels, has been gravely handicapped by the lack of adequate facilities and staff, which has made it unduly dependent on the industry and on state regulatory agencies.[87]

It is perhaps useful to note, although without further elaboration, some of the more specific suggestions and criticisms that have been made:

1. The United States should develop added facilities for the conduct of its own surveys and for preliminary exploration.
2. The conduct of lease sales should be on a more regular and predictable basis so as to facilitate industry planning.[88]
3. Properties that may be rich in gas should be identified and given emphasis in the leasing program.
4. A fixed-bonus system of leasing would provide greater opportunity to smaller companies to participate in exploration and development.
5. Royalties on gas production might be fixed at a lower level than on oil to encourage directional drilling and accelerated development of gas-producing properties.[89]
6. Gas royalties might be fixed on a sliding scale, with the rate ascending over the course of time, in order to stimulate more rapid development of the leasehold.[90]
7. The federal government should require more detailed plans and undertakings from the lessee and should administer with greater rigor the requirements for extension of leases.
8. The administration of the federal leaseholds should be divorced from state regulatory schemes, in particular those aspects of the state system that tend to cause overdrilling and that impose artificial restrictions on the rate of production.[91]

More fundamental than these specific issues, however, is the need for governmental machinery adequate to the task of policy making and administration. This has two aspects. The first and most

[85] See Krueger et al., *Study of the Outer Continental Shelf*, vol. 1, pp. 619–23. It has recently been reported that a Senate judiciary subcommittee is preparing to examine state prorationing control over federal leases off Louisiana and Texas (*Oil and Gas Journal*, April 20, 1970, p. 74).

[86] 30 C.F.R. 250.35; 43 C.F.R. 3385a.1.

[87] See, generally, Krueger et al., *Study of the Outer Continental Shelf*; and Krueger, "Development and Administration of the Outer Continental Shelf."

[88] In a press release, March 11, 1968, the Department of the Interior (Bureau of Land Management) recognized the desirability of a more advanced scheduling of sales. Thereafter, delays in the scheduling of sales were caused by fear of spillage.

[89] This suggestion has been advanced by the present chairman of the FPC. See "Remarks of Chairman Nassikas to the American Petroleum Institute," FPC release, November 11, 1969, p. 10.

[90] A proposal along these lines is discussed in the prepared testimony of J. Daniel Khazoom of the FPC Office of Economics, submitted in the reopened Southern Louisiana proceedings, AR 69-1, May 1970.

[91] See text accompanying fn. 85.

obvious need is to develop within the responsible department an expert staff with adequate facilities and manpower at its command. At present, Interior's ability to gather and assess information and to engage in effective administration and enforcement seems woefully inadequate. The second aspect is to bring into the administration other agencies and officials of government with a significant role to play in the making of policy. To take a clear example, the FPC, which has comprehensive responsibility for the regulation of natural gas in interstate commerce, should be heard in the councils that plan the government's leasing policies. One mechanism would be to create a joint board, drawing its membership from the interested departments and agencies,[92] that would function as the secretary's advisor on major issues of policy and program.

A Comment on Incentives

It has recently become fashionable for members of the FPC to make regular obeisance to the precept that there should be no reduction in the incentives extended to the gas and oil industry.[93] During the pendency of the proposals that culminated in the Tax Reform Act of 1969, the commission supported the maintenance of gas and oil percentage depletion at the 27.5 percent rate.[94] Earlier, the FPC chairman, dissenting from the Cabinet Task Force's strenuous criticism of the immensely costly[95] oil import control program, defended that program on the ground that an increase in oil imports would result in a lessening of domestic oil exploration and that this would adversely affect gas supply to the extent that gas is found as an incident of the search for oil.[96]

Even if one unreservedly accepts the commissioner's underlying premises—that energy requirements and gas demands are growing; that exploration and development should be encouraged; and that the risks and uncertainties of the industry warrant special recognition—it does not follow that the incentives provided by existing measures are worth anything like the price we pay nor does it follow that the incentives operate efficiently.

The gigantic cost of the oil import control program to American consumers has been noted earlier.[97] It should be added that percentage depletion is an inefficient device for inducing the search for minerals and the replenishment of reserves. The depletion deduction is derived by applying the statutory percentage (now 22 percent) to the gross income from mining (but not to exceed 50 percent of the net income therefrom). The subsidy is therefore related to current profits rather than costs, increasing the after-tax return on production. A company that has discovered a single producing field will receive the allowance provided by percentage depletion as long as production remains profitable. The benefits are entirely unrelated to the amount of the company's investment in the producing property (indeed, the tax benefits may come to exceed that investment many times over) and their availability is in no way dependent on the initiation of any further exploratory effort.[98]

A far more efficient device would be an allowance strictly confined to exploratory costs, for example, a provision authorizing an immediate write-off of such expenses (or, if one wished to heighten

[92]This would include the Departments of State, Defense, Commerce, and Transportation. See Krueger et al., *Study of the Outer Continental Shelf*, vol. 1, chap. 2.

[93]E.g., "Remarks of Chairman Nassikas to the American Petroleum Institute," p. 9.

[94]See the chairman's testimony in *Supplies of Natural Gas*, Hearings before the Subcommittee on Minerals, Materials and Fuels of the Senate Committee on Interior and Insular Affairs, 91 Cong. 1 sess. (1969), pp. 127, 153.

[95]See fn. 73.

[96]Cabinet Task Force, *The Oil Import Question*, pp. 375–81. Mr. Nassikas's comments take no account of the fact that cheaper oil would reduce the demand for gas, particularly from users in the industrial segment of the market.

[97]See fn. 73.

[98]For an elaboration of the argument that percentage depletion is an inefficient device for stimulating new discovery, see Canadian Royal Commission on Taxation (Carter Commission), *Report* (1966), vol. 4, chap. 23. See also *Tax Reform Studies and Proposals, U.S. Treasury Department*, Joint Publication, House Ways and Means Committee and Senate Finance Committee (February 5, 1969), p. 427, in which the Treasury presented studies indicating that percentage depletion has "a rather low payoff [in terms of additions to reserves] for the revenue foregone."

the incentive, a multiple thereof).[99] It has also been suggested that, at the least, percentage depletion should be modified so as to make it a condition of the allowance that the company "plow back" an equivalent amount in further exploration.[100]

One other somewhat neglected aspect of present tax policy warrants notice: the grant of special tax treatment to the production of oil and gas in a foreign territory. American companies engaged in overseas gas and oil enterprises obtain, in addition to foreign tax credits, the full measure of benefits (including percentage depletion and the expensing of intangible well drilling costs) applicable to domestic mineral extraction.[101] Yet, the lion's share of this production is destined for foreign markets and makes no significant contribution to the preservation of the domestic resources base. More than that, the availability of substantial subsidy for the overseas development of mineral resources inevitably diverts capital that oil and gas companies might otherwise direct to exploration within the continental United States. The magnitude of this diversion is suggested by the following testimony of John G. Winger, vice president of the Chase Manhattan Bank and manager of its Energy Economics Division: "The industry is international in character. And, of all the capital spending throughout the Free World, American companies alone account for 70 per cent. Of the spending devoted to the search for new reserves, they are responsible for as much as 85 percent."[102]

The efficacy and the social cost of these incentives pose issues of substantial and continuing moment. To say that there is an interest in stimulating the finding of new sources of gas marks the beginning, not the end, of the inquiry.

Import Policy

The volume of undiscovered natural gas in Canada is estimated to be at least equal to that potentially available in the "lower forty-eight" states.[103] Canada is currently selling more than 40 percent of its annual production to American transmission companies.[104] Although this represents less than 4 percent of total American consumption, the Canadian industry is at a relatively early stage of development, and Canadian production will doubtless become an increasingly important supplementary source of supply for the American market.[105] There is a mutual interest in a development of this trade. American demand is strong and Canadian supply far exceeds Canada's domestic needs. The United States is of course the natural market for that excess.

There is also an underlying conflict of interest on the subject of price—essentially the same conflict that exists within the United States between producing and consuming states. Questions of import policy have thus come to turn primarily on the matter of accommodation between Canada's natural interest in obtaining a maximum return on Canadian exports and the interest of the United States in keeping prices down and in preventing erosion of the system of domestic controls erected on the authority of the Gas Act.

Although the Federal Power Commission has no jurisdiction over Canadian producers or pipelines operating within Canada, it does exercise a veto power over importation. Section 3 of the Natural Gas

[99] Existing provisions authorizing the expensing of intangible well drilling costs are not so limited.

[100] This suggestion was advanced by Senator Jack Miller of Iowa, among others, in *Tax Reform Act of 1969*, Hearings before the Senate Finance Committee on H.R. 13270, 91 Cong. 1 sess. (1969), pt. 1, pp. 702 ff. In the Canadian government's *White Paper on Tax Reform*, issued in November 1969, it is proposed that depletion allowances would have to be "earned" by continuing exploration and/or development of Canadian minerals. See "CCH Canadian Tax Reports," *Tax Topics*, November 12, 1969.

[101] In the 1969 deliberations on tax reform, the Treasury supported the continuation of percentage depletions on foreign oil and gas production and the subject received only cursory consideration by Congress. See *Tax Reform Act of 1969*, pp. 761 ff.

[102] Prepared testimony submitted in May 1970 to the FPC in the reopened Southern Louisiana proceedings, AR 69-1.

[103] BNG, *Staff Report on National Gas*, pp. 54-55.

[104] Ibid.

[105] The National Energy Board of Canada has predicted that Canadian exports of gas to the United States will increase tenfold by 1990. (*New York Times*, November 24, 1969, p. 70, col. 7.)

Act proscribes any exportation or importation of natural gas to or from a foreign country without an order of authorization from the commission. This prohibition carries with it the power to fix the conditions on which approval will be granted. A further requirement was imposed by an executive order issued in 1953.[106] Finding that the proper conduct of foreign relations requires that "executive permission be obtained for construction and maintenance at the borders of the United States of facilities for the exportation or importation of electric energy and natural gas," the president directed that a presidential permit be obtained for such border facilities. The FPC is authorized to issue the permit on the president's behalf after obtaining the favorable recommendations of the secretaries of state and defense. If the FPC and the two secretaries are not in agreement, the matter is to be referred to the president for decision.

The unfolding of the American-Canadian relationship is revealed by a series of proceedings involving the delivery of gas by Westcoast Transmission Company, a Canadian pipeline, to El Paso Natural Gas Company at the international border near Sumas, Washington. The original arrangements for importation, made pursuant to a 1954 contract, were approved without serious controversy. Subsequently the contract was renegotiated to provide for additional volumes, an extended term, and an increase in price. In an opinion handed down in 1967 when the revised contract came before it, the commission stated, "It would only be fair to alert both the United States and Canadian interests at an early time that gas imports into the United States will not be approved at unjustifiably high prices.[107] Comparing the proposed price terms under the new contract with prices at which Westcoast was selling to Canadian customers, the FPC concluded that the prices to be charged El Paso were out of line. The approach was not unlike that which the commission was following in deciding whether to certificate sales by domestic producers to domestic pipelines.[108] In keeping with its conclusion, the commission conditioned its approval of the importation of additional gas on the submission of a substitute agreement modifying the price terms.

At this point, however, Canada's National Energy Board, exercising its statutory control over the exportation of gas, balked. Passing on Westcoast's application to vary the terms on which the board had previously approved the exportation of added volumes of gas, the Canadian agency concluded that the conditions imposed by the American commission were unacceptable. It stated that the FPC's "in-line" approach failed to take account of "fair market value" and implied that the FPC was seeking to impose its regulatory precepts on companies beyond its jurisdiction. Approval of Westcoast's application, it observed, "would seem to carry with it, despite any protestations we might make, a de facto acceptance of the in-line method of pricing our gas exports, with its implication that the only way to improve export prices in future would be to increase prices charged by the transmission company to its Canadian distributor customers."[109] The board indicated that there were three tests that it would normally apply in determining whether the price at which an applicant proposed to export gas was "just and reasonable in relation to the public interest":[110] (1) whether the price covers the properly allocable share of costs incurred by the Canadian transmission company; (2) whether the price is not less than that available to comparably situated Canadian customers; and (3) whether the delivered price of Canadian gas in the market area approximates the cost of obtaining an alternative supply of energy from American sources. The first two tests, the board stated, establish a floor; the third a ceiling. It concluded that the prices rejected by the FPC fell within this range.

Caught between the millstones, the parties thereupon worked out a further agreement, which they duly submitted to the FPC in 1968 as a settlement proposal. The price terms embodied in the settlement were lower than those set by the parties in the earlier renegotiation but higher than those that

[106] Executive Order 10485, September 3, 1953, 3 C.F.R. 1949–53.
[107] *El Paso Natural Gas Co.*, 38 FPC 311, 319 (1967).
[108] See the discussion of the CATCO case in fn. 31.
[109] *Westcoast Transmission Co.* (December 1967, mimeo.), 8–14 ff., 8–26.
[110] The standard is that set forth in the National Energy Board Act, 7–8 Eliz. II c. 46, par. 83 (1959).

the FPC had specified in its conditioned approval of 1967. This time, the commission approved the amended agreement[111] and the board in turn granted approval to Westcoast's renewed application.

For its part, Canada has recognized that the American authorities can ill afford to approve any arrangement that would make Canadian gas more expensive in the market area than an added increment of American gas—an approval that, if it were given, would create overwhelming pressures to increase the ceiling prices on domestic gas. At the same time, the Canadian board has preserved its position that it is not within the prerogative of American agencies to monitor the costs or profits of Canadian exporters.

The two agencies, in short, have apparently arrived at a tacit understanding as to what falls within the range of political acceptability, while preserving ample room for maneuver. There has been no confrontation since the Westcoast affair. A more serious test of regulatory policy may come if at some point in the future we become heavily dependent on Canadian supply and find it necessary to negotiate for purchases of gas from sources substantially more distant (and hence more expensive) than the Alberta fields from which we are currently drawing.

Beyond this, it should be observed that, as is so often the case, gas issues and oil issues are far from separable. The extent of Canada's willingness, over the long run, to pursue a gas export policy solicitous of American interests may well depend in considerable measure on U.S. readiness to relax the oil import controls that severely restrict Canada's opportunity to sell its oil in the American market.[112]

Liquid Natural Gas

Through processes that involve cooling and application of pressure, natural gas can be converted to a liquid (LNG).[113] In liquid form it occupies only 1/632 of its volume as vapor. Although liquefaction is an expensive process requiring large capital investment, a number of gas distribution companies have found it advantageous to build facilities to liquefy, store, and reconvert limited quantities of gas. The compensating factor is that storage capacity enables the distributor to buy gas during off-peak periods and to satisfy peaks of demand without incurring the large demand charges that it would be required to pay the pipeline if it preempted a larger allotment of the latter's transmission capacity. LNG can be regasified quickly and relatively cheaply for such peakshaving purposes.

Present technology does not make it commercially feasible to build an LNG pipeline (such a line would require very costly insulation). LNG can be transported, however, by marine tankers, rail tank cars, or tank trucks specially designed for the purpose. Gas distributors that have facilities for liquefaction commonly sell some LNG to other nearby distributors or to industrial customers having use for LNG. These transactions are currently of small magnitude and the movements are ordinarily made by cryogenic tank trucks. In a few instances, transportation has been across the Canadian border.

Although the sales and transportation of LNG within the United States have been of only peripheral significance to date, they have already raised for the FPC new issues of regulatory authority. Moreover, as indicated below, LNG is becoming increasingly important on the international scene and there are prospects of a marked increase in its use in the United States. Before seeking to identify the various questions of regulatory policy to which these developments may give rise, there is the threshold question of the FPC's jurisdiction in this area.

The basic statutory issue is whether liquid natural gas is natural gas within the meaning of the Natural Gas Act.[114] The FPC is proceeding on the assumption that it is,[115] although the commission has

[111] *El Paso Natural Gas Co.*, 39 FPC 162 (1968). Since 1968 the FPC has approved importation of additional increments. FPC Docket no. G–8932, orders of January 20 and May 12, 1970.

[112] For expressions of the Canadian view, see *New York Times*, August 7, 1969, p. 47, col. 4; February 16, 1970, p. 53, col. 5.

[113] *LNG Information Book* (July 1968), published by the American Gas Association, contains an excellent discussion of LNG technology.

[114] Natural gas is simply defined as "natural gas unmixed, or any mixture of natural and artificial gas" (sec. 2).

[115] In a press release, January 14, 1963, the FPC first indicated its intention to assert jurisdiction.

never asserted that the various liquid hydrocarbons commonly stripped from natural gas (butane, propane, etc.) and sold or transported in interstate commerce constitute "jurisdictional" gas.[116]

The question of LNG's status has not yet been tested in the courts. It seems very probable, however, that the agency's position will be sustained. If the Gas Act had been construed to apply exclusively to the relationship between pipelines and their customers, there would be no basis for regulating transactions in gas moved by means other than pipe. As observed earlier, however, the Phillips decision in 1954[117] took a more expansive view, holding that the statute also authorizes the economic regulation of producers. It is thus open to the commission to argue that LNG is still natural gas (different in form, but chemically the same and readily reconvertible to its original state) and hence subject to the act whenever it is transported or sold for resale in interstate commerce. The fact that LNG was unknown when the act was passed is certainly not conclusive. The statutory definitions are broad enough to comprehend it and, in light of the Phillips decision, the same may be said of the statutory scheme. Certainly, if LNG were produced in significant volumes and distributed to the conventional users of gas by any of the available media of transportation, the existing regulatory scheme would be directly affected. In a recent case involving the power of the Federal Communications Commission (FCC) to regulate community antenna television (CATV), the Supreme Court upheld the agency's authority, despite the absence of directly applicable statutory language, on the ground that regulation was needed in order to protect existing FCC licensees.[118] Only a few months earlier, in the *Permian Basin Area Rate Cases*, the Court had ruled, "We are, in absence of compelling evidence that such was Congress' intention, unwilling to prohibit administrative action imperative for the achievement of an agency's ultimate purpose."[119]

If LNG is treated as a species of natural gas, it would seem to follow that the FPC may exercise the same powers in relation to LNG transactions as it now exercises in the case of "gaseous" gas. A sale for resale in interstate commerce would require certification. While a direct sale, for example, a sale of LNG to an electric utility for boiler fuel purposes, would not itself be subject to regulation, the commission would have a jurisdictional handle if consummation of the transaction required transportation in interstate commerce.[120] To the extent, however, that jurisdiction is predicated on transportation, one immediately encounters problems of dual regulation. Subject to certain exceptions, the carriage of LNG by railroad, truck, barge or ship already falls within the general regulatory authority of the Interstate Commerce Commission (ICC) or the Federal Maritime Commission.[121]

The FPC has not sought congressional resolution of this issue but has initiated a rule-making proceeding that may largely avoid interagency conflict or overlap.[122] The proposed rule would (1) exempt from the certification requirements of the Gas Act (sec. 7) the transportation of LNG by motor carrier, rail, or barge; (2) exempt from certification limited and incidental sales of LNG (that is, sales of less than 625,000 gallons annually) effectuated by motor carriage, rail, or barge; and (3) impose reporting requirements on those undertaking transportation or sale under the provisions for exemption.

The view that LNG is "jurisdictional" gas also means that exportation and importation require a Section 3 authorization and that such transactions may be disapproved if found not "consistent with the national interest." The commission has issued orders of authorization,[123] but the jurisdictional issue has yet to be litigated.

[116] See Henry F. Lippitt II, "Regulatory Problems in the Use and Development of Liquid Methane," 39 *Texas Law Review* 601 (1961).

[117] *Phillips Petroleum Co. v. Wisconsin.*

[118] *U.S. v. Southwestern Cable Co.*, 392 U.S. 157 (1968).

[119] 390 U.S. 747, 780 (1968).

[120] See *F.P.C. v. Transcontinental Gas Corp.*, 365 U.S. 1 (1961).

[121] Private, as distinguished from common or contract, motor carriers are exempt from ICC jurisdiction. See Interstate Commerce Act, pt. 2, sec. 203(a)(17), 49 U.S.C. 303(a)(17). Moreover, water carriers of commodities in bulk are exempt from regulation under pt. 3 of the act in certain instances. See sec. 303(b), (c), and (d), 49 U.S.C. 903(b), (c), and (d).

[122] 35 Fed. Reg. 3076 (February 17, 1970).

[123] See, e.g., *Phillips Petroleum Co., Marathon Oil Company*, 37 FPC 677 (1967).

If the role of LNG in the U.S. fuel economy were to remain what it is today, the legal questions noted above would be of no great practical moment. LNG is a growing factor in the international fuel market, however, and may come to play a much more prominent role in the United States as well.[124] Large quantities of LNG have been shipped from North Africa to France, Italy, Spain, and the United Kingdom.[125] A substantial project for the movement of LNG from Alaska to Japan is well under way.[126] Various companies with interests in overseas gas production have announced plans for shipping LNG to U.S. ports.[127] A group that includes the Philadelphia Gas Works is actively exploring the possibility of satisfying a portion of base load utility requirements by the importation of Venezuelan liquid gas.[128] There are limiting factors. The construction of facilities for liquefaction takes time and substantial investment. A more serious bottleneck is the scarcity of cryogenic tankers. Nonetheless, the FPC's Bureau of Natural Gas has prognosticated that by 1980 LNG could account for as much as 5 percent of U.S. gas consumption.[129]

Whether this in fact takes place will depend in considerable measure on price and regulatory policy. The Philadelphia group believes that there is a realistic possibility in the near term of importing LNG at prices close to those that would be involved in obtaining added increments of domestic pipeline gas. Among other things, however, this is contingent on an agreement with Venezuela to sell wellhead gas at prices much more favorable than those that obtain in the American gas fields. Moreover, even if the Philadelphia plan proves feasible, it cannot be assumed that LNG for base load will prove to be generally available to American gas distributors at competitive prices. The cost of marine transportation is one significant element. LNG brought from North Africa, as contemplated in certain proposals, would involve considerably higher shipping costs than gas carried from Venezuela. Also, while it might prove practicable to bring LNG to a coastal city enjoying deep-water port facilities, the costs might be prohibitive for a city some distance inland since the ultimate consumers would have to bear the added burden of transportation or transmission costs from the port of entry.

In a study conducted by the Arthur D. Little Company,[130] it is estimated that LNG liquefied in Venezuela could be delivered to New York for a cost approximating 40 cents per Mcf exclusive of purchased gas costs (estimated at 10–15 cents per Mcf). The weighted average wholesale price of pipeline gas in the New York City area is approximately 42 cents. The estimate for shipments from Cook Inlet, Alaska, to Los Angeles is approximately 38 cents (again, exclusive of purchased gas costs), which is about the average wholesale cost of pipeline gas in that area. On the basis of these figures it would appear that the cost of LNG would run 20–25 percent higher than the weighted average cost of pipeline gas. However, the differential is somewhat less, since new increments of pipeline gas cost more than the weighted average price. Thus, it would seem that the price of imported LNG for favorably located port cities would be somewhat higher than the cost of pipeline gas, but that the figures are not very far apart.[131]

Some pipeline and distribution companies would doubtless be prepared to break the domestic price line in order to acquire a commitment of substantial additional supplies in the form of LNG. The policy question that this would pose for the FPC is essentially the same as that noted in connection with the importation of pipeline gas from Canada, and the answer would appear to be the same: that the regulatory agency could hardly expect to maintain for long the prevailing ceilings on domestic

[124] See, generally, T. J. Joyce, "LNG–Key to an International Natural Gas Industry?" *American Gas Journal*, June 1969, p. 28.

[125] BNG, *Staff Report on National Gas*, pp. 60–64; AGA, *LNG Information Book*, pp. 9–10.

[126] The sale is by the Phillips-Marathon group and the contract price is 52 cents per Mcf. See fn. 123.

[127] See BNG, *Staff Report on National Gas*, p. 62; *New York Times*, November 30, 1969, pp. 111–12, col. 5; *Oil and Gas Journal*, January 26, 1970, p. 100.

[128] Interview with Charles H. Frazier, National Economic Research Associates, consultant to Philadelphia Gas Works.

[129] BNG, *Staff Report on National Gas*, p. 64.

[130] Ibid. The results of the study are reproduced in the BNG report, p. 63.

[131] In a cost study conducted by Purvin & Gertz, Inc., Dallas, estimates are made that are close to those in the Arthur D. Little study and point to the same conclusion. See *Oil and Gas Journal*, May 20, 1968, p. 70.

production if it permitted the importation of substantial quantities of foreign gas at significantly higher delivered prices.

On these assumptions, it would seem a fair prediction that some imported LNG will be sold to American gas utilities for base load, as well as for peakshaving, purposes, but that sales for base load will be largely confined to certain favorably located areas. In short, the prospect—barring a major development in the technology or a drastic change in regulatory policy—is that LNG sales will increase substantially in volume, but that the role of LNG will nonetheless remain a purely supplementary one in terms of supplying gas to gas distribution systems.

In considering LNG prospects, Alaskan gas calls for special mention. Alaska's potential reserves are great,[132] but the physical conditions and the distances involved would inevitably make it costly to transmit gas to the "lower forty-eight" by construction of a trans-Canadian pipeline. LNG shipments to the West Coast are a promising alternative.[133] The prospects, however, would be greatly enhanced if Congress could be persuaded to modify the provisions of the Jones Act[134] so as to permit LNG to be transported from Alaska to the other states by tankers of foreign registry. The Arthur D. Little estimate of the cost of shipping LNG from Alaska to Los Angeles, as given above, was based on the assumption that foreign tankers would be used. Shipment by U.S.-flag tankers built in the United States, according to that study, would involve costs amounting to an additional 8 cents per Mcf. The Jones Act has a particularly harsh impact on Hawaii, which has no indigenous source of natural gas on which to draw and is unable to import LNG from the logical source of supply because there are no American tankers equipped to carry it.

There is another important dimension to LNG. It not only is subject to regasification and use in conventional gas distribution systems, but also has potential uses as a liquid commodity, principally as a motive fuel. Attractive because it is relatively free of pollutants, LNG can be used to propel aircraft, railroads, ships, and motor vehicles. The Institute of Gas Technology has predicted that its consumption for these purposes will, one day, far outstrip its use by the gas utilities.[135] If this prediction proves accurate, there will be a new set of regulatory problems.

First, there is the question of import controls. Should the FPC undertake to monitor the importation of quantities of LNG destined for use as a liquid commodity? Certainly the prices that airlines are prepared to pay for jet fuel is a subject far removed from the original purposes of the Gas Act.[136] A limitation on such imports might be much to the liking of domestic oil producers and refiners, but it is difficult to see how it would serve any protective purpose from the standpoint of the consumers on whose behalf the statute was enacted.

Second, there is the question of whether the FPC should seek to assert authority if domestic producers of natural gas begin to liquefy gas for use as motive fuel. If such use becomes widespread, domestic producers will certainly wish to enter the market rather than yield it to imported LNG. It is equally clear that if a new and attractive unregulated market develops, gas supplies will be diverted from the regulated market, with possible adverse effects for the gas utilities and their customers.

At the present juncture, one can only guess at the pace of the anticipated development and the extent to which LNG will be used as a liquid commodity. The need for responsive measures turns both on this and on other future developments. If, for example, oil import controls should be greatly re-

[132] In addition to the major discoveries recently made in the North Slope area, there is great potential in the continental shelf adjoining Alaska.

[133] The cost of shipping LNG is far less per mile than the cost of transmitting gas by pipelines. However, one must add to the LNG cost computation the heavy cost of liquefaction, estimated at about 20 cents per Mcf. See BNG, *Staff Report on National Gas*, pp. 51, 63. See also Nicholas Blederman, "Economics of Gas from the Far North," *American Gas Journal*, March 1970, p. 26.

[134] 46 U.S.C. 883.

[135] This prediction is set forth in BNG, *Staff Report on National Gas*, p. 64. See also James P. Lewis, "LNG as a Vehicle Fuel—Fact or Fancy?," *American Gas Journal*, June 1969.

[136] It may be noted parenthetically that LNG may already be able to compete quite effectively with jet fuels currently in use. See J. R. Lofstrom and T. J. Joyce, "LNG: Fuel with a Future," *American Gas Journal*, June 1968, p. 27.

laxed (as the President's Task Force has vigorously urged) and the price of fuel oil reduced, the pressure of the existing demand for natural gas would be substantially relieved. Again, if synthetic gas should become commercially available within the next few years (see below), the diversion of natural gas to the market for motive fuel would provide far less cause for concern. The one conclusion that can be asserted with some confidence is that decisions as to needs and priorities within the fuel economy can only be made intelligently on the basis of a national energy policy—a policy that at this point is conspicuously absent. These decisions are not within the competence of an agency that, even if it were wiser than there is present reason to suspect, has only parochial responsibility and authority.

Synthetic Gas

In the long run, probably the most promising means of supplementing the supply of gas is by manufacturing gas of pipeline quality from coal. Coal is of course available in virtually limitless quantities and at widely scattered locations. Processes for the conversion of coal to gas with high British thermal units (Btu) have proved technologically feasible and the question now, as with LNG, is primarily one of commercial feasibility.[137]

The BNG has estimated that one process (the hydrogasification process currently in use at the Chicago pilot plant operated by the Institute of Gas Technology) under optimal conditions, could produce gas of pipeline quality at about 33 cents per million Btu.[138] Although this figure is approximately 70 percent higher than the cost of an Mcf of natural gas in the field,[139] that is not the critical comparison. Coal fields are advantageously located in relation to many of the nation's heavily populated and highly industrialized areas where the consumption of energy runs high. Gasification plants located near such centers of demand, producing gas at the projected 33 cent figure and in a position to deliver it at low transmission cost, would be able to compete with pipelines carrying natural gas from far distant points. Moreover, it is reasonable to expect that natural gas prices will increase during the next few years, while further experimentation with the gasification of coal might point the way to further economics in that productive process.

The research and development that have occurred to date have been financed in large measure by the federal government through the Office of Coal Research, Department of the Interior. Yet the expenditure has been meager compared with the importance of the undertaking. The favorable results that have been achieved fully warrant a greatly intensified developmental program.[140]

The position of the natural gas industry vis-à-vis synthetic gas also deserves notice. Recognizing that significant competition may ultimately come from this source, oil and gas companies have been persistently engaged in acquiring control of major coal companies. As pointed out in a study recently submitted by a group of energy economists to a congressional committee, eleven of the largest U.S. petroleum companies (in order of their rank by assets, Standard Oil of New Jersey, Texaco, Gulf, Shell, Atlantic Richfield, Continental Oil, Sun Oil, Occidental, Standard Oil of Ohio, Ashland, and Kerr-McGee) have obtained substantial positions in coal. The group further stated: "It could well be that the self-interest of certain companies with dominant positions, if not of the industry as a whole, would call for delaying the inauguration of a synthetic fuels industry in order to protect existing investments in crude oil and natural gas."[141] To date, however, the Antitrust Division and the Federal

[137]It is also possible to obtain gas of high heating value from oil shale. That technology, however, is not well developed, and investor interest in an oil shale industry has waned in the last several years. See BNG, *Staff Report on National Gas*, p. 73. Views of the Department of the Interior on the prospects of gas-from-coal are expressed in *Supplies of Natural Gas*, pp. 15 ff. See also K. W. Bennett, "Energy Scramble Leads to Coal Gas," *Iron Age*, February 12, 1970, pp. 51–52.

[138]BNG, *Staff Report on National Gas*, pp. 68–70.

[139]One million Btu is the norm for an Mcf of natural gas.

[140]The Institute of Gas Technology, which has made major contributions to gas-from-coal technology, received a $10.8 million five-year contract from the government in 1968. Federal support of research and development in the energy field, however, has been overwhelmingly devoted to nuclear energy.

[141]Netschert, Gerber, and Stelzer, "Competition in the Energy Markets," p. 26; app. B, table 6; and p. 31.

Trade Commission have stood idle, raising no objection to the penetration of the coal industry by oil and gas companies.

Pipelines carrying synthetic gas from coal mines to cities are perhaps five to ten years away. Hence, no regulatory issue is imminent. It should be noted, however, that existing federal agencies would lack jurisdiction over an interstate synthetic gas pipeline. The FPC's authority is confined to natural gas or a mixture of natural and artificial gas. The ICC, for its part, is excluded from the regulation of pipelines transporting either natural or artificial gas.[142] However, the basic considerations that prompted enactment of the Natural Gas Act would appear to be no less applicable to the interstate transmission of synthetic gas.[143] On the other hand, there is insufficient basis for assuming at this stage that effective competition will be lacking at the producing end of the business.

CONCLUSIONS

While further regulatory proceedings may well reveal that there have been recent increases in costs that warrant modification of gas prices, there are persuasive reasons for believing that a significant factor in the reluctance of producers to commit additional supplies to the market is the conviction that the commission is in the process of beating a massive retreat from existing price ceilings. It appears likely that the claims of shortage have been exaggerated. Certainly, the evidence of a serious decline in the ratio of reserves to production is suspect, resting as it does on the self-serving testimony of an industry that has been permitted by the regulatory authorities to draw a veil over the underlying data.

Price regulation would inevitably break down if the FPC should depart from the principle that prices are to be anchored to costs. However, the commission, for its part, must find means of accelerating its proceedings and of making periodic adjustments promptly in the light of changing conditions that affect industry costs. Since the industry characteristically sells gas under long-term contracts, a serious time lag in the agency's response to higher costs acts as a brake on the commitment of future supplies.

It remains true, of course, that the demand for gas, intensified by environmental concerns, grows apace and that the supply is not inexhaustible. The commission, in exercising its powers of certification, must accordingly give increasing attention to the proposed end use of the gas. By the same token, there is a strong interest in stimulating additional sources of supply.

While the United States has long followed a policy of offering large tax incentives to those who engage in the search for gas, the method it has employed is wasteful and inefficient. Moreover, the development of our extensive offshore resources has undoubtedly suffered from the lack of an effective federal leasing policy and from woefully inadequate administration.

For some submarkets the importation of Canadian gas is commercially feasible. Imported LNG may also become a supplementary source of supply for distributors situated at advantageous locations. Nonetheless, in the absence of extraordinary circumstances, the commission should deny import authority if the delivered price is significantly higher than the cost of gas delivered from domestic sources. To do otherwise would cause irresistible pressure on domestic price controls.

Perhaps the most promising alternative source of supply in the long run is synthetic gas of pipeline quality derived from coal, a resource that the United States possesses in immense quantities. This suggests two points: (1) the desirability of a more intensive federal effort to speed further development of the technology, and (2) a recognition by the antitrust authorities that the rapid penetration of the coal industry by gas and oil interests has serious portent for interfuel competition.

[142] Interstate Commerce Act, pt. 1, sec. 1 (1) (b), 49 U.S.C. 1 (1) (b).

[143] See discussion under "The Regulatory Scheme" above.

6 STRUCTURE OF THE NATURAL GAS PRODUCING INDUSTRY

*Clark A. Hawkins**

Examining the structure of an industry enables one to say something about the extent of competition in the market served by the industry—to determine if it is of the competitive, oligopolistic, or monopolistic type. Oligopoly or monopoly may imply antitrust action to promote more competitiveness, or government regulation or ownership if the market represents a natural monopoly or something akin to it. There are all shades of oligopoly and monopoly, however, and the textbook example, especially of monopoly, seldom exists. Instead, "monopoly power" is a term substituted for monopoly and oligopoly to describe structural characteristics of an industry that presumably should be exorcised.

In the natural gas producing industry the possibility of government ownership or antitrust action is greater than in industry generally, because most sellers are large integrated oil producers. Assuming that ownership by government is out of the question, even an antitrust approach (were it warranted) against gas producers would require consideration of the impact on oil operations. Conversely, if antitrust action were taken against oil companies due to monopoly power in their oil markets, there would be implications for natural gas. Both the oil and gas producers are regulated in large part, although in the case of oil the regulation is disguised. Natural gas regulation is for the protection of the gas consumer, and oil regulation is for the protection of the nation (ostensibly because of defense considerations).

An analysis of monopoly power, or market power, is usually made in terms of market structure, conduct, and performance, and these factors are difficult to measure. Market structure is defined by Kaysen and Turner as "those stable features of the environment of a business firm which determine or condition the firm's decision." They maintain that the elements of market structure can be divided into five classes: "(1) the breadth of the market and the character of demand, (2) the number and size distribution of sellers and buyers, (3) the conditions of entry for new sellers and buyers, (4) the character and importance of product differentiation, and (5) the degree of independence of action among sellers and buyers." They define market power as follows: "A firm possesses market power when it can behave persistently in a manner different from the behavior that a competitive market would force on a firm, facing otherwise similar cost and demand conditions. When the 'can' is emphasized, we see that the concept of market power is basically a structural concept."[1] Here, then, is the key argument for an examination of structure, although it should be noted that Kaysen and Turner are primarily interested in antitrust policy rather than in regulation.

The conditions of entry for new sellers and expansion for existing sellers, as well as the character and importance of product differentiation, are very often categorized as "barriers to entry." These

*Professor of Finance and Economics, College of Business and Public Administration, University of Arizona.
The author acknowledges the helpful criticism given this paper by Helmut Frank and Joseph Walka and also the assistance of graduate student Alan Frankle.

[1]Carl Kaysen and Donald F. Turner, *Antitrust Policy* (Harvard University Press, 1959), pp. 71, 75.

arise typically from minimum scale efficiency, capital requirements for entry, absolute cost advantages, and product differentiation. There are no apparent economies of scale, as usually considered, in discovering and producing natural gas. However, from the risk standpoint, there may be economies if it is recognized that, although risk may not be an obstacle to entry, it may be inversely related to size. A large company can drill more wells than a small company without facing bankruptcy if the wells are unsuccessful. Thus, although there may be modest capital requirements to entry, risks may be greater for the small firm. New firms may also be at a disadvantage due to existing firms having many of the best deposits–best in the sense of being easiest and/or cheapest to find. Some good leases and luck may be important requirements for entry as a producer. As for product differentiation, a barrier to entry is practically nonexistent apart from quality differences that are largely nature induced. Except for possibly the risk factor, the barriers to entry for natural gas production are substantially less than in most manufacturing. Even the risk case is not clear, since it may be an inducement to certain types of venture capital that would be in a high marginal tax bracket. The rise of mutual-type funds selling pieces in drilling ventures in recent years is some indication that the high risks themselves do not deter entry.[2]

Real ambiguity exists in a discussion of structure in the natural gas producing industry. Analysis and discussion of structure normally take place in the context of antitrust possibilities–reducing barriers to entry, preventing or breaking up monopoly, and so on. But this sort of approach does not really apply for natural gas production, except insofar as one wishes to approach the entire petroleum industry on these grounds. Such a discussion is outside the purview of this paper, but this does not necessarily imply that the antitrust approach is irrelevant or less desirable than regulation.

The usual reason for regulation in the textbook economic sense is the existence of natural monopoly. It should be obvious, however, that regulation has been imposed in situations other than natural monopoly–for exemple, trucking, milk, and insurance–and with all sorts of licensing laws. The disadvantage of the regulatory approach to monopoly power compared with the antitrust approach is the almost invariable propensity of the former to introduce artificialities into the market. The original intention of regulation as a substitute for a competitive market gets confused with government assurance and protection of the profits of certain firms, certain industries, or certain segments of industries, as well as with the prosecuting of goals, such as redistribution of income. In many cases, instead of fostering competition, regulation is imposed for the benefit of the members of an industry, to protect them from outside competition or to block entry into the industry.

Largely on the economic grounds of efficiency and allocation of resources, economists may debate whether regulation or antitrust action, or neither, should be applied in a particular situation. But it should be recognized as a practical matter that the disposition of the argument very often will ultimately rest more on political than on economic judgments. Nevertheless, to consider regulation on economic grounds, structure should be examined, and one of the main elements of market structure is "the number and size distribution of sellers and buyers," or concentration. This characteristic for the natural gas producing industry is examined next.

CONCENTRATION

The purpose of measuring industry concentration is to assess its various possible effects, only some of which may be economic, such as its effects on income distribution, resource allocation, distribution of power, efficiency of the firm, and rate of technological progress.[3] However, these effects are

[2]Clinton Oil Company, founded in 1963, has become one of the largest independent oil and gas operators in the Mid-Continent and Gulf Coast areas. It has been responsible for some significant oil and gas discoveries in Kansas and Texas. In addition, the company claims to have "enjoyed a success ratio in its drilling operations significantly higher than the overall industry average." This information is from its *Annual Report, 1969.*

[3]See Tibor Scitovsky, "Economic Theory and the Measurement of Concentration," *Business Concentration and Price Policy* (Princeton: National Bureau of Economic Research, 1955).

difficult to measure and represent a large order. Stigler stated that "the purpose of a measure of concentration is to predict the extent of the departure of price (or, alternatively, of rate of return) from the competitive level."[4] Ultimately, of course, such a determination is an economic judgment, because there is no formula for pinpointing the extent of this departure. Our first step here is to determine whether or not a large part of the industry's production is accounted for by large producers.

There are two general types of concentration indexes: one type measures the percentages of industry assets, sales, employment, and the like, which are accounted for by a certain number of firms (absolute measures of concentration); the other type measures the dispersion or inequality of firm sizes (relative measures of concentration). A practical advantage of the first over the second is that the first does not require a knowledge of the total number of firms in the industry. A theoretical advantage is that it is the dominant position of a few firms in an industry that has the more serious potential consequences for competition, rather than the number of firms and their dispersion. A complete study of structure, however, should include relative measures as well as absolute. Assets and employment are not very useful measures in the natural gas producing industry, because many large producers are integrated petroleum companies, and these attributes mostly are not (and in large part cannot be) segregated by product lines. This general problem occurs throughout the analysis of this industry. Therefore, in discussing concentration in the natural gas producing industry, absolute measure and sales figures are relied on to the exclusion of other variables.

The concentration curve shows the cumulative percentage of industry assets, sales, or employment accounted for by the firms in the industry. The firms are shown on the horizontal axis, beginning with the largest. The cumulative percentage of the industry variable is given on the vertical axis. The area under the curve is taken as a measure of concentration; the greater the area, the higher the concentration. It gives a relative measure graphically as well as numerically and can be used to compare concentration in several industries. The higher the curve is at a given point, the higher the concentration, except that ambiguity results when the curves cross one another. Figure 6-1 shows the concentration curve for interstate sales for 1963 and 1968. Figure 6-2 shows a Lorenz curve for the same years. This is a relative measure that can be made by using the firms reporting to the Federal Power Commission (FPC) and their sales. The 45-degree line from the origin to the point of 100 percent sales and 100 percent producers represents an industry composed of firms of equal size. That is, a given percentage of the industry will hold the corresponding percentage of sales of the industry. Since the percentage of firms in the industry is cumulated from the smallest to the largest, the more the curve bends down away from the 45-degree line the more unequal the size distribution of firms. There is considerable inequality of size in the natural gas producing firms. With over 3,000 producers, and the top eight having one-third or more of the sales, this is no surprise.

The idea of taking the percentages of output, employees, or total value attributable to the largest four, eight, and twenty producers originated in the 1930s[5] and is still widely used as the concentration measure to examine market power. Kaysen and Turner have taken this kind of an approach to analyzing what they refer to as "structurally oligopolistic markets." They define this type of market as follows:

> A structurally oligopolistic market is one in which the few largest sellers in the market have a share of the market sufficient to make it likely that they will recognize the interaction of their own behavior and their rival's response in determining the values of the market variables.[6]

They take as the dividing point a market share of one-third of the total sales for the eight largest sellers and, within this class, distinguish two types of structural oligopoly. A Type I oligopoly occurs

[4] George J. Stigler, *The Organization of Industry* (Homewood: Richard D. Irwin, Inc., 1968), p. 30.

[5] See Gardner C. Means, *The Structure of the American Economy* (National Resources Planning Board, 1939); and W. L. Thorp and W. F. Crowder, *The Structure of Industry*, Monograph no. 27 (Washington, D.C.: Temporary National Economic Committee, 1941).

[6] Kaysen and Turner, *Antitrust Policy*, p. 27.

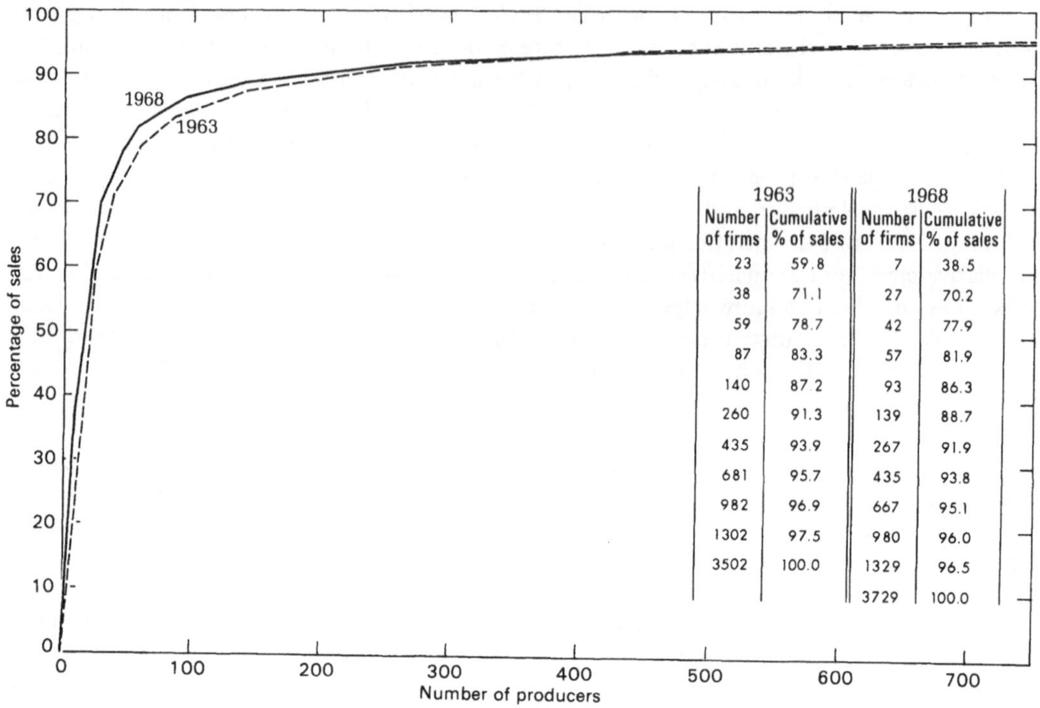

1963 Number of firms	1963 Cumulative % of sales	1968 Number of firms	1968 Cumulative % of sales
23	59.8	7	38.5
38	71.1	27	70.2
59	78.7	42	77.9
87	83.3	57	81.9
140	87.2	93	86.3
260	91.3	139	88.7
435	93.9	267	91.9
681	95.7	435	93.8
982	96.9	667	95.1
1302	97.5	980	96.0
3502	100.0	1329	96.5
		3729	100.0

FIGURE 6-1. Concentration curve for interstate natural gas sales, 1963 and 1968.

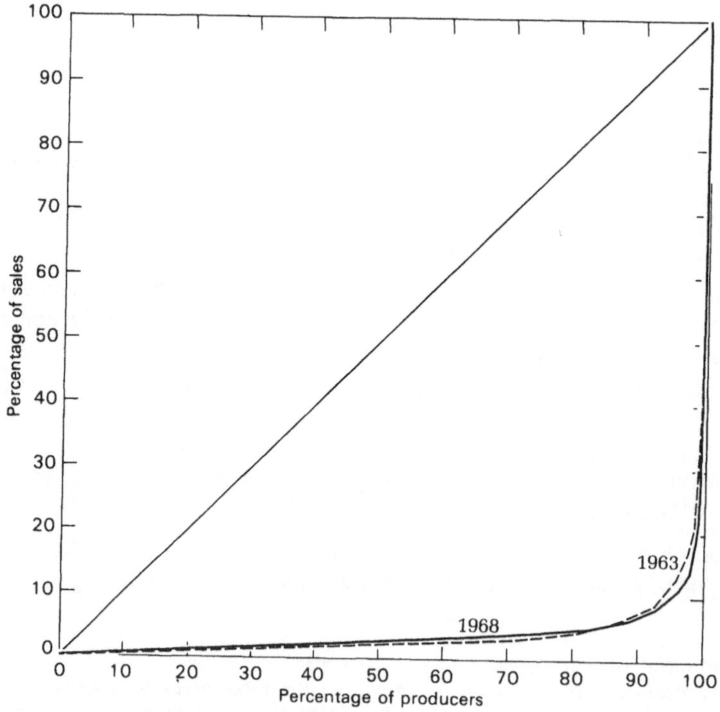

1963 % of firms	1963 % of sales
62.8	2.5
71.9	3.1
80.5	4.3
87.5	6.1
92.5	8.7
96.0	12.8
97.5	16.7
98.3	21.3
98.9	28.9
99.3	40.2
100.0	100.0

1968 % of firms	1968 % of sales
64.3	3.5
73.7	4.0
82.1	4.9
88.3	6.2
92.8	8.1
96.2	11.3
97.5	13.7
98.4	18.1
98.8	22.1
99.2	29.8
99.8	61.5
100.0	100.0

FIGURE 6-2. Lorenz curve for interstate natural gas sales, 1963 and 1968.

when the first eight firms have at least 50 percent of the market and the first twenty have at least 75 percent of the market. A Type II oligopoly exists when the largest eight firms account for one-third of sales, and the rest of the market is relatively unconcentrated. The distinguishing feature between the two is the large unconcentrated sector in Type II that could be considered a competitive constraint on the behavior of the largest eight firms.

It is interesting to note that if interstate sales by producers are examined, the producing industry falls almost exactly into the second category. The percentage of sales accounted for by the top eight firms ranges from about one-third in 1961 to about 40 percent in 1968. Tables 6-2 through 6-9 give the volume and percentage of sales for the top twenty-four producers for this time period. Table 6-1 gives the top twenty in 1955. A slight trend toward more concentration is apparent in the figures. The top twenty held 53 percent of sales in 1961 and 63.1 percent in 1968. The top eight held 32.4 percent in 1961 and 41.8 percent in 1968. It should be noted that the sales figures cannot be taken as the volumes of output of the various producers because, aside from the omission of all intrastate sales, those sales reported to the FPC include some gas purchased by other producers. These intra-industry sales occur among at least the largest fifty producers, with the possible exception of the Canadian producers. Such sales are attributed to the first-named producer so that double counting does not occur. Although the practical effect of this is unknown in identifying concentration in output, it is unlikely to significantly bias the sales concentration data.[7] Production concentration might be slightly less due to the sales being attributed to the first company named. The top twenty-four companies are given, rather than only the usual twenty, because of an identification of the top twenty-four as major supply companies from 1963 (twenty-three in 1965) by the FPC.[8]

Table 6-1. Natural Gas Sales by Producers to Interstate Pipelines, 1955

Rank	Producer	Volume (*Mcf*)	% of total	Cumulative %
1	Phillips	511,113,955	8.74	8.74
2	Standard (Indiana)	324,296,232	5.54	14.28
3	Humble	268,144,970	4.58	18.86
4	Union Producing	231,320,866	3.95	22.81
5	Cities Service	214,508,599	3.67	26.48
6	Shell	190,547,320	3.26	29.74
7	Magnolia	163,906,059	2.80	32.54
8	Chicago Corp.	150,237,142	2.57	35.11
9	Gulf	147,643,569	2.52	37.63
10	Texas Co.	131,512,727	2.25	39.88
11	Atlantic	124,790,611	2.13	42.01
12	Shamrock	89,595,951	1.53	43.54
13	Pure	87,971,532	1.50	45.04
14	Sun	82,351,101	1.41	46.45
15	Ohio	78,834,690	1.35	47.80
16	Tidewater	77,796,361	1.33	49.13
17	Sunray	72,341,319	1.24	50.37
18	Skelly	69,544,744	1.19	51.56
19	Southern Production	64,553,341	1.10	52.66
20	Superior	62,184,420	1.06	53.72
Total: top 20		3,143,195,509		
Total: other		2,707,714,263		
Grand total		8,339,356,000		

SOURCE: Leslie Cookenboo, Jr., "Competition in the Field Market for Natural Gas," *The Rice Institute Pamphlet*, Monograph in Economics, vol. 44, no. 4 (January 1958), p. 48.

[7]In 1968 Humble, the largest producer, sold approximately 1,109 billion cubic feet, of which about 97 billion cubic feet, or 8.7 percent, was in association with other producers. Since Humble is the first-named producer, this should constitute the highest absolute figure and indicate the magnitude of the highest percentage figure of this type for the largest producers.

[8]FPC, Bureau of Natural Gas, *The Gas Supplies of Interstate Natural Gas Pipeline Companies, 1968* (January 1970), p. 9.

Table 6-2. Natural Gas Sales by Producers to Interstate Pipelines, 1961

Rank	Producer	Volume (*Mcf*)	% of total	Cumulative %
1	Phillips	661,328,762	7.9	7.9
2	Pan American Petroleum	461,448,960	5.5	13.5
3	Humble Oil	358,481,858	4.3	17.8
4	Shell Oil	318,808,523	3.8	21.6
5	Texaco	273,957,034	3.3	24.9
6	Socony Mobil Oil	224,113,059	2.7	27.6
7	Gulf Oil	223,711,587	2.7	30.2
8	Atlantic Refining	181,846,018	2.2	32.4
9	Superior Oil	178,985,618	2.1	34.6
10	Sinclair Oil & Gas	164,629,233	2.0	36.5
11	Union Producing	161,347,284	1.9	38.5
12	Union Oil of California	161,016,019	1.9	40.4
13	Sun Oil	158,337,777	1.9	42.3
14	Cities Service	154,909,902	1.9	44.2
15	Champlin Oil & Refining	145,238,628	1.7	45.9
16	Continental Oil	140,002,636	1.7	47.6
17	Union Texas National Gas	133,723,696	1.6	49.2
18	Tidewater Oil	112,442,443	1.3	50.5
19	Calif. Co. Div. of Chevron	105,877,983	1.3	51.8
20	Warren Petroleum	102,617,430	1.2	53.0
21	Skelly Oil	99,031,546	1.2	54.2
22	Sunray Mid Continent Oil	97,755,161	1.2	55.4
23	Pure Oil	97,241,802	1.2	56.6
24	Ohio Oil	90,932,739	1.1	57.7
Total: top 24		4,807,785,698		
Total: other		3,531,570,302		
Grand total		8,339,356,000		

SOURCE: FPC, *Sales by Producers of Natural Gas to Interstate Pipeline Companies* (1962).

Table 6-3. Natural Gas Sales by Producers to Interstate Pipelines, 1962

Rank	Producer	Volume (*Mcf*)	% of total	Cumulative %
1	Phillips	617,985,842	7.0	7.0
2	Pan American Petroleum	497,530,561	5.6	12.6
3	Shell Oil	402,940,484	4.5	17.1
4	Humble Oil	390,001,122	4.4	21.5
5	Texaco	321,154,332	3.6	25.1
6	Socony Mobil Oil	311,767,894	3.5	28.6
7	Gulf Oil	265,288,747	3.0	31.6
8	Superior Oil	216,247,671	2.4	33.9
9	Union Oil of California	199,938,375	2.2	36.0
10	Atlantic Refining	184,574,529	2.1	38.0
11	Sinclair Oil & Gas	181,508,997	2.0	42.0
12	Continental Oil	175,320,490	2.0	42.0
13	Sun Oil	171,435,951	1.9	43.9
14	Union Producing	169,633,996	1.9	43.9
15	Cities Service	143,770,708	1.6	47.4
16	Champlin Oil & Refining	140,424,783	1.6	49.0
17	Westcoast Transmission (Canadian)	132,626,196	1.5	50.5
18	Calif. Co. Div. of Chevron	128,336,413	1.4	51.9
19	Alberta & Southern Gas (Canadian)	126,211,326	1.4	53.3
20	Union Texas Petroleum Div. Allied	121,350,464	1.4	54.6
21	Tidewater Oil	121,033,959	1.4	56.0
22	Skelly Oil	115,990,453	1.3	57.3
23	Warren Petroleum	100,203,316	1.1	58.4
24	Marathon Oil	99,272,444	1.1	59.5
Total: top 24		5,334,948,953		
Total: other		3,629,376,047		
Grand total		8,964,325,000		

SOURCE: FPC, *Sales by Producers of Natural Gas to Interstate Pipeline Companies* (1963).

Table 6-4. Natural Gas Sales by Producers to Interstate Pipelines, 1963

Rank	Producer	Volume (*Mcf*)	% of total	Cumulative %
1	Phillips	613,624,000	6.5	6.5
2	Pan American Petroleum	535,237,000	6.5	12.2
3	Shell Oil	441,733,000	4.7	16.9
4	Humble Oil	438,622,000	4.7	21.6
5	Socony Mobile Oil	358,115,000	3.8	25.4
6	Texaco	335,817,000	3.6	29.0
7	Gulf Oil	296,633,000	3.2	32.1
8	Sinclair Oil & Gas	224,653,000	2.4	34.4
9	Union Oil of California	223,595,000	2.4	36.8
10	Superior Oil	213,853,000	2.3	39.1
11	Continental Oil	205,143,000	2.2	41.3
12	Atlantic Refining	183,440,000	1.9	43.2
13	Sun Oil	182,203,000	1.9	45.1
14	Union Producing	170,914,000	1.8	46.9
15	Cities Service	158,406,000	1.7	48.6
16	Calif. Co. Div. of Chevron	147,565,000	1.6	50.2
17	Westcoast Transmission (Canadian)	138,169,000	1.5	51.7
18	Champlin Oil & Ref.	136,735,000	1.5	53.2
19	Union Texas Petroleum Div. Allied	134,418,000	1.4	54.6
20	Skelly Oil	130,125,000	1.4	56.0
21	Alberta & Southern Gas (Canadian)	128,064,000	1.4	57.4
22	Tidewater Oil	125,470,000	1.3	58.8
23	Marathon Oil	104,093,000	1.1	59.9
24	Ashland Oil	99,842,000	1.1	61.0
Total: top 24		5,726,469,000		
Total: other		3,689,833,000		
Grand total		9,416,302,000		

SOURCE: FPC, *Sales by Producers of Natural Gas to Interstate Pipeline Companies* (1964).

Table 6-5. Natural Gas Sales by Producers to Interstate Pipelines, 1964

Rank	Producer	Volume (*Mcf*)	% of total	Cumulative %
1	Phillips	619,938,889	6.2	6.2
2	Pan American Petroleum	615,071,099	6.1	12.3
3	Humble Oil	503,780,689	5.0	17.3
4	Shell Oil	479,866,953	4.8	22.1
5	Socony Mobil Oil	410,906,934	4.1	26.2
6	Texaco	356,457,117	3.5	29.7
7	Gulf Oil	352,580,568	3.5	33.2
8	Sinclair Oil & Gas	272,500,230	2.7	35.9
9	Continental Oil	249,477,950	2.5	38.4
10	Union Oil of California	240,628,333	2.4	40.8
11	Superior Oil	226,830,004	2.3	43.1
12	Sun Oil	203,758,259	2.0	45.1
13	Atlantic Refining	202,288,810	2.0	47.1
14	Union Producing	177,609,757	1.8	48.9
15	Cities Service	171,703,634	1.7	50.6
16	Calif. Co. Div. of Chevron	156,686,800	1.6	52.2
17	Alberta & Southern Gas (Canadian)	146,067,886	1.5	53.7
18	Westcoast Transmission (Canadian)	143,199,042	1.4	55.1
19	Skelly Oil	133,336,889	1.3	56.4
20	Tidewater Oil	133,216,161	1.3	57.6
21	Champlin Petroleum	132,529,862	1.3	58.9
22	Union Texas Petroleum Div. Allied	128,044,219	1.3	60.2
23	Marathon Oil	111,324,327	1.1	61.3
24	Ashland Oil	110,621,613	1.1	62.4
Total: top 24		6,278,426,025		
Total: other		2,783,109,975		
Grand total		10,061,535,000		

SOURCE: FPC, *Sales by Producers of Natural Gas to Interstate Pipeline Companies* (1965).

Table 6-6. Natural Gas Sales by Producers to Interstate Pipelines, 1965

Rank	Producer	Volume (*Mcf*)	% of total	Cumulative %
1	Humble Oil	656,856,220	6.4	6.4
2	Pan American Petroleum	631,271,629	6.1	12.5
3	Phillips	611,540,672	5.9	18.4
4	Shell Oil	493,906,694	4.8	23.2
5	Gulf Oil	425,514,121	4.1	27.3
6	Socony Mobil Oil	413,090,344	4.0	31.3
7	Texaco	332,439,459	3.2	34.5
8	Continental Oil	331,244,000	3.2	37.7
9	Union Oil of California	283,387,275	2.7	40.4
10	Sinclair Oil & Gas	266,739,907	2.6	43.1
11	Atlantic Refining	227,188,368	2.2	45.3
12	Superior Oil	220,212,942	2.1	47.4
13	Sun Oil	211,992,960	2.1	49.5
14	Union Producing	171,000,689	1.7	51.2
15	Cities Service Oil	169,854,414	1.6	52.8
16	Alberta & Southern Gas (Canadian)	152,298,574	1.5	54.3
17	Westcoast Transmission (Canadian)	143,991,086	1.4	55.7
18	Champlin Petroleum	143,320,208	1.4	57.1
19	Skelly Oil	143,090,778	1.4	58.5
20	Union Texas Petroleum Div. Allied	136,423,849	1.3	59.8
21	Calif. Co. Div. of Chevron	130,915,214	1.3	61.1
22	Tidewater Oil	130,704,007	1.3	62.4
23	Ashland Oil	107,786,085	1.0	63.3
24	Warren Petroleum	102,000,689	1.0	64.3
Total: top 24		6,637,070,184		
Total: other		3,678,329,816		
Grand total		10,315,400,000		

SOURCE: FPC, *Sales by Producers of Natural Gas to Interstate Pipeline Companies* (1966).

Table 6-7. Natural Gas Sales by Producers to Interstate Pipelines, 1966

Rank	Producer	Volume (*Mcf*)	% of total	Cumulative %
1	Humble Oil	770,612,919	6.9	6.9
2	Pan American Petroleum	685,819,131	5.2	13.1
3	Phillips	665,450,028	5.0	19.1
4	Shell Oil	539,052,933	4.8	23.9
5	Gulf Oil	493,762,882	4.4	28.3
6	Mobil Oil	446,779,174	4.0	32.2
7	Union Oil	365,918,247	3.6	35.7
8	Continental Oil	333,319,245	3.0	38.6
9	Texaco	320,416,463	2.9	41.5
10	Sinclair Oil & Gas	273,553,969	2.5	44.0
11	Superior Oil	233,225,431	2.1	46.1
12	Atlantic Richfield	229,419,742	2.1	48.2
13	Sun Oil	210,002,983	1.9	50.1
14	Cities Service Oil	175,174,883	1.6	51.7
15	Calif. Co. Div. of Chevron	171,356,282	1.5	53.2
16	Union Producing	170,573,276	1.5	54.7
17	Alberta & Southern Gas (Canadian)	165,812,840	1.5	56.2
18	Union Texas Petroleum Div. Allied	152,513,781	1.4	57.6
19	Champlin Petroleum	152,341,957	1.4	59.0
20	Westcoast Transmission (Canadian)	152,341,957	1.4	60.3
21	Skelly Oil	149,933,887	1.3	61.6
22	Tidewater Oil	133,145,095	1.2	62.8
23	Marathon Oil	122,587,518	1.1	63.9
24	Warren Petroleum	119,110,195	1.1	65.0
Total: top 24		7,232,341,309		
Total: other		3,900,324,420		
Grand total		11,132,665,729		

SOURCE: FPC, *Sales by Producers of Natural Gas to Interstate Pipeline Companies* (1967).

Table 6-8. Natural Gas Sales by Producers to Interstate Pipelines, 1967

Rank	Producer	Volume (*Mcf*)	% of total	Cumulative %
1	Humble Oil	895,342,366	7.5	7.5
2	Pan American Petroleum	723,118,958	6.1	13.6
3	Phillips	677,293,687	5.7	19.3
4	Gulf Oil	613,139,638	5.1	24.4
5	Shell Oil	560,265,171	4.7	29.1
6	Mobil Oil	472,222,091	4.0	33.1
7	Texaco	435,747,483	3.7	36.8
8	Union Oil of California	396,612,915	3.3	40.1
9	Continental Oil	346,771,098	2.9	43.0
10	Sinclair Oil and Gas	271,171,286	2.3	45.2
11	Superior Oil	229,119,973	1.9	47.1
12	Atlantic Richfield	226,710,168	1.9	49.0
13	Sun Oil	215,633,355	1.8	50.8
14	Calif. Co. Div. of Chevron	207,012,019	1.7	52.5
15	Alberta & Southern Gas (Canadian)	206,106,980	1.7	54.2
16	Westcoast Transmission (Canadian)	188,028,293	1.6	55.8
17	Union Producing	177,587,326	1.5	57.3
18	Cities Service Oil	169,478,854	1.4	58.7
19	Union Texas Petroleum Div. Allied	163,989,267	1.4	60.1
20	Champlin Petroleum	156,791,886	1.3	61.5
21	Getty Oil	153,322,709	1.3	62.8
22	Skelly Oil	145,643,271	1.2	64.0
23	Marathon Oil	130,830,722	1.1	65.1
24	Warren Petroleum	119,614,615	1.0	66.1
	Total: top 24	7,881,554,941		
	Total: other	4,042,981,059		
	Grand total	11,924,536,000		

SOURCE: FPC, *Sales by Producers of Natural Gas to Interstate Pipeline Companies* (1968).

Table 6-9. Natural Gas Sales by Producers to Interstate Pipelines, 1968

Rank	Producer	Volume (*Mcf*)	% of total	Cumulative %
1	Humble Oil	1,109,232,556	8.6	8.6
2	Pan American Petroleum	758,163,249	5.9	14.5
3	Gulf Oil	730,324,434	5.7	20.2
4	Phillips	661,242,535	5.2	25.4
5	Shell Oil	613,560,965	4.8	30.2
6	Mobil Oil	539,680,271	4.2	34.4
7	Texaco	523,539,875	4.1	38.5
8	Union Oil	425,897,887	3.3	41.8
9	Continental Oil	383,334,018	3.0	44.8
10	Sinclair Oil and Gas	284,040,864	2.2	47.0
11	Sun Oil	256,807,518	2.0	49.0
12	Alberta & Southern Gas (Canadian)	241,363,664	1.9	50.9
13	Superior Oil	235,737,400	1.8	52.7
14	Cities Service	228,864,475	1.8	54.4
15	Atlantic Richfield	226,625,424	1.8	56.2
16	Calif. Co. Div. of Chevron	209,308,057	1.6	57.8
17	Westcoast Transmission (Canadian)	199,280,313	1.6	57.8
18	Union Texas Petroleum Div. Allied	165,454,246	1.3	60.7
19	Union Producing	156,378,923	1.2	61.9
20	Getty Oil	155,041,233	1.2	63.1
21	Champlin Petroleum	154,183,703	1.2	64.3
22	Skelly Oil	148,378,679	1.2	65.5
23	Marathon Oil	144,758,329	1.1	66.6
24	Warren Petroleum	122,075,580	1.0	67.6
	Total: top 24	8,673,184,198		
	Total: other	4,150,836,949		
	Grand total	12,824,021,147		

SOURCE: FPC, *Sales by Producers of Natural Gas to Interstate Pipeline Companies* (1969).

Although the interstate sales figures given in table 6-1 are not total figures, they are about the best available for getting an idea of concentration. It would not be meaningful, as mentioned above, to examine assets or employees of natural gas producing companies, and at the same time two other variables that it would be desirable to examine are not available. Total national sales or production by the top companies would be a more precise measure of concentration than interstate sales. Neither sales nor production can be used, however, because the companies do not report them on a comparable basis. For example, Esso reports sales, Mobile reports gross production, and Standard Oil of Indiana reports net production. The other variable that is not available is reserves. Since reserves correspond to the amounts of future sales controlled, they represent a measure of concentration with potential that may be more important than current sales. Uncommitted reserves have been held by some (such as National Economic Research Associates) to be the most crucial concentration for an assessment of competition in field markets. This now seems to be largely a moot question, because current evidence indicates little uncommitted reserves available. Statements by pipeline and distribution officials about their inability to obtain more gas, and the possibility of curtailing customers, FPC hearing testimony, and producer statements all point in this direction.[9] In any case, very few companies make information public on reserves,[10] whether total or uncommitted reserves, and thus it is impossible to examine producer concentration from this viewpoint.

The obvious thing that can be said about reserves is that they are declining relative to production, as measured by the life index. In 1968 and 1969, when fewer reserves were added to the proved category than production (something claimed imminent for at least ten years by producers) there was some indication of worry at the FPC, and certain gas distributors reassessed their view of regulation of field prices. In tables 6-10 and 6-11 there are figures for reserves of interstate pipelines. Table 6-10 gives that part of reserves owned by interstate pipelines and dedicated to them by contract. It can be seen from this table that about 70 percent of proved reserves are committed to the pipelines. The life index for pipelines, although just slightly higher, seems to be falling monotonically along with the nationwide average. The absolute amounts under contract to pipelines rise, but the amounts owned by pipelines stay nearly constant or fall slightly. This is no doubt due to the disincentive (until changed in 1969) for pipelines to drill by placing all pipeline gas in the rate base at "cost," rather than permitting a market or area ceiling price. This problem stems all the way back to the Hope case, of course. Permission now to allow new pipeline gas the relevant producer ceiling may encourage pipeline exploration.

The concentration of buyers is also of importance in characterizing the market. Purchases of natural gas by interstate pipelines are given in tables 6-12 through 6-19 for the top twenty purchasers from 1961 to 1968. The top four buyers took 40 percent of output and the top twenty about 90 percent, a share that declines slightly during this period. A simple examination of market shares on both sides of the market indicates more evidence of monopsony than of monopoly. To substantiate that

[9] In "Power Crisis," *Wall Street Journal*, June 2, 1970, the following appeared: "A number of gas suppliers have told industrial customers they're clamping a lid on new business. Some utilities are actually cutting back. . . . This winter Pittsburgh Plate Glass's glass division had gas service reduced at various plants on two dozen occasions." In testimony introduced in the reopened South Louisiana hearing, as reported in "Virtually No Uncommitted Gulf Gas Left," *Oil and Gas Journal*, May 18, 1970, the following statement was made: "Of 2588 gas completions shut in as of December 31, 1969, all but 128 were committed to contract. Of the uncommitted group, only 29 represented gas reserves which could economically be made available to the interstate market. And all of these wells were the subject of contract negotiations." In the 1969 *Annual Report* of Standard Oil of New Jersey, there is this statement: "Humble has the largest proved gas reserves of any single company in the United States. Virtually all of these reserves, except those in Alaska and those required for its own production and manufacturing operations, have now been committed to sale."

[10] Some do, however. For example, Continental Oil in its 1969 *Annual Report* states: "At the end of the year, the (Western Hemisphere) Division's estimated reserves of natural gas totaled 7711 billion cubic feet, 22 billion cubic feet less than at the end of the previous year. All of these gas reserves were located in the United States and Canada." In Standard Oil of Indiana's 1969 *Annual Report*, the company's U.S. and Canadian natural gas reserves for the years 1960 through 1969 are given: 20,379 billion cubic feet in 1969; 20,464 billion cubic feet in 1968; and 15,358 billion cubic feet in 1960.

Table 6-10. Natural Gas Reserves of Interstate Pipelines, 1963-68

Year	(1) Proved reserves[a]	(2) Owned by interstate pipelines[b]	(3) Under contract to interstate pipelines[b]	(4) Total committed to interstate pipelines[b]	(5) 4 ÷ 1
	(- - - - - -	- - - - - - *billion Mcf*	- - - - - -	- - - - - -)	*(percent)*
1963	276.2	23.4	168.6	192.0	69.5
1964	281.3	22.9	169.6	192.5	68.4
1965	286.5	23.0	171.8	194.8	68.0
1966	289.3	22.2	176.2	198.4	68.6
1967	292.9	22.2	179.9	202.1	68.9
1968	287.4	21.7	177.0	198.7	69.1

[a] AGA, *Gas Facts* (1963 through 1968).
[b] FPC, *The Gas Supplies of Interstate Natural Gas Pipeline Companies, 1968* (January 1970).

Table 6-11. Natural Gas Life Indexes, Nationwide and Interstate Pipelines, 1963-68

Year	Nationwide pipelines[a]				Interstate pipelines[b]			
	Reserves	Production	R/P ratio	F/P ratio	Reserves	Production	R/P ratio	F/P ratio
	(- - - - - *billion Mcf*	- - - - -)			(- - - - - *billion Mcf*	- - - - -)		
1963	276.2	14.5	19.0	1.26	192.0	9.6	20.0	–[c]
1964	281.3	15.3	18.3	1.33	192.5	10.3	18.7	1.04
1965	286.5	16.3	17.3	1.31	194.8	10.5	18.6	1.21
1966	289.3	17.5	16.5	1.16	198.4	11.3	17.6	1.31
1967	292.9	18.4	15.9	1.19	202.1	12.0	16.8	1.30
1968	287.4	19.3	14.9	0.71	198.7	12.8	15.5	0.73

[a] AGA, *Gas Facts* (1963 through 1968).
[b] FPC, *The Gas Supplies of Interstate Natural Gas Pipeline Companies, 1968* (January 1970).
[c] Not available.

Table 6-12. Natural Gas Purchases by Interstate Pipelines, 1961

Rank	Pipeline	Volume *(Mcf)*	% of total	Cumulative %
1	United Gas Pipeline	1,239,441,531	14.9	14.9
2	El Paso Natural Gas	1,107,707,926	13.3	25.2
3	Tennessee Gas Transmission	894,661,417	10.7	38.9
4	Northern Natural Gas	528,796,873	6.3	45.2
5	Transcontinental Gas Pipeline	474,014,712	5.7	50.9
6	Cities Service Gas	370,229,794	4.4	55.3
7	Texas Eastern Transmission	357,612,637	4.3	59.6
8	Natural Gas Pipeline of America	324,361,136	3.9	63.5
9	Lone Star Gas	293,475,635	3.5	67.0
10	United Fuel Gas	280,828,242	3.4	70.4
11	Texas Gas Transmission	277,654,224	3.3	73.7
12	Southern Natural Gas	272,107,398	3.3	77.0
13	Panhandle Eastern Pipeline	222,529,562	2.7	79.8
14	Trunkline Gas	213,120,003	2.6	82.4
15	Arkansas Louisiana Gas	208,760,712	2.5	84.9
16	Michigan Wisconsin Pipeline	181,106,863	2.2	87.1
17	Colorado Interstate Gas	166,558,002	2.0	89.1
18	Olin Gas Transmission	136,267,385	1.6	90.7
19	American Louisiana Pipeline	132,752,766	1.5	92.2
20	Transwestern Pipeline	110,428,894	1.3	93.5
Total: top 20		7,783,415,712		
Total: other		555,939,797		
Grand total		8,339,355,509		

SOURCE: FPC, *Sales by Producers of Natural Gas to Interstate Pipeline Companies* (1962).

Table 6-13. Natural Gas Purchases by Interstate Pipelines, 1962

Rank	Pipeline	Volume (*Mcf*)	% of total	Cumulative %
1	United Gas Pipeline	1,272,350,332	14.2	14.2
2	El Paso Natural Gas	1,135,832,441	12.6	26.8
3	Tennessee Gas Transmission	906,388,527	10.1	36.9
4	Northern Natural Gas	576,222,587	6.4	43.3
5	Transcontinental Gas Pipeline	479,532,923	5.3	48.6
6	Texas Eastern Transmission	384,491,703	4.3	52.9
7	Cities Service Gas	371,572,895	4.1	57.0
8	Natural Gas Pipeline of America	340,102,685	3.8	60.8
9	Lone Star Gas Company	308,258,936	3.4	64.2
10	Southern Natural Gas	290,731,562	3.2	67.4
11	United Fuel Gas	285,696,771	3.2	70.6
12	Texas Gas Transmission	285,655,957	3.2	73.8
13	Panhandle Eastern Pipeline	267,200,812	3.0	76.8
14	Trunkline Gas	259,314,302	2.9	79.7
15	Arkansas Louisiana Gas	214,893,847	2.4	82.1
16	Michigan Wisconsin Pipeline	185,626,436	2.1	84.2
17	Colorado Interstate Gas	154,131,060	1.7	85.9
18	Humble Gas Transmission	141,555,112	1.6	87.5
19	American Louisiana Pipeline	130,271,913	1.5	89.0
20	Pacific Gas Transmission	126,211,326	1.4	90.4
Total: top 20		8,106,042,127		
Total: other		858,282,958		
Grand total		8,964,325,085		

SOURCE: FPC, *Sales by Producers of Natural Gas to Interstate Pipeline Companies* (1963).

Table 6-14. Natural Gas Purchases by Interstate Pipelines, 1963

Rank	Pipeline	Volume (*Mcf*)	% of total	Cumulative %
1	United Gas Pipeline	1,341,211,836	14.2	14.2
2	El Paso Natural Gas	1,114,967,391	11.8	26.0
3	Tennessee Gas Transmission	956,445,814	10.2	36.2
4	Northern Natural Gas	580,245,651	6.2	42.4
5	Transcontinental Gas Pipeline	525,913,868	5.6	48.0
6	Texas Eastern Transmission	395,560,475	4.2	52.2
7	Cities Service Gas	378,301,254	4.0	56.2
8	Natural Gas Pipeline of America	370,234,386	3.9	60.1
9	Lone Star Gas	333,564,104	3.5	63.6
10	United Fuel Gas	324,057,527	3.4	67.0
11	Texas Gas Transmission	320,277,209	3.4	70.4
12	Panhandle Eastern Pipeline	316,116,037	3.4	73.8
13	Southern Natural Gas	308,816,230	3.3	77.1
14	Trunkline Gas	283,150,181	3.0	80.1
15	Arkansas Louisiana Gas	241,662,586	2.6	82.7
16	Michigan Wisconsin Pipeline	192,365,137	2.0	84.7
17	Colorado Interstate Gas	152,291,928	1.6	86.3
18	Humble Gas Transmission	142,058,402	1.5	87.8
19	Transwestern Pipeline	140,446,398	1.5	89.4
20	American Louisiana Pipeline	135,141,762	1.4	90.8
Total: top 20		8,552,828,176		
Total: other		863,474,791		
Grand total		9,416,302,967		

SOURCE: FPC, *Sales by Producers of Natural Gas to Interstate Pipeline Companies* (1964).

monopsony power existed, or that it prevailed over market power on the seller's side, would require more careful analysis, however, and in fact seems somewhat unlikely at present, because of the lack of uncommitted reserves and the evidence of increasing demand relative to supply at current prices. All that can be noted is the obvious point that the buying side is more concentrated than the selling side.

Table 6-15. Natural Gas Purchases by Interstate Pipelines, 1964

Rank	Pipeline	Volume (*Mcf*)	% of total	Cumulative %
1	United Gas Pipeline	1,424,623,616	14.2	14.2
2	El Paso Natural Gas	1,248,384,712	12.4	26.6
3	Tennessee Gas Transmission	989,772,235	9.8	36.4
4	Northern Natural Gas	615,766,095	6.1	42.5
5	Transcontinental Gas Pipeline	558,202,954	5.5	48.0
6	Cities Service Gas	413,805,290	4.1	52.1
7	Texas Eastern Transmission	405,860,380	4.0	56.1
8	Natural Gas Pipeline of America	405,047,639	4.0	60.1
9	United Fuel Gas	360,649,195	3.6	63.7
10	Lone Star Gas	334,336,694	3.3	67.0
11	Panhandle Eastern Pipeline	328,138,478	3.3	70.3
12	Texas Gas Transmission	324,225,935	3.2	73.5
13	Southern Natural Gas	319,730,082	3.2	76.7
14	Trunkline Gas Company	299,010,228	3.0	79.7
15	Arkansas Louisiana Gas	261,834,667	2.6	82.2
16	Michigan Wisconsin Pipeline	206,527,561	2.1	84.3
17	Transwestern Pipeline	186,527,497	1.9	86.2
18	Colorado Interstate Gas	178,097,211	1.8	88.0
19	Pacific Gas Transmission	146,067,886	1.5	89.5
20	American Louisiana Pipeline	140,371,466	1.5	90.9
Total: top 20		9,147,069,821		
Total: other		914,465,375		
Grand total		10,061,535,196		

SOURCE: FPC, *Sales by Producers of Natural Gas to Interstate Pipeline Companies* (1965).

Table 6-16. Natural Gas Purchases by Interstate Pipelines, 1965

Rank	Pipeline	Volume (*Mcf*)	% of total	Cumulative %
1	El Paso Natural Gas	1,291,779,219	12.5	12.5
2	United Gas Pipeline	1,277,593,186	12.4	24.9
3	Tennessee Gas Transmission	969,568,924	9.4	34.3
4	Northern Natural Gas	638,187,123	6.2	40.5
5	Transcontinental Gas Pipeline	574,875,637	5.6	46.1
6	Texas Eastern Transmission	473,841,262	4.6	50.7
7	Natural Gas Pipeline of America	443,075,956	4.3	55.0
8	Cities Service Gas	418,332,348	4.1	59.1
9	United Fuel Gas	413,810,680	4.0	63.1
10	Texas Gas Transmission	357,888,768	3.5	66.6
11	Lone Star Gas	348,502,032	3.5	70.0
12	Panhandle Eastern Pipeline	332,720,638	3.2	73.2
13	Southern Natural Gas	325,314,682	3.2	76.4
14	Trunkline Gas	323,984,158	3.1	79.4
15	Arkansas Louisiana Gas	269,382,097	2.6	82.0
16	Michigan Wisconsin Pipeline	201,554,388	2.0	84.0
17	Transwestern Pipeline	199,730,489	1.9	85.9
18	American Louisiana Pipeline	186,344,092	1.8	86.7
19	Colorado Interstate Gas	166,240,266	1.6	89.3
20	Pacific Gas Transmission	152,598,574	1.5	90.8
Total: top 20		9,365,333,520		
Total: other		950,066,066		
Grand total		10,315,399,586		

SOURCE: FPC, *Sales by Producers of Natural Gas to Interstate Pipeline Companies* (1966).

In his important study of competition in interstate pipeline markets for new reserves in the 1950s prior to the introduction of regulation, Paul MacAvoy found evidence indicating either monopsonistic or competitive behavior. His approach to this investigation was based on pricing patterns: "The differences in theoretical pricing patterns, for competition versus monopoly or monopsony markets, can be traced to particular conditions of development costs and of maximum offer prices." He postulated

Table 6-17. Natural Gas Purchases by Interstate Pipelines, 1966

Rank	Pipeline	Volume (*Mcf*)	% of total	Cumulative %
1	United Gas Pipeline	1,323,513,367	11.9	11.9
2	El Paso Natural Gas	1,306,070,269	11.7	23.6
3	Tennessee Gas Pipeline	977,725,915	8.8	32.4
4	Northern Natural Gas	698,348,800	6.3	38.7
5	Transcontinental Gas Pipeline	618,084,120	5.6	44.3
6	Natural Gas Pipeline of America	536,320,184	4.8	49.1
7	Texas Eastern Transmission	520,509,090	4.7	53.8
8	United Fuel Gas	466,378,360	4.2	58.0
9	Cities Service Gas	439,332,233	3.9	61.9
10	Michigan Wisconsin Pipeline	436,852,856	3.9	65.8
11	Texas Gas Transmission	382,846,430	3.4	69.2
12	Southern Natural Gas	375,318,613	3.4	72.6
13	Lone Star Gas	372,727,651	3.3	75.9
14	Panhandle Eastern Pipeline	369,620,158	3.3	79.2
15	Trunkline Gas Company	341,732,970	3.1	82.3
16	Arkansas Louisiana Gas	295,467,447	2.7	85.0
17	Transwestern Pipeline	230,151,708	2.1	87.0
18	Colorado Interstate Gas	174,002,474	1.6	88.6
19	Pacific Gas Transmission	165,812,840	1.5	90.1
20	Humble Gas Transmission	155,494,562	1.4	91.5
Total: top 20		10,186,310,047		
Total: other		946,355,682		
Grand total		11,132,665,729		

SOURCE: FPC, *Sales by Producers of Natural Gas to Interstate Pipeline Companies* (1967).

Table 6-18. Natural Gas Purchases by Interstate Pipelines, 1967

Rank	Pipeline	Volume (*Mcf*)	% of total	Cumulative %
1	United Gas Pipeline	1,429,848,627	12.0	12.0
2	El Paso Natural Gas	1,410,977,856	11.8	23.8
3	Tennessee Gas Pipeline Div. of Tenn.	1,044,130,687	8.8	32.6
4	Northern Natural Gas	772,882,008	6.5	39.1
5	Transcontinental Gas Pipeline	696,172,270	5.8	44.9
6	Natural Gas Pipeline of America	610,385,264	5.1	50.0
7	Texas Eastern Transmission	534,183,614	4.5	54.5
8	Michigan Wisconsin Pipeline	494,143,864	4.1	58.6
9	United Fuel Gas	492,278,436	4.1	62.7
10	Cities Service Gas	434,995,157	3.6	66.3
11	Texas Gas Transmission	534,183,614	4.5	54.5
12	Panhandle Eastern Pipeline	412,961,216	3.5	73.3
13	Southern Natural Gas	401,633,406	3.4	76.8
14	Lone Star Gas	378,025,376	3.2	80.0
15	Trunkline Gas	359,055,446	3.0	83.0
16	Arkansas Louisiana Gas	304,642,826	2.6	85.6
17	Transwestern Pipeline	261,543,054	2.2	87.8
18	Pacific Gas Transmission	206,106,980	1.7	89.5
19	Colorado Interstate Gas	183,070,952	1.5	91.0
20	Humble Gas Transmission	158,627,543	1.3	92.3
Total: top 20		11,007,460,631		
Total: other		917,075,249		
Grand total		11,924,535,880		

SOURCE: FPC, *Sales by Producers of Natural Gas to Interstate Pipeline Companies* (1968).

various patterns of price with respect to variations in contract volume, distance to point of sale, term length of contract, and the presence or absence of renegotiation or escalation clauses, depending on whether monopoly power, monopsony power, or competition dominated. Where monopoly power existed, prices could be expected to be set differently for each buyer without necessary regard for the variables. For monopsony, prices should be uniform enough to result in an absence of any pattern. In a competitive situation, price levels should be uniform for all buyers, with differences related to the

Table 6-19. Natural Gas Purchases by Interstate Pipelines, 1968

Rank	Pipeline	Volume (*Mcf*)	% of total	Cumulative %
1	United Gas Pipe Line	1,513,267,815	11.8	11.8
2	El Paso	1,478,894,488	11.5	23.3
3	Tennessee Natural Gas Lines	1,120,145,760	8.7	32.0
4	Northern Natural Gas	828,613,272	6.5	38.5
5	Transcontinental Gas Pipeline	755,462,122	5.9	44.4
6	Natural Gas Pipeline of America	679,678,593	5.3	49.7
7	Texas Eastern Transmission	611,252,613	4.8	54.5
8	Michigan Wisconsin Pipeline	529,949,440	4.1	58.6
9	United Fuel Gas	526,354,335	4.1	62.7
10	Southern Natural Gas	463,933,874	3.6	66.3
11	Texas Gas Transmission	437,668,611	3.4	69.7
12	Panhandle Eastern Pipe Line	433,885,929	3.4	73.1
13	Cities Service Gas	422,951,324	3.3	76.4
14	Lone Star Gas Company	422,433,152	3.3	79.7
15	Arkansas Louisiana Gas	381,200,529	3.0	82.7
16	Trunkline Gas	371,488,911	2.9	85.6
17	Transwestern Pipeline	288,205,176	2.2	87.8
18	Pacific Gas Transmission	241,363,664	1.9	89.7
19	Colorado Interstate Gas	186,227,189	1.5	91.2
20	Humble Gas Transmission	167,925,296	1.3	92.5
Total: top 20		11,860,902,093		
Total: other		963,119,054		
Grand total		12,824,021,147		

SOURCE: FPC, *Sales by Producers of Natural Gas to Interstate Pipeline Companies* (1969).

variables hypothesized to influence price. MacAvoy set up multiple regression equations with price as the dependent variable and the hypothesized variables to explain on price. He examined the pre-1960 major markets and drew the conclusion that "gas markets were diverse in structure and behavior, and were generally competitive or were changing from monopsony toward competition."[11]

Two previous studies, covering roughly the same time but employing different methodology, were done by Cookenboo and Neuner.[12] The purpose of Cookenboo's study was to analyze the economic issues within the industry and to decide if workable competition or if monopoly predominated. He discussed the conflicting views held by various persons in the mid-1950s about the competition in the natural gas producing industry, including those of M. L. Lindahl and J. W. McKie (who believed competition was effective) and M. G. Glaeser and J. K. Galbraith (who believed it was not). After a brief history of the development of the industry, the structure of the industry in the mid-1950s is examined from the standpoint of U.S. interstate sales and of all Texas sales (table 6-1 shows Cookenboo's figures for U.S. interstate sales). When comparing the United States as a whole with Texas, the following summarizes the result:[13]

Sales held by:	All U.S. interstate sales, 1955	All Texas sales, 1953
	(------------------ *percent* ------------------)	
Top 4 producers	23	21
Top 8 producers	35	33
Top 12 producers	44	42
Top 16 producers	49	49
Top 20 producers	54	54

[11]Paul W. MacAvoy, *Price Formation in Natural Gas Fields: A Study of Competition, Monopsony, and Regulation* (Yale University Press, 1962), pp. 83, 265.

[12]Leslie Cookenboo, Jr., "Competition in the Field Market for Natural Gas," *The Rice Institute Pamphlet*, Monograph in Economics, vol. 44, no. 4 (January 1958); and Edward J. Neuner, *The Natural Gas Industry* (University of Oklahoma Press, 1960).

[13]Cookenboo, "Competition for Natural Gas," pp. 48, 51.

Cookenboo also found that new contract sales in the early 1950s and the sales in the Gulf Coast region were very close in concentration to all contract sales during the same period. His general conclusion was that the field market was one of low concentration, with effective competition prevailing.

In his study, Neuner looked at the time period of the main postwar expansion of the industry (1945-53), examining concentration, price levels, and new contracts in the major fields. The Cookenboo and Neuner studies are similar, except that Neuner's constitutes a more complete analysis of new contracts. Neuner's conclusions were similar to Cookenboo's: "It is evident that the level of concentration in natural gas production is not excessive by comparison with manufacturing industry. The structural condition of natural gas field markets, by these tests, does little to support a claim of gas producer monopoly."[14] In addition, Neuner concluded that concentration in submarkets was insufficient to generate monopolistic pricing and behavior there.

The submarket problem is important; concentration measures may not appear high at the national level but could at the regional level. The basic question here is how the market should be defined. Because the courts have been very impressed with market shares when ruling on antitrust cases, it is important that the market or markets be carefully delineated.

If the idea of potential competing products constituting a "product market" is brushed aside (which is usually done in considering natural gas market structure, even though we know oil and coal do provide competition for gas), a market is then usually defined as the area in which a product is sold. Due to an extensive natural gas pipeline system, the market for natural gas is generally considered to be a national one.[15] However, it is interesting to conjecture to what extent the submarket of intrastate sales differs from the national interstate market. These two markets may have singular differences due to the absence of regulation in submarket intrastate sales. It is known, for example, that an important intrastate market exists in Texas, but size and size distribution of sellers and buyers are difficult to ascertain. Interstate sales from Texas production are given in tables 6-20 to 6-22 for 1964, 1966, and 1968 for the top twenty producers. It can be seen that the concentration of interstate sellers in Texas is higher then in the national market. This points to the possibility that the Texas intrastate market is more concentrated than the national market, but it would be necessary to examine Texas intrastate sales figures to sustantiate this conclusion. These figures are not readily available, however, and it would be interesting to compare the entire intrastate market, as an unregulated market, with the interstate regulated market, especially since 1960.

It is very difficult to get analytically reliable figures on the intrastate market. Tables 6-23 through 6-26 are attempts to get an idea of the intrastate market's approximate size. In column 8 in table 6-23, estimates of intrastate sales are given. Column 2 is net marketed production (according to the Bureau of Mines), which, as defined, is to exclude line losses, field use, and so forth. Since the figures in column 3 of table 6-23 are for purchases from producers by pipelines, they should be net of interproducer sales and presumably are before interpipeline sales. Some of these sales are intrastate (column 4) and are so designated for 1965 through 1968. (Such sales for the top twenty-four producers are given in tables 6-27 through 6-30.) They are not identified in 1961 through 1964, however, and for these years are estimated at the approximate ratio of intrastate to total sales for 1965-68. Subtracting these intrastate sales from total producer sales to interstate pipelines leaves interstate sales to pipelines (column 5). Adding this to interstate pipeline production gives total interstate sales, which, when subtracted from net marketed production, give an estimate of intrastate sales as a residual.

There are three reasons why these figures in column 8 probably underestimate true intrastate sales:

1. If the pipelines' own production is not all marketed (for example, some is sold intrastate or some is lost), then interstate sales will be overstated.

[14] Neuner, *The Natural Gas Industry*, p. 245.

[15] Although this was not accepted in the early stages of the gas regulation controversy, especially by the proponents of the regulation, by the time of the area rate hearings it was generally conceded to be the case. In the examiner's decision in the Permian Basin Area hearing, for example, he made the point of the market being a national one numerous times. The FPC's opinion also acknowledged this, as did the Supreme Court's final decision.

Table 6-20. Texas Interstate Gas Sales by Producers, 1964

Rank	Producer	Volume (*Mcf*)	% of total	Cumulative %
1	Phillips	492,370,673	14.5	14.5
2	Pan American	196,674,454	5.8	20.3
3	Humble	156,734,532	4.6	24.9
4	Texaco	143,768,297	4.2	29.1
5	Shell	136,661,076	4.0	33.1
6	Gulf	114,365,440	3.4	36.5
7	Socony Mobil	113,571,610	3.4	39.9
8	Sun Oil	112,147,038	3.3	43.2
9	Champlin	111,621,842	3.3	46.5
10	Sinclair	104,950,977	3.1	49.6
11	Atlantic Refining	86,925,641	2.6	52.2
12	Union Producing	83,336,881	2.5	54.7
13	Shamrock	80,765,084	2.4	57.1
14	West Texas Gathering Co.	78,469,634	2.3	59.4
15	Calif. Co. Div. of Chevron	74,268,888	2.2	61.6
16	Continental	73,849,615	2.2	63.8
17	H. R. Cullen et al.	64,199,837	1.9	65.7
18	Lone Star	63,278,557	1.9	67.6
19	Cities Service	61,647,345	1.8	69.4
20	Sunray DX	57,792,334	1.7	71.1
Total: top 20		2,407,399,755		
Total: other		977,541,417		
Grand total		3,384,941,172		

SOURCE: FPC.

Table 6-21. Texas Interstate Gas Sales by Producers, 1966

Rank	Producer	Volume (*Mcf*)	% of total	Cumulative %
1	Phillips	500,221,836	14.4	14.4
2	Humble	302,814,543	8.7	23.1
3	Pan American	220,456,764	6.3	29.4
4	Shell	203,438,761	5.9	35.3
5	Mobil	147,690,368	4.3	39.6
6	Champlin	118,713,212	3.4	43.0
7	Continental	113,598,921	3.3	46.3
8	Sun	111,919,859	3.3	49.6
9	Gulf	109,619,733	3.2	52.8
10	Sinclair	105,717,094	3.0	55.9
11	Union Producing	88,303,819	2.5	58.4
12	Texaco	87,669,815	2.5	60.9
13	Atlantic Richfield	84,451,467	2.4	63.3
14	Shamrock	78,023,311	2.2	65.5
15	Calif. Co. Div. of Chevron	77,195,419	2.2	67.7
16	Sunray DX	66,278,245	1.9	69.6
17	Cities Service	60,663,873	1.7	71.4
18	Lone Star	60,507,324	1.7	73.1
19	Mitchell Geo. & Associates	60,208,375	1.7	74.8
20	Warren	57,148,669	1.6	76.4
Total: top 20		2,654,638,408		
Total: other		818,360,454		
Grand total		3,472,998,862		

[a]Not available.

SOURCE: FPC.

2. The calculation of intrastate sales to interstate pipelines is only for producers selling 2 billion cubic feet or more. Leaving out the small producers would tend to understate intrastate sales.
3. It is very difficult to capture direct industrial sales, many of which are intrastate, in the figures publicly reported by the Bureau of Mines or the American Gas Association (AGA). Thus,

Table 6-22. Texas Interstate Gas Sales by Producers, 1968

Rank	Producer	Volume (*Mcf*)	% of total	Cumulative %
1	Phillips	505,921,934	13.6	13.6
2	Humble	491,210,196	13.2	26.8
3	Pan American	226,540,086	6.1	32.9
4	Shell	210,829,527	5.7	38.6
5	Mobil	156,686,914	4.2	42.8
6	Gulf	151,455,218	4.1	46.9
7	Champlin	123,672,606	3.3	50.2
8	Sun Oil	120,192,833	3.2	53.4
9	Texaco	113,318,352	3.0	56.4
10	Sinclair	100,522,456	2.7	59.1
11	Continental	98,808,144	2.7	61.8
12	Calif. Co. Div. of Chevron	86,737,837	2.3	64.1
13	Atlantic Richfield	86,587,439	2.3	66.4
14	Union Producing	70,677,010	1.9	68.3
15	Lone Star	67,209,456	1.8	70.1
16	Cities Service Oil	62,137,644	1.7	71.8
17	Warren Petroleum	60,687,947	1.6	73.4
18	Sunray DX	59,530,246	1.6	75.0
19	Mitchell Geo. & Associates	54,501,000	1.5	76.5
20	Skelly Oil	43,399,948	1.2	77.7
Total: top 20		2,890,626,793		
Total: other		831,441,231		
Grand total		3,722,068,024		

SOURCE: FPC.

although the interstate sales are presumably quite accurate as reported to the FPC, the totals and hence intrastate sales (as a residual) are understated.

Three other attempts to estimate intrastate sales or production for the intrastate market are made in tables 6-24 through 6-26. Percentages of production attributable to interstate and intrastate sales, prepared by the Bureau of Natural Gas (BNG), are given for 1963-68 and are applied to the AGA production figures for these years in table 6-24. Intrastate production is given as one-third or slightly more of the total. The intrastate production figures in this table are nearly twice the size of the figures for intrastate sales in table 6-30.

In table 6-25 figures compiled by the Future Requirements Committee (FRC) were used to make estimates of intrastate sales. Since Texas and Louisiana export large amounts of gas to other states, the amounts consumed in these two states should give a conservative estimate of intrastate sales. These figures are given in table 6-25 and are higher than the intrastate sales figures estimated in table 6-23. An objection to these figures might be raised with regard to the FRC definition of consumption. The FRC defines "actual requirements" as "the aggregate usage of gas in the past as reported by the respondent companies based on their records of gas sold, used in their own operations, lost in transmission, or unaccounted for."[16] Field use is excluded from this definition, and the "total less other" figures given represent sums of firm residential, commercial, and industrial and interruptible as listed by the committee. Hence, it seems reasonable to accept these sums as representing sales, or at least a good proxy.

Table 6-26 has been prepared from Bureau of Mines information. Column 1 is the 1968 consumption in states with net deliveries. Net delivery means that more goes out of the state than comes in—much of what comes in over the borders is not used in the state but is passed through to other states. The states in column 1, although passing large amounts of gas through (as in the case of Louisiana), are themselves net exporters. Hence, the amounts consumed in those states represents an estimate of intrastate consumption, which comes to about 38.6 percent of total U.S. consumption. These figures

[16]FRC, *Future Natural Gas Requirements of the United States* (Denver Research Institute, University of Denver, September 1969), p. 46.

Table 6-23. Interstate and Intrastate Gas Sales, 1961-68

Year	(1) Total net marketed production[a]	(2) Producer total sales to interstate pipelines[b]	(3) Producer intrastate sales to interstate pipelines[b]	(4) Producer interstate sales to interstate pipelines[b] (3 - 4)	(5) Interstate pipeline own production[b]	(6) Total interstate sales (4 + 5)	(7) Intrastate sales (1 - 7)	(8) Intrastate sales as % of total net marketed production
	(------------	------------	------------	thousand Mcf	------------	------------	------------)	
1961	10,992,393	8,339,356	583,754[c]	7,755,602	882,729	8,638,331	2,354,062	21.4
1962	11,511,281	8,964,325	617,502[c]	8,346,823	861,596	9,208,419	2,302,862	20.0
1963	12,169,894	9,416,303	659,141[c]	8,757,162	930,402	9,687,564	2,482,330	20.4
1964	12,948,529	10,061,535	704,307[c]	9,357,228	951,999	10,309,227	2,639,302	20.4
1965	13,693,230	10,315,400	808,587	9,506,813	982,340	10,489,153	3,204,077	23.4
1966	14,963,862	11,132,666	745,855	10,386,811	1,004,228	11,391,039	3,572,823	23.9
1967	15,764,778	11,924,536	735,322	11,189,214	1,027,909	12,217,123	3,547,655	22.5
1968	16,836,791	12,824,021	837,738	11,986,283	1,064,435	13,050,718	3,786,073	22.5

[a] Bureau of Mines information in *Gas Facts.*

[b] FPC, *Sales by Producers of Natural Gas to Interstate Pipeline Companies* (1961 through 1968).

[c] Not identified as such in FPC, *Sales by Producers of Natural Gas*. Estimated from the ratio of intra to total in years 1965-68 (a difference of approximately 7 percent).

Table 6-24. Natural Gas Production Attributable to Interstate and Intrastate Sales, 1963-68

Year	Total production[a] *(1,000 Mcf)*	Interstate production *(1,000 Mcf)*	Interstate production as a % of total[b]	Intrastate production *(1,000 Mcf)*	Intrastate production as a % of total[b]
1963	14,546,025	9,614,923	66.1	4,931,102	33.9
1964	15,347,028	10,282,509	67.0	5,064,519	33.0
1965	16,252,293	10,547,738	64.9	5,704,555	35.1
1966	17,491,073	11,334,215	64.8	6,156,858	35.2
1967	18,380,838	12,039,449	65.5	6,341,389	34.5
1968	19,373,428	12,805,836	66.1	6,567,592	33.9

[a] AGA, *Gas Facts* (1963 through 1968).
[b] BNG, *A Staff Report on National Gas Supply and Demand* (September 1969), table 21, p. 89.

Table 6-25. Natural Gas Used in Louisiana and Texas, 1964-68

(thousand Mcf)

Year	Louisiana	Texas	Total	Total less other
1964	869,764 (43,494)	2,411,032 (170,031)	3,280,796	3,067,271
1965	903,808 (47,380)	2,708,096 (190,781)	3,611,904	3,373,743
1966	1,068,334 (50,645)	2,934,620 (192,386)	4,002,954	3,759,923
1967	1,207,679 (55,658)	3,067,580 (189,306)	4,275,259	4,030,295
1968	1,366,211 (63,100)	3,307,928 (197,291)	4,674,139	4,413,748

NOTE: The figures contain firm, interruptible, and "other." The figures in parentheses are the "other," which have been subtracted in the last column.

SOURCE: Future Requirements Committee, *Future Natural Gas Requirements of the United States*, vol. 3 (Denver Research Institute, University of Denver, September 1969).

Table 6-26. Estimated Intrastate Consumption of Natural Gas, 1968

(million cubic feet)

Consumption of states with net deliveries		Marketed production of states with net receipts	
Kansas	525,592	Alabama	230
Louisiana	1,661,782	Alaska	17,343
North Dakota	38,481	Arkansas	156,627
New Mexico	309,409	Arizona	881
Oklahoma	552,189	California	714,893
Texas	4,153,513	Colorado	121,424
West Virginia	189,027	Florida	108
Wyoming	90,191	Illinois	4,380
Total	7,520,184	Indiana	234
% of total consumption	38.6	Kentucky	89,024
		Maryland	864
		Michigan	40,480
		Mississippi	135,051
		Missouri	14
		Montana	19,313
		Nebraska	8,129
		New York	4,632
		Ohio	42,673
		Pennsylvania	87,987
		Tennessee	48
		Utah	46,151
		Virginia	3,389
		Total	1,493,875
		% of total consumption	7.6

Total U.S. consumption: 19,459,939
Total U.S. marketed production: 19,322,400

SOURCE: U.S. Bureau of Mines, "Natural Gas Production and Consumption: 1968," *Mineral Industry Surveys*, August 25, 1969.

Table 6-27. Intrastate Gas Sales by Producers to Interstate Pipelines, 1965

Rank	Producer	Volume (*Mcf*)
1	Humble Oil	182,985,585
2	Pan American Petroleum	27,526,339
3	Phillips	13,390,588
4	Shell Oil	19,978,937
5	Gulf Oil	49,674,222
6	Socony Mobil Oil	7,989,857
7	Texaco	72,903,729
8	Continental Oil	12,452,722
9	Union Oil	4,226,614
10	Sinclair Oil & Gas	8,288,568
11	Atlantic Refining	3,804,188
12	Superior Oil	246,119
13	Sun Oil	39,937,389
14	Union Producing	36,709,138
15	Cities Service	16,800,777
16	Alberta & Southern Gas (Canadian)	000
17	Westcoast Transmission (Canadian)	000
18	Champlin Petroleum	4,345,445
19	Skelly Oil	675,349
20	Union Texas Petroleum Div. Allied	8,862,652
21	Calif. Co. Div. of Chevron	4,261,616
22	Tidewater Oil	35,364,687
23	Ashland Oil	000
24	Warren Petroleum	10,866,458
Total: top 24		561,290,979
Total: other		247,295,774
Grand total		808,586,753

SOURCE: FPC, *Sales by Producers of Natural Gas to Interstate Pipeline Companies* (1966).

Table 6-28. Intrastate Gas Sales by Producers to Interstate Pipelines, 1966

Rank	Producer	Volume (*Mcf*)
1	Humble Oil	171,367,937
2	Pan American Petroleum	32,356,956
3	Phillips	27,410,443
4	Shell Oil	23,629,238
5	Gulf Oil	47,886,627
6	Mobil Oil	9,068,579
7	Union Oil	9,048,855
8	Continental Oil	5,413,768
9	Texaco	25,151,767
10	Sinclair Oil and Gas	9,204,659
11	Superior Oil	000
12	Atlantic Richfield	6,701,156
13	Sun Oil	25,473,340
14	Cities Service	22,119,295
15	Calif. Co. Div. of Chevron	3,625,017
16	Union Producing	33,366,570
17	Alberta & Southern Gas (Canadian)	000
18	Union Texas Petroleum Div. Allied	7,181,412
19	Champlin Petroleum	4,556,596
20	Westcoast Transmission (Canadian)	000
21	Skelly Oil	1,462,648
22	Tidewater Oil	28,594,040
23	Marathon Oil	4,653,834
24	Warren Petroleum	26,853,622
Total: top 24		525,117,359
Total: other		220,737,098
Grand total		745,854,457

SOURCE: FPC, *Sales by Producers of Natural Gas to Interstate Pipeline Companies* (1967).

Table 6-29. Intrastate Sales by Producers to Interstate Pipelines, 1967

Rank	Producer	Volume (*Mcf*)
1	Humble Oil	182,812,659
2	Pan American Petroleum	30,908,292
3	Phillips	26,972,721
4	Gulf Oil	54,493,130
5	Shell Oil	13,653,823
6	Mobil Oil	7,503,064
7	Texaco	33,110,006
8	Union Oil of California	7,619,000
9	Continental Oil	4,112,113
10	Sinclair Oil and Gas	8,881,163
11	Superior Oil	000
12	Atlantic Richfield	6,424,200
13	Sun Oil	29,813,378
14	Calif. Co. Div. of Chevron	3,775,879
15	Alberta & Southern Gas (Canadian)	000
16	Westcoast Transmission (Canadian)	000
17	Union Producing	33,150,885
18	Cities Service	23,764,685
19	Union Texas Petroleum Div. Allied	7,195,466
20	Champlin Petroleum	5,689,596
21	Getty Oil	12,347,223
22	Skelly Oil	1,297,676
23	Marathon Oil	724,483
24	Warren Petroleum	26,192,321
Total: top 24		520,441,763
Total: other		239,407,794
Grand total		759,849,557

SOURCE: FPC, *Sales by Producers of Natural Gas to Interstate Pipeline Companies* (1968).

Table 6-30. Intrastate Sales by Producers to Interstate Pipelines, 1968

Rank	Producer	Volume (*Mcf*)
1	Humble	241,374,389
2	Pan American Petroleum	21,782,404
3	Gulf Oil	56,040,315
4	Phillips	25,447,900
5	Shell	10,365,395
6	Mobil	2,102,345
7	Texaco	49,626,275
8	Union	8,291,676
9	Continental	5,348,045
10	Sinclair	7,207,059
11	Sun Oil	35,050,989
12	Alberta & Southern Gas Co. (Canadian)	000
13	Superior Oil	000
14	Cities Service	27,259,463
15	Atlantic Richfield	12,686,797
16	Calif. Co. Div. of Chevron	7,041,180
17	Westcoast Transmission (Canadian)	000
18	Union Texas Petroleum Div. Allied	5,993,986
19	Union Producing	30,141,267
20	Getty Oil	19,244,262
21	Champlin Petroleum	5,291,256
22	Skelly Oil	1,195,960
23	Marathon Oil	6,176,585
24	Warren Petroleum	25,933,470
Total: top 24		603,601,018
Total: other		234,136,628
Grand total		837,737,646

SOURCE: FPC, *Sales by Producers of Natural Gas to Interstate Pipeline Companies* (1969).

could overstate intrastate sales, however, because some amounts moving into the states could be consumed there. In other words, all interstate movements into these states need not pass through to other states, and hence all that is consumed by these states need not have been produced intrastate. However, if these figures were not overstated by more than 7.6 percent of the total, they would still be representative. Column 2 gives figures of marketed production in states that are net importers. It would seem that few sales made in these states would be for use outside the state. Hence, these figures represent mostly intrastate sales excluded from the column 1 estimate. The figure of 7,520,184 million cubic feet in column 1 may not be a good estimate of intrastate sales but may give an idea of intrastate consumption relative to the total.

The above figures seem to indicate that the intrastate market is probably between 33 percent and 40 percent of the total, rather than less than 33 percent. The intrastate market is increasing in absolute size along with the national market, but it is difficult to ascertain a trend relative to the national market. Threatened energy shortages in northeastern parts of the country, however, may make plant locations develop faster within intrastate gas markets than they otherwise would.[17]

The FPC estimates the intrastate market to be one-third of the total national market (table 6-24). However, for industrial sales, the Bureau of Mines reports the industrial and electric sales for Texas and Louisiana to be:[18]

	Industrial	Electric utilities
	(- - - - - - - - - - *thousand Mcf* - - - - - - - - - -)	
Louisiana	864,848	252,831
Texas	1,696,231	852,173
	2,561,079	1,105,004
Total	3,666,083	

These figures are considerably below those given in table 6-25 for Texas and Louisiana. However, the figures in table 6-25 are as defined above by the FRC, whereas, these are "delivered to consumers." These industrial deliveries are over 40 percent of total U.S. industrial sales and must be largely intrastate.

There is evidence that producers attempt to stay in the intrastate market where possible. Advertisements have been made by producers to induce industry to relocate in Texas and use intrastate gas, and there have been sales by producers of reserves to intrastate pipelines and/or distributors.[19] Sales of reserves by producers to interstate pipelines have, of course, been held to be equivalent to interstate gas sales and thus subject to FPC jurisdiction.[20] Aside from the possibility of a higher price in the intrastate market because it is not covered by FPC regulation, there is another reason for preference for the intrastate market. This is because a contract sale intrastate may represent a better return on investment, even if the price were below that which could be obtained interstate, if the reserves could be produced faster for the intrastate sale. The rational firm wishes to maximize the present value of total revenues, not simply total revenues, for a given capital investment. The faster the reserves are produced, the sooner the revenues are obtained and hence the higher the present value. In addition, the uncertainty associated with regulation imparts a potential risk to selling in the regulated

[17]In "Power Crisis," there is a reference to Armco Steel's decision to locate a steel plant in Texas rather than in the Midwest due to assurance of gas supply in Texas.

[18]U.S. Department of the Interior, Bureau of Mines, *Mineral Industry Surveys* (August 25, 1969).

[19]For example "Houston Natural Gas Corp.'s Intratex Gas Co. subsidiary said it contracted with Humble Oil and Refining Co., the chief domestic subsidiary of Standard Oil Co. (New Jersey), to purchase substantial gas reserves in the West Comex field of Pecos County and Block 16 of Ward County, Texas. . . . As previously announced, the subsidiary entered similar contracts to purchase reserves in the West Texas area from Texaco Inc., Superior Oil Co., and Union Oil Co." (*Wall Street Journal*, April 29, 1970.)

[20]*United Gas Improvement Co. v. Continental Oil Co.*, 381 U.S. 392 (1965).

market that is not attached to selling in the unregulated market. This perceived risk difference would also give a preference for the intrastate market.

The current weighted average prices paid for gas by an intrastate (Texas) pipeline company, Coastal States Gas Producing Company, for the months of October, November, and December 1969 were 13.5 cents, 14.87 cents, and 15.06 cents per Mcf respectively.[21] This is comparable to the Bureau of Mines average wellhead price for the United States of 16.4 cents (1968). The weighted average new contract price for the United States as calculated from data reported by the FPC (column 5, table 6-31) was 19.4 cents for 1968. During the quarter in which Coastal States purchases were examined, there were some sales as high as 19.5 cents, 20.8 cents, 20.9 cents, and 21 cents per Mcf. During 1970 it was recorded that intrastate sales in the Permian Basin were as high as 24 cents per Mcf, and in the Texas-Gulf Coast region for up to 25.7 cents per Mcf—with a 1 cent escalation every four years.

Table 6-31. Prices and Additions to Natural Gas Reserves, 1955-69

Year	(1) New discoveries[a]	(2) Increase in proved reserves over previous year[a]	(3) Average wellhead price[b]	(4) Weighted average new contract price[c]
	(- - - - mil. cu. ft.	- - - -)	(- - - - ¢/Mcf	- - - -)
1955	5,688,009	11,921,613	10.4	19.0
1956	5,605,864	14,000,671	10.8	17.7
1957	8,950,119	8,746,922	11.3	18.5
1958	5,580,624	7,531,655	11.9	19.1
1959	5,769,245	8,408,639	12.9	17.9
1960	6,600,963	1,155,895	14.0	18.5
1961	6,907,729	3,947,316	15.1	17.5
1962	6,299,164	6,005,216	15.5	17.9
1963	5,577,934	3,872,375	15.8	16.8
1964	6,909,301	5,100,221	15.4	16.8
1965	6,543,709	5,217,469	15.6	16.8
1966	6,057,725	2,863,882	15.7	17.8
1967	5,695,171	3,574,898	16.0	18.6
1968	2,922,041	(5,557,851)	16.4	19.4
1969	3,813,000	(12,241,000)		

[a] AGA.
[b] Bureau of Mines.
[c] FPC.

The economic studies conducted in the late 1950s led to the conclusion that monopoly did not exist in field markets. The structure of the industry does not seem to have changed substantially since then, with the exception that the intrastate market may now be more important. Regulation may have caused some changes in the industry structure that are not readily apparent but that might be more observable if it were possible to compare the regulated interstate market with the unregulated intrastate market. However, the intrastate data are very difficult to obtain, and such a comparison would require an extended research study.

One thing that regulation has changed with respect to interstate contracts is the status of some of the clauses that at the outset of regulation were felt to be objectionable.[22] These clauses are usually placed under the categories: (1) fixed price escalation, (2) tax sharing, (3) inflation adjustment and/or spiral escalation, (4) favored nation, and (5) price redetermination. For the first, the prevailing practice is 1 cent every four years. Tax sharing is permitted, and the usual contract provides for reimburse-

[21] This information was obtained from "Purchasers' Monthly Gas Tax Report," Form 60-1.50 (files, Texas Controller).

[22] Permissible price changes are set forth in FPC, *Regulations under the Natural Gas Act* (August 1, 1967), especially sec. 154.93. I would like to thank Mr. R. E. Johnson of El Paso Natural Gas Company for information about current practices in the industry relative to these clauses.

ment for seven-eighths of additional taxes. The third and fourth categories are not permitted. Although price redetermination is permitted, the general practice is not to use it as such but to incorporate a so-called area rate clause that essentially permits negotiating a higher price up to any higher relevant area rate that is instituted.

RENT, WINDFALL, AND SUPPLY

Despite the consensus of those who have studied the natural gas producing industry and have concluded that it exhibits effective or workable competition, rather than monopoly, the proponents of field price regulation have held that, even if competition is present, regulation is justified on other grounds. The argument for regulation has shifted to the view that, in the absence of regulation, producers have the ability to earn a type of excess profit.[23] It should be noted, however, that as an empirically testable proposition it is quite impossible to decide if this is true–or was true in the preregulation period–because of the ubiquitous problem of the integrated nature of the large producers. In other words, differences in returns cannot be determined for different-sized firms, or differently located firms, or firms in the now regulated interstate market versus the unregulated intrastate market, because it is impossible to obtain returns on the natural gas business of most companies.

The economic justification claimed for regulating field prices was to prevent the earning of rents and windfalls by producers. The argument for regulation on rent and windfall grounds carried considerable weight in the implementation of producer regulation, and this argument was put forward most clearly by Professor Alfred Kahn in the Permian Basin hearing. The consulting group (National Economic Research Associates) with which he was associated no doubt had the most impact on the hearing and ultimately on the commission, and Professor Kahn himself probably had the most significant single influence on the outcome. His views and recommendations about how the area rate concept of price regulation should be prosecuted in large part set the tone for the configuration of current regulation. His proposal of the two-price system, which was adopted, rested in part on his rent and windfall argument, a summary of which is given in the following quotations from his testimony:

Kahn: . . . Because of these circumstances–the long production cycle and the historical by-product relationship of gas to oil–the windfalls conferred on producers of gas by the increase of market price on all sales even if only to the level of recent or replacement costs have been unusually large.

Question: Your use of the term windfall suggests that this additional flow of revenues to producers under old contracts is in some economic sense unnecessary. Do you mean to convey this impression?

Kahn: Yes. Even a competitive price system produces all sorts of income that economists classify as "economic rents," that is to say, differential returns to individuals who happen to be advantageously situated compared with their competitors for reasons that may have nothing to do with their own particular merit or effort. Economic rents are obtained by individuals who have possession of some resource that is unique, or nonreproducible, for which there is a demand. Rents do not fulfill the same economic function as other costs, which must be reimbursed if supply is to be forthcoming. The supply of these nonreproducible resources would be available, in the presence of competition, at any positive price, thus rents do not determine price; they are instead determined by price. If, for example, the demand for a nonreproducible good increases, the price will go up and the owners will receive increased rents. The classic example is the high rental income accruing to the owner of land in the middle of a city. These rents, which are traditionally referred to as the "unearned increment," do not necessarily justify price regulation: most economists who object to private parties earning them–and there is a long tradition to this effect–would prefer to remove them by taxation; however, in the absence of such a tax, it is conceivable that price regulation, by curtailing the "unearned increment," might add more to the sum total of economic satisfactions of the community at large than it would diminish them by interfering with the allocative functions of the pricing system. This is precisely the character of the windfalls conferred by the combination of circumstances I have described above. The higher price extended to old gas contracts in consequence of the rising price of new reserve commit-

[23] See, for example, the difference in the thrust of the arguments for regulation in Senator Paul Douglas's article, "The Case for the Consumer of Natural Gas," *Georgetown Law Journal* (June 1956); and in Professor Alfred Kahn's testimony in the Permian Basin hearing, *In the Matter of Area Rate Proceeding*, Docket no. AR 61-1.

ments is and was unnecessary to elicit those supplies already discovered, developed and committed. I might add that natural gas is by no means the only industry in which governments have intervened to regulate prices in order to hold down the economic rents that would otherwise be earned by producers of goods or services whose supply was inelastic or short.

Question: Is it only because of the possible presence of monopolistic elements that you feel regulation with reference to cost will be required?

Kahn: No. It will serve also to diminish or prevent the earning of rents, to which I have already referred. The most familiar instances of economic rents are those accruing to the owners of completely nonreproducible goods like land. But in economic terms nonreproducibility is a matter of degree—of the degree of elasticity of supply. A completely nonreproducible good will have a completely inelastic supply. An increase in demand for a product like gas, with a supply that was inelastic although certainly not completely so, will similarly, in the absence of regulation, have conferred large amounts of economic rent on producers with average costs of production considerably below those of the marginal supplies that the market required. Cost-based regulation, if intelligently applied, can diminish these windfalls, while interfering little or not at all with the flow of supply.[24]

From an economic standpoint the argument of the ability of producers to earn rents must carry the burden of the position for regulation. The desirability of redistributing income from stockholders of producing companies to consumers of natural gas as a welfare improvement is a separate area of discussion. Professor Kahn was careful to point out above that interfering with the allocative function of the price system may increase satisfaction. One can also say that it may not, and with that, nothing further can be said by way of a welfare pronouncement. Despite Professor Kahn's well-stated testimony, which was persuasive to those concerned with implementation of producer regulation (the hearing examiner and FPC commissioners), under more rigorous economic analysis it is anything but clear that producers would earn rents in the absence of regulation.

Rents and windfalls must be carefully defined. A casual usage of these terms typically implies the "getting of something for nothing" or "unearned increment," but this is true only under restrictive conditions. The theory of economic rent is dependent on alternative cost and alternative use. Economic rent is payment in excess of that which is necessary to bring a factor into production. In the long run, rents are payments in excess of the costs of production, and production costs include the opportunity costs of factors in alternative employments. If a factor has more than one use, then the excess of its return in the best use over its possible return in other uses will be rent, or it may be profit. George Stigler defines the difference:

Rents are the returns to productive services in excess of what they can earn in alternative employments. When these returns accrue to the owners of resources hired or purchased by the firm they are called rents, but when they accrue to the entrepreneur they may be called profits. Firms in a competitive industry may receive profits because of differences among firms or because of a state of disequilibrium.[25]

In the short run the capital input to a production process is fixed, and hence the supply of output for the short time period is fixed as long as price is greater than the variable cost of output. No price above variable cost is required to bring the factor produced into the market. In this sense supply is fixed, and does not enter into the determination of price. Payments above variable cost can therefore be considered rents, but are more correctly called (after Marshall) quasi-rents. Marshall claimed that the hire (or quasi-rent) of a factor can diverge only for short periods from normal profits on its cost of production. This can occur on the down side only during a period when the capital input cannot be withdrawn and variable cost is covered. It can occur under net positive investment or accumulation of capital in an industry during the time when profits above normal (which may be thought of as rents as per Stigler's definition) exist to attract net capital. But this conception of rent or profit need not arise either as scarcity rents or as monopoly rents. Scarcity rent, which is the concept of pure economic rent as defined by Marshall, is income derived from holding free gifts of nature. While lease

[24] Kahn, *Matter of Area Rate Proceeding*, pp. 33–35, 50–51.

[25] George J. Stigler, *The Theory of Price* (2nd rev. ed., Macmillan, 1952), p. 180.

and royalty payments are in the category of pure economic rent because the landowner is fortunate enough to have oil or gas deposits beneath his property, the same situation does not apply for the producer. And if we accept the position that competition exists in field markets, then monopoly profits or rents are ruled out. Thus, although profits higher than normal may exist in an industry if there is workable competition and if barriers to entry and/or expansion by existing firms are weak, the higher profits will attract capital and by so doing will prevent them from continuing.

In the correct economic usage of the term, it is improper to use windfall to discuss the possibility of unearned increment or excess profit arising in natural gas production. Pigou defined windfalls as follows:

By windfalls I mean accretions to the real value of people's property that are not foreseen by them and are not in any degree due to efforts made, intelligence exercised, risks borne, or capital invested by them.[26]

In the context of the definition, he was discussing taxation of windfalls, which he maintains should take place only on real increments rather than on apparent increments resulting from a rise in the general price level or from a drop in interest rates. Notice that Pigou used windfalls in regard to capital assets or investments. Keynes does the same.[27] He used the term windfall loss in discussing an unforeseen drop in the capital value of an asset as having an effect on the propensity to consume. If we wish to stick with the traditional usage of windfall, then it would be better to reserve it for unanticipated changes in the value of a capital asset and to use rent or economic rent for discussion of a particular type of return on a commodity or service.

The decision to spend money in drilling for oil or gas is a capital investment decision. This is so regardless of the accounting treatment (especially for income tax purposes) of such expenditures. The benefits to be derived from such expenditures are expected to accrue over some period of time into the future when the capital is committed. Money spent on unsuccessful wells is also a part of capital investment, because it is expected to be recovered in the course of obtaining hydrocarbon deposits. Capital will be shifted away from exploration if there is the expectation of a higher return in other investments; the decision to drill for oil or gas is no different from any other type of investment decision. Tables 6-32 and 6-33 give at least an indication of a move away from investment in drilling during the decade of the 1960s. Exploratory wells, successful new well completions, and a number of drilling rigs in operation all have fallen from 20 percent to 40 percent. Since the percentage of gas well completions to total oil and gas has remained about the same over the period, this is some indication that the relative decline in drilling activity has been proportional in the two areas.

Table 6-32. Successful New Well Completions, 1960-69

Year	Oil	Gas[a]	Total oil and gas	Gas as % of oil and gas
1969	14,368	4,083	18,451	22.1
1968	14,331	3,456	17,787	19.4
1967	15,320	3,619	18,939	19.1
1966	16,780	4,377	21,157	20.2
1965	18,761	4,724	23,485	20.1
1964	20,620	4,855	25,475	19.1
1963	20,288	4,751	25,039	19.0
1962	21,249	5,848	27,097	21.6
1961	21,101	5,664	26,765	21.2
1960	21,186	5,258	26,444	19.9

[a]Includes condensate wells.

SOURCE: AGA, *Gas Facts* (1960 through 1969).

[26] A. C. Pigou, *A Study in Public Finance* (3rd. ed., London: Macmillan, 1947), p. 156.

[27] John Maynard Keynes, *The General Theory of Employment Interest and Money* (London: Macmillan, 1936), p. 57.

Table 6-33. Exploratory and Development Wells and Drilling Rigs, 1960-69

Year	New field wildcats	Exploratory other than new field wildcats	Development	Total exploratory and development	Drilling rigs in operation (yearly average)
1969	5,956	3,745	22,486	32,187	2,074
1968	5,205	3,674	21,720	30,599	2,095
1967	5,271	3,788	23,103	32,162	2,207
1966	6,158	4,155	26,071	36,384	2,514
1965	6,182	3,284	30,044	39,510	2,783
1964	6,632	4,115	32,216	42,963	3,066
1963	6,570	4,094	30,722	41,386	2,952
1962	6,794	4,003	32,982	43,779	3,089
1961	6,909	4,083	32,879	43,871	3,464
1960	7,320	4,384	32,314	44,018	3,543

SOURCES: Various issues of *Gas Facts, Oil and Gas Journal, Bulletin of American Association of Petroleum Geologists, World Oil.*

A relatively small investment in exploration can yield a very high return, or a large investment can yield a minuscule return. This is an obvious way of saying that the production function for obtaining natural gas has an important random component. However, the ex ante expectation of the profit from such investment on some risk-adjusted basis must be comparable to the expectations of competing investment projects or there will be a net flow of funds away from such activity. The risk in exploration is in the production function, contrary to the traditional business risk of firms, which usually lies in marketing. Because of the possibility of large returns in oil and gas discovery from a small investment there will continue to be risk takers who will speculate in such ventures. This is borne out by the increasing importance of the independents who raise money through fund selling. On net, however, there must be movement away from commitment to drilling if average returns fall.

The attraction of capital to drilling for natural gas if returns rose on this type of investment is predicted on the assumption that the supply of natural gas in the United States is not price inelastic. Given (1) the present concern about supply being inadequate in the near future, (2) estimates of the potential amounts of gas to be found, and (3) the fact that less than one-third of existing supply is associated with oil, there is strong weight to the conclusion that the bulk of potential supply is not insensitive to price.

The idea that natural gas supply is price inelastic is vaguely appealing on the basis of two facts that generate incorrect inferences: (1) there is only so much natural gas existing, and it is impossible to change this amount; and (2) a lot of natural gas is associated with oil production and is in a by-product category. One of the inferences is that the price of natural gas should be "low," that gas should even be given away,[28] because before the existence of the long-distance pipelines most gas was discovered in the search for oil. It is because of this historical relationship of gas to oil (and its low price relative to oil) that Kahn made his rent and windfall argument. Why should buyers be expected to pay as much for gas as for oil on some comparable basis, when the companies that found gas were, after all, looking for oil? Anything over the variable cost of producing gas or any price permitted on existing gas reserves that did not reflect its low value prior to its access to the long-distance interstate market should be considered a sort of windfall to producers.

This kind of thinking about supply has influenced the posture of regulation since its inception, abetted by the distributors' desire to keep down the price they pay for gas to help them to continue to expand. It has also been responsible for the view that on a British thermal unit (Btu) basis, or other comparable basis, the "value" of gas is less than that of oil. In fact, it is quite likely that on a heat content basis a Btu of natural gas should sell for more than a Btu of oil. This is now unprovable one way or the other but is based on the opinion that if all governmental interventions in the market for

[28] The classic statement in this direction is probably Joel Dirlam, "Natural Gas Cost, Conservation and Pricing," *American Economic Review*, May 1958: ". . . as long as states prohibit flaring of gas, producers can afford to sell at a negative price to permit oil production to get under way."

both oil and natural gas were removed (ceiling prices on gas, import quotes on oil, and prorationing of oil production) gas prices would rise and oil prices would fall.

There is the possibility of an excess demand for gas that is different from that usually defined as excess demand. In the usual case, excess demand indicates a disequilibrium, which in the absence of price control and rationing will not persist. Some adjustment of schedules will occur or, for positive excess demand, price will simply rise to equate the quantity holders wish to hold (Wicksteed's reservation demand) plus the quantity nonholders wish to hold with the quantity holders wish to dispose of. In other words, given a short time period, fixed supply, and a free market, an equilibrium price will emerge that will cause excess demand to be zero. Now with gas, consider a short time period, say, a year. New demand takes the form of new long-term contracts plus the expansion of sales that existing contracts permit. As long as sufficient proven reserves exist, either uncommitted from prior time periods or discovered during the current year to meet these new requirements, no excess demand will exist in the conventional sense, despite the possibility of the fixed ceiling price being imposed below what the equilibrating price would be for zero excess demand. However, there will be disguised excess demand if the ceiling price is below equilibrium.

Suppose hypothetically that the field market had no regulation, but sellers were faced with buyers who were part of a monopsonistic cartel. For obvious reasons the supply of natural gas is probably relatively price inelastic *on the down side*, that is, with respect to existing supply. The buyer, if a monopsonist cartel, may lower price with little short-run effect on the quantity supplied, but new buyers who are not part of the agreement may then enter the market, because the elasticity of the excess demand created by the lower price is larger than anticipated by the cartel. If after some time supply became short and it were not possible by some type of nonmarket behavior to eliminate some buyers, price would rise to eliminate the excess demand and the cartel arrangement would have proved unstable. Now with regulation, something analogous happens. Price is not lowered as in the textbook example of the monopsonist, but rather a ceiling is imposed, which, if below equilibrium, will create a nonobservable excess demand as long as uncommitted reserves exist and/or discovery keeps far enough ahead of production. If the ceiling is below equilibrium this situation will not continue, and the excess demand will manifest itself. If the market were free, the price would rise. If the market could not price adjust, the excess demand would persist and eventually there would be unsatisfied buyers. This may well be what is now occurring. If the market were allocating supply, excess demand would be zero; if not, then the allocative responsibility must be through some other mechanism. In this case it would be the FPC, which may soon find itself performing a rationing function on gas consumption.[29]

If the FPC does not want this function, which would be subject to some rather severe political pressures, then there seems no escape from a price rise. An increase in price would not only tend to elicit more supply but would reduce demand. No one knows what the equilibrium price would be, although the FPC will be required by events to raise average prices if it does not assume the role of rationer. The legally required way to set price is based on cost, and it is quite beyond the FPC or anyone else to determine the cost of natural gas for an integrated producer, let alone for a segment of the industry. The way those presenting data and calculations of average cost of an Mcf of gas in area hearings switch back and forth among sources, the way they make joint cost allocations, and the way they apply "factors" represent pseudo-science to a degree it would be difficult to equal. At the final determination, the examiner's fumbling among the numbers and making a cafeteria-style selection from those presented, plus the commission adding a few delicate adjustments of its own, make the whole thing nothing short of ludicrous. And sad to say, it appears that all of this ritual will be duplicated in the reopened South Louisiana and Permian hearings.

As FPC Chairman John Nassikas said before the Senate Subcommittee on Minerals, Materials, and Fuels on November 13, 1969: "We have not yet developed a supply curve that will show us the rela-

[29]The New York Public Service Commission has already taken the position that if there is a supply problem, the FPC should ration gas to industrial consumers and save it for residential users.

tionship between the price level and discovery of reserves." To economists this is no surprise nor a necessary indication of dereliction on anyone's part, since such a relationship is difficult to obtain even for a commodity or service whose market is relatively perfect. It is even more difficult in a market, part of which is regulated and part not. Besides the aggregation problem, regulation can introduce distortion in supply and demand patterns that would not exist in an unregulated situation. Despite the difficulty of obtaining such a relationship, however, it must be granted that one exists, which regulated prices that are below equilibrium cannot hope to measure.

If the market for natural gas were free of regulation, it might be easier to obtain some supply-price relationship from empirical observation than is true for some other commodities. This is so because of the way the supply of gas can be considered over a time series. Whereas with the supply and price of, say, standard Chevrolet automobiles one can observe only the price and the number cleared from the market at that price (a supply and demand intersection), with gas, the additions to "inventory" might well be the supply price response variable rather than amount sold. Inventory of manufactured goods is often in the one- to two-month range, or less, but with gas the traditionally desired level has been twenty years.

From the figures in table 6-31 a little naive statistical exercise was performed. Four simple regressions and correlations were calculated over the years 1955-68, with new discoveries (D) and increase in proved reserves (I) regressed on average wellhead (W) and weighted average new contract prices (C). These equations came out:

$$D = 8{,}325{,}523 - \underset{(165{,}553)}{159{,}805\,W} \qquad R^2 = 0.0721$$

$$D = 15{,}520{,}583 - \underset{(398{,}507)}{523{,}902\,C} \qquad R^2 = 0.1259$$

$$I = 19{,}060{,}380 - \underset{(1{,}570{,}682)}{753{,}295\,C} \qquad R^2 = 0.0188$$

$$I = 30{,}417{,}410 - \underset{(382{,}888)}{1{,}773{,}652\,W} \qquad R^2 = 0.6413$$

The last equation is the only one showing a statistically significant relationship, and it is negative. There is a ready explanation for this, however. The average wellhead price was gradually rising over this period, with the expiration of older contracts at very low prices. At the same time, consumption has been increasing faster than increases in new reserves, so these values are falling. The other equations show no relationship between the variables. If there existed an "incentive" price to search for gas, one would hypothesize a positive relationship between weighted average new contract price and new discoveries, since these prices are what can be expected for new gas discovered. The sign on the regression between the two is negative, but not statistically significant. Hence, the impact of existing price in the market on additions to inventory of gas supply has been random during this period. Although there are other determinants of new discoveries, such as the price of oil, with most gas found being nonassociated, there should be an aggregate positive relationship between gas price and new discoveries. One could interpret this lack of relationship as simply showing that supply is not responsive to price, and hence we are back at the old stand that supply is price inelastic. The rejoinder to this could well be that supply is not responsive at *these* prices. The aggregate supply curve might well have a kink in it, as shown in figure 6-3. A part of supply, in the sense of additions to inventory, that is price inelastic consists of associated gas and a random element of gas discovery, and it can be argued that the ceiling prices have never gotten into the elastic range. If the equilibrium price (P_e) occurs in the elastic portion of the supply curve (as shown in figure 6-3), then excess demand (AB) created by an average ceiling price in the inelastic portion of the supply curve (P_c) would constitute a serious problem for the regulatory agency to resolve. And this may be exactly what the FPC is faced with today.

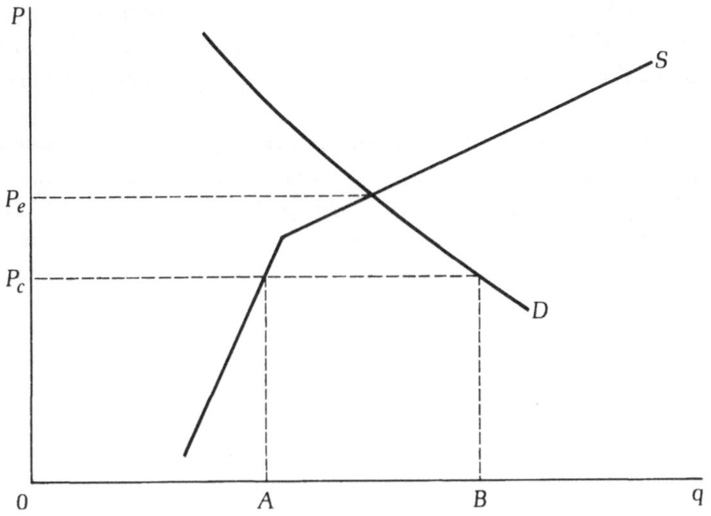

FIGURE 6-3. Potential supply curve for natural gas.

If one does not believe that the supply of natural gas (within some limits of the 1,000 trillion cubic feet to 2,000 trillion cubic feet still to be found in the United States) is price elastic, then the arguments of Dirlam and Kahn of a decade ago must be accepted. Kahn's position that the raising of price for old gas would confer windfalls and for new gas would constitute the paying of rents led him to argue that higher prices would not influence supply. Less than ten years ago his view was that there existed a burden on consumers of carrying (in the price they had to pay) an over-supply of gas in the form of too high a life index.[30]

This is scarcely the concern now, if it ever really was. The immediate problem for the FPC is how high to raise ceiling prices to get into the elastic portion of the supply curve, that is, to induce investment into drilling for gas. What rise in the ceiling would accomplish this is unknown, and its determination would require careful study; but because of the lack of clean data and the demand-supply curve identification problem it may well be impossible. However, industry spokesmen and producer witnesses have recently indicated that 5 cents to 10 cents per Mcf is the range of necessary increase.[31]

That substantial further gas supply is available from liquefied natural gas (LNG) imports is readily apparent from the contracts already signed. That it is considerably higher priced is also apparent. There are huge coal reserves in the United States that also provide extremely large latent supply. Based on the current demand for industrial coal and the recent flap over attempted implementation of new mine safety standards, there is some evidence that it will cost more and more to produce coal in the near future for whatever purpose. Since no one has built a full-scale plant to gasify coal, the costs of gas from this source must be estimated.[32] Casual empirical evidence gives an indication that the vector of primary supplemental supply for the United States is likely to be pipeline imports from Canada and Alaska, LNG imports, and coal gasification, in that order. In any event, it would be safe to say that the long-run supply curve will be elastic when we must rely mainly on these alternatives

[30] Kahn, *Matter of Area Rate Proceeding*, pp. 106-08.

[31] That a wellhead price rise of 5-10 cents per Mcf is needed to reverse the decline in reserves through increased drilling was stated by LeRoy Culbertson, head of Phillips Petroleum's natural gas and gasoline department, to the AGA, May 25, 1970. In a report on the testimony of Radford Shantz, vice president of Foster Associates, in the South Louisiana hearing, an actual figure of the magnitude necessary is not given, but price increases of several cents per Mcf were indicated ("Gas Supply Would Rise with Price Hike," *Oil and Gas Journal*, May 4, 1970, p. 94).

[32] See estimates of R. Bruce Foster in chap. 3.

and that there will be discontinuities in the supply curve as the primary basis shifts from the less costly to the more costly. What we would now like to know is the configuration of the supply curve in the region of obtaining those domestic natural gas supplies for the interstate market that are as yet undiscovered.

Since there is considerable room between current ceilings on gas and comparable volume and heat content cost estimates for LNG and coal gas, the FPC may simply have to experiment with ceiling price increases between the present ones and LNG costs to test supply response.[33] It should really do something along these lines, because if it does not there is going to be a large social cost associated with the failure to obtain those domestic supplies that could be obtained at less than the cost of the alternative sources.

[33]The FPC must use some imagination and depart from its current practices. One possibility is for some type of "investment tax credit" approach. The FPC could raise ceilings by, say, 5 cents for producers, which would step up their search for gas. For example, out of the increase in revenues in year one from the 5-cent increase on gas contracts, which would rise by the 5-cent amount, a gas producer would be required to invest a minimum percentage, say, 25 percent in gas drilling in year two over the amount he invested in gas drilling in year one. This sort of approach would not be without its problems, such as in determining exactly how much is spent on gas drilling. Something like this would be a radical departure from traditional regulation but should not be excluded solely on these grounds, since surely by now it is recognized that natural gas field price regulation is not a traditional regulatory effort.

7 THE REGULATION-INDUCED SHORTAGE OF NATURAL GAS

*Paul W. MacAvoy**

Since its inception by Supreme Court decree in 1954,[1] the regulation of prices at the gas wellhead by the Federal Power Commission (FPC) has been "welfare" policy in the sense of redistributing income. As with local rent controls or maximum charges for medical services paid for by the state, gas price restrictions were to "keep prices down for the consumer." Problems were encountered because of the large number of sources of supply of gas reserves, not because a few firms were able to restrict supply. The attention of the FPC centered on holding the price line on hundreds of transactions for differing qualities of gas sold by different producers to almost as many industrial buyers, local retail gas utilities, or long-distance interstate pipelines. The regulatory problem of first priority has been to find uniform price ceilings that would apply equitably to heterogeneous sellers and buyers.

The problem brought the FPC to a standstill in the 1950s. In its experience with natural gas companies, it had legislative and judicial mandates to fix specific prices on the sales of each company in keeping with accounting costs of service on these sales, because these companies were interstate gas pipelines with local monopoly or oligopoly power to set prices to retail distributing utility companies. The FPC attempted to deal with gas producers in the same fashion, first with the Phillips Petroleum Company in the remand from the Supreme Court of the original case for field price regulation. This was to attempt the impossible, given that costs of service were joint costs of oil and gas exploration, and joint costs of exploration and development, and that neither theoretical nor empirical procedures exist for finding them. Finding specific costs of service was abandoned, however, not for analytical reasons, but because the time and FPC resources taken up in making a single finding were needed to process hundreds of cases. Also the decades-long time delays that were going to be encountered in maintaining the procedure would have imposed prohibitive costs on the industry.

In the 1960s the FPC devised regional price ceilings to apply to all sales in large producing areas whether these sales were made by companies encountering high or low exploration and development costs. The problems of "fairness to all" were those of finding an areawide ceiling price to apply to gas reserves not only costing different amounts to develop, but also having different demands because of particular advantages of location or of size. The ceiling prices had to strike a compromise between the gains of some producers and the losses of others and between the gains of buyers and the losses of producers. Here, the regulatory process worked out differences in the way that local rent ceilings did

*Professor of Management and Economics, Sloan School of Management, Massachusetts Institute of Technology.
This paper was completed in conjunction with work being done by the author on an operations analysis of the Federal Power Commission, sponsored by the Brookings Institution under its program of Studies in the Regulation of Economic Activity. Acknowledgment is made of this support and of that of Resources for the Future and of the wise counsel of Franklin Fisher, Edwin Kuh, and Lester Thurow. The comments and criticisms of M. A. Adelman, Keith Brown, Edward Erickson, Richard L. Gordon, Clark Hawkins, John R. Rebman, and Haskell Wald greatly improved the results—at least according to the author.

[1] In *Phillips Petroleum Co. v. Wisconsin*, 347 U.S. 672 (1954).

in various parts of the United States. The higher or highest prices set in competitive regional markets in the late 1950s were made the interim regulatory ceiling prices, and proceedings were initiated to show whether average costs of recent gas development and production were so high as to require further price increases. The proceedings were massive and prolonged, and they produced voluminous documentation for different levels of average total costs that were used as the basis for FPC judgment. The FPC judgment was defined as equitable because it involved due process according to the U.S. Court of Appeals.[2]

The FPC finding of just prices to cover all transactions in a broadly defined region and time period raises far more problems than it solves. The resulting redistribution of income should be investigated and should be in keeping with general practice on moving toward income equality (whatever that may be). Entirely new problems are encountered related to the supply of new reserves and additions to demands for these reserves brought forth by the ceiling prices themselves. Market-clearing prices—those bringing forth more reserves from producers equal to the buyers' demands for more reserves—are set by the marginal costs for more gas, not the average costs used in finding the ceiling prices. If the FPC judgmental procedure were to result in an accurate forecast of future average total costs for additional supply and were to set prices equal to these average costs, prices would be too low to clear the market. The marginal costs of additional supply exceed the average costs in a competitive market for a wasting resource so that market-clearing marginal cost prices exceed the regulatory ceiling prices. The problem of the 1960s was that the ceiling prices added to demand and reduced supplies of new reserves so as to create excess demand.

This problem was recognized early in the regulatory process. It was minimized by some on the grounds that the reduction in supply would be very small and that the increase in demand also would be very small.[3] It would appear that this problem has been one of the highest magnitude and that it has brought about substantial misallocations of new supplies by causing supplies to be denied to those the FPC was seeking to favor.

In the last few years there have been a number of indications of a serious shortage of natural gas reserves. Some have been statements by industry experts on new evidence of a shortage or on what ought to be done. In many declarations on the subject, it has been stated that the FPC area rates have been an important cause of the supply-demand difference and that new area rate policies are needed to reduce the excess demand.

The sources of these statements have been so diverse as to suggest a consensus. Not only producers but also buyers have claimed that there was substantial regulation-induced excess demand and that ceiling prices under new contracts were too low. The buyers have shown their own shortage in the Midwest and on the eastern seaboard by announcing restrictions on sales to new customers in Chicago, New Jersey, and Pennsylvania. The chairman of the Columbia Gas (transmission) System recently announced that the company would limit its new sales commitments, "because customers have recently made requests for next winter far exceeding [previous] requirements . . . [and] we cannot obtain the additional gas to meet these demands." The chairman laid the shortage to "unrealistic area pricing policies on gas production," combined with unforeseen increases in demand.[4]

M. D. Borger, chief executive officer of Consolidated Natural Gas Company (Pittsburgh), in his testimony in the Southern Louisiana area rate proceeding, described that company's 1967–70 attempts to secure more gas for retail delivery as unsuccessful: "We were unable to contract with our [pipeline]

[2]See *Permian Base Area Rate Cases*, 390 U.S. 747 (1968); *Southern Louisiana Area Rate Proceedings*, AR 61–2, 41 FPC 301 (March 20, 1969) affirmed in *Southern Louisiana Area Rate Area Cases v. FPC*, 428 F2d 407 (5th Cir. 1970).

[3]See Alfred E. Kahn, "Economic Issues in Regulating the Field Price of Natural Gas," 50 *American Economic Review* (1960), pp. 504, 509, where reference is made to "thin" and "inelastic" supply with increased demands during the 1950s.

[4]"Columbia Gas Will Limit New Commitments to Sell Natural Gas: Short Supply Cited," *Wall Street Journal*, April 10, 1970.

suppliers for any additional long term supplies . . . [and] were routinely met with the response that there was a shortage of gas for purchase in the gas fields in Texas, Louisiana and the midcontinent areas of the United States." Borger's own investigations led to some strong conclusions on cause and effect: "My conclusion now is that the volumes of undedicated reserves which may be available for purchase by the pipelines are insufficient to meet current increases in utility demand . . . in large part because of the cumulative effects of regulation under the Natural Gas Act during the period since the *Phillips* decision in 1954. In my opinion, price reductions and price rollbacks and instability have stifled exploration operations while the resulting low rates have created an artificially high demand for natural gas."[5] Other distributors and some of the state regulatory commissions have disagreed with these reasons for gas shortages, but the statements of Columbia and Consolidated represent a new and strong trend of policy views of those buying gas under long-term contracts for home delivery. The pipelines to the North and East have found the regulated prices below what they would be willing to pay for additional production rather than go without, and they have joined the producers in seeking higher prices.

The FPC has also acknowledged the existence of an important shortage and has begun to consider new policy in light of this condition. The staff dealing with natural gas regulation has issued reports in recent years showing that additions to gas reserves of the pipelines have been insufficient to meet "market requirements for extended periods of time."[6] Most recently, the staff seemed to be saying that the pipelines could have met demands in 1963 for fifteen years thereafter, while they likely could meet 1968 demands for no more than seven years.[7] The staff view has been shared by the commission, at least to the point of considering it likely that ceiling prices have been far too low to clear gas field markets. In the opinion issued on the rehearing of Southern Louisiana area rates, the commission concluded that "the vital importance of future additional gas supply from offshore warrants . . . further proceedings . . . looking towards a possible revision of the area rate proceedings for such gas."[8] And to some extent a commission's policy reflects the views of the chairman. John Nassikas, FPC chairman, has recently taken a stance on a national policy issue outside FPC jurisdiction on the basis of the "natural gas shortage" alone. Because supplies of new gas reserves have not been sufficient, Chairman Nassikas has urged that oil import quotas be maintained to provide every possible incentive for finding gas in the quota-induced search for domestic oil.[9] This position implies that prices of oil should be kept too high because FPC-regulated prices of gas are too low. Also, it is very likely an inexpedient policy within interrelated markets for energy sources, since higher prices for oil increase the demand for gas as a substitute boiler fuel, adding to excess demand. The Nassikas view can be rationalized only as emergency policy—given that the shortage of gas is acute, that raising natural gas field prices will not have the immediate effect of relieving this shortage, and that the oil quota price adds more to gas supply than demand.

All these expressions of opinion are incomplete. They lead to the conclusion that a natural gas shortage exists, partly at least as a result of regulated prices having been set too low to clear field markets of increased demands. They do not show the magnitude of shortage, nor the effects on different consumer groups. The sources of excess demand are as important as the magnitude, because those doing without might be especially disadvantaged in using substitute fuels. Yet only the FPC staff reports show measures of deficiency, and these are based on "requirements" rather than on price-related demands so that the effects of the price ceiling are not shown.

[5] Testimony before the FPC, *Southern Louisiana Area Rate Proceeding*, AR 69–1 (1970), pp. 5, 6, 8, and 10.

[6] FPC, Bureau of Natural Gas, *The Gas Supplies of Interstate Natural Gas Pipeline Companies, 1968* (1970), p. 14.

[7] Assuming that in ibid., chart 3, p. 15, 2 trillion cubic feet of "normal replacement" reserves are forthcoming from extension and revisions of present fields.

[8] *Southern Louisiana Area Rate Proceedings*, Opinion no. 546–A.

[9] See "Supplementary Views of the Chairman, Federal Power Commission: Impact on the Natural Gas and Electric Utility Industries" in Cabinet Task Force on Oil Import Control, *The Oil Import Question: A Report on the Relationship of Oil Imports to the National Security* (February 1970), pp. 369–89.

The size and source of excess demands will be assessed in the sections that follow. First, the size of the shortage in the last few years is estimated by recourse to an inventory analysis, in which the inflow of newly discovered gas reserves is put against new production in order to find the extent of reserve backing for that new production. The critical element here is the judgment on the amount of reserves required to provide assurance of continued supply to a new customer. If this backing should have been as extensive as that for new deliveries in the 1950s, then more than half again as much reserves should have been available in the last five years. If the backing had to be twenty years' production, then one and one-fifth again as much new reserves should have been committed to the pipelines.

In the second section an attempt is made to construct measures of the role played by the price ceiling in the shortage. The measures are from two econometric models of gas supply and demand constructed on the basis of market-clearing conditions in the 1950s but used to simulate behavior without regulation by extrapolation of these conditions to the 1960s. The simulations of newly discovered reserves and prices are compared with actual new reserves and prices under area rate regulation, and the differences are attributed to the regulation. Here, reserves seem to be substantially less than they would have been if prices had been higher–and the higher prices would have cleared markets by increasing supply and reducing demand proportionately.

The last section is a survey of those who received gas in these shortage markets–the home consumer at regulated resale prices or the industrial user at unregulated prices. Regulation seems to have played a large role in this decision as well, and as a consequence the home consumer does not seem to have done very well.

AN "INVENTORY" APPROACH TO EXCESS DEMAND

The field markets for natural gas are not similar to those for other fuels, because there are no large-volume spot transactions for the purchase and delivery of gas at one location in one day. Rather, reserves of gas are dedicated to pipeline buyers, and these are depleted by production into pipelines over an extended period of time. The dedications are sales by contracts that specify initial prices, delivery period, and the property rights to inground gas reserves. The contract markets can be said to clear if the amount of newly dedicated reserves meets the inventory requirements for new production going by pipeline to final industrial and home consumers.

These inventory needs certainly exceed the amount actually produced under a new contract in the first year of that contract. A pipeline would not make a connection with an industrial firm for a single year's deliveries, nor would it sign a ten- to twenty-year contract for delivery to a retail gas distributing company without reserve backing. The question is how much backing is optimal, given that more reserves provide more certainty for the pipeline against default on its promised deliveries, but also more reserves require greater pipeline expenditures.[10] To some extent the regulatory process provides the answer: the FPC has considered the proper amount of reserves to be twenty times initial production so that (regulated) demands for new reserves should be based on "the assumption that each new market commitment is backed by a twenty year supply."[11] But the average buyer–pipeline and nonpipeline–has sought less extensive security than that required by the commission. Additional reserves ΔR_t have been added by all buyers on average each year since 1947–about the time interstate delivery began on a large scale–equivalent to 19.5 years of additional deliveries ΔQ_t each year. For national reserves and production, $\Delta R_t/\Delta Q_t = a + \beta T$ has been $\Delta R_t/\Delta Q_t = 13.5 + 1.0T$ ($R^2 = 0.21$) with

[10] In theory, at least, a longer waiting period for production imposes higher costs on the supplier, necessitating higher contract prices to the pipelines. This cost increase was not reflected in significantly higher prices on longer-term contracts, however, during the period just before area rate regulation. See Paul MacAvoy, *Price Formation in Natural Gas Fields: A Study of Competition, Monopsony and Regulation* (Yale University Press, 1962), pp. 262–65.

[11] FPC, Bureau of Natural Gas, *A Staff Report on National Gas Supply and Demand* (September 1969), p. 18. Twenty years reserve "backing" is equivalent to twelve years "deliverability"–that is, twelve years production at well or field capacity.

T=1 to T=11 for the years 1947-57,[12] or 14.5 years reserve backing at the beginning of the period and 24.5 years at the end. The inference might be that it was optimal over the 1950s for most buyers to keep less reserve backing than required under regulation.

The demand for new reserves by interstate pipelines might well be at the rate providing reserve backing in keeping with a continuation of 1947 conditions. This backing on average may have been sufficient for final buyers in unregulated intrastate and industrial markets, whether they were retail gas distributors or power companies. It could be sufficient under regulation–what is secure enough in competitive intrastate and industrial markets is secure enough in interstate markets. Alternatively, twenty years backing is required, as one of the guarantees of the "public interest" protected by the FPC–the optimal reserve backing is that judged so by the commission. Applying the rate $\Delta R_t/\Delta Q_t$ given by $(a + \beta T)$ to actual production on new contracts, ΔQ results in an estimate D_1 of the expansion demands for new reserves, where these demands are in keeping with continued 1947 national patterns. Applying the twenty-year ratio to ΔQ results in the "commission-approved" reserve backing for new demands–the amount that would provide the backing sought by the regulator for the level of security deemed "necessary" for final consumers.

At the same time that some interstate pipelines are seeking reserves for expansion, others are demanding new reserves to replace amounts depleted by completion of earlier contracts. These new replacements are not expected to be complete. A number of old contracts are not likely to be continued, because prices for replacement gas have to be greater than the original prices; the buyers taking gas at the earlier prices change to oil or coal rather than pay the higher prices required by new field purchase agreements. Replacement demand D_2 is equal to $\gamma Q_t\,(a + \beta T)$ where γ is the proportion of established deliveries Q_t that are now depleted and are going to be replenished. Alternatively, $D_2 = 20\gamma Q_t$ for the commission-desired twenty-year reserve backing.

Total new reserve demands $D_1 + D_2 = (\Delta Q_t + \gamma Q_t)\,(a + \beta T)$, or twenty times $(\Delta Q_t + \gamma Q_t)$ at the commission-approved inventory level. Finding these demands requires estimates of ΔQ, the amount of deliveries to new customers in the first year, and γ, the proportion of established deliveries to be replenished; but both require detailed studies of sales contracts and of the elasticities of demand of old and new customers that are not possible given present data conditions. It is possible to estimate the total $(\Delta Q + \gamma Q_t)$ directly, however, because $(\Delta Q + \gamma Q_t) = (Q_t - Q_{t\text{-}1}) + (\gamma + \epsilon)Q_t$, where $(\gamma + \epsilon)$ is the rate of depletion of reserves, γ is replacement, and ϵ is nonreplacement, since additional production takes place *only* for either expansion or replacement purposes. Production under new contracts $(\Delta Q + \gamma Q_t)$ by definition must either expand production $(Q_t - Q_{t\text{-}1})$ or replace depleted production $(\gamma + \epsilon)Q_t$.[13] Estimation is confined to finding $(Q_t - Q_{t\text{-}1})$, the year-to-year change in the total, and $(\gamma + \epsilon)$, the proportion of any year's production that depletes reserve holdings.

The estimates of new pipeline production times the required pipeline reserve backing show the reserve shortage as in table 7-1. For new production, the value of $(\gamma + \epsilon)$ in each year was taken to be $1/(a + \beta T)$, the rate of depletion of national reserves for T=1 for 1947,[14] and the rates Q_t, $Q_{t\text{-}1}$ were as shown in the FPC staff study *The Gas Supplies of Interstate Pipelines, 1968*. To find the required reserves, the new production $[Q_t - Q_{t\text{-}1} + (\gamma + \epsilon)Q_t]$ was either multiplied by $(a + \beta T)$, the

[12] The production and reserves series are from American Gas Association, American Petroleum Institute, and Canadian Petroleum Association, *Reserves of Crude Oil, Natural Gas Liquids, and Natural Gas in the United States*, vol. 23 (May 1969). In order to find *additional* reserves dedicated to *additional* production, the year-to-year changes in reserves are calculated first. These equal $(R_t - R_{t\text{-}1} + Q_t)$. Second, an amount, Q_{47}, is subtracted, given that this equals the required replacement for existing 1947 commitments that have not been cancelled. Then $\Delta R_t = (R_t - R_{t\text{-}1}) + (Q_t - Q_{47})$.

[13] Here $(\gamma + \epsilon)$ is the proportion of previous years' production that depletes reserves, γ is that proportion that is replaced by new contracts written with the same established customers, and ϵ is that proportion not replaced by new contracts.

[14] This is the rate of depletion of new national reserves in 1947. It is taken to be representative of depletion of old interstate pipeline reserves in 1964-68, because the period of large-scale development of the interstate lines began in 1947, and the initial reserves were probably being depleted at the national rate.

Table 7-1. Estimates of Supply and Demand for Natural Gas Reserves, 1964-68

(trillion cubic feet)

Year	Supply (additions to reserves)[a]	"Unregulated" reserve demand[b] (continuation of previous national average reserves accumulation)	"Regulated" reserve[c] demand (20-year reserve lifetime)
1964	10.7	[0.7 + .07 (9.6)] [14.5] = 19.8	27.4
1965	12.8	[0.2 + .07 (10.3)] [14.5] = 13.2	18.4
1966	14.9	[0.8 + .07 (10.5)] [14.5] = 22.1	30.7
1967	15.7	[0.7 + .07 (11.3)] [14.5] = 21.4	29.8
1968	9.4	[0.8 + .07 (12.0)] [14.5] = 23.6	32.8

[a] "Additions to reserves" are $(R_t - R_{t-1}) + Q_t$, as shown by pipeline company responses to FPC Form 15 and 15A.

[b] "Unregulated" demand is $[Q_t - Q_{t-1}) + (\gamma + \epsilon)Q_t] (a + \beta T)$, as explained in the text. Here $(\gamma + \epsilon) = (a + \beta T)^{-1}$.

[c] "Regulated" demand is $[(Q_t - Q_{t-1}) + (\gamma + \epsilon) Q_t]$ 20, in keeping with the commission-expected twenty-year reserve backing.

SOURCE: FPC, Bureau of Natural Gas, *The Gas Supplies of Interstate Pipelines, 1968* (January 1970).

trend of pre-regulation marginal reserve-production ratios, or by the FPC's "twenty times production" factor. On these bases, the amount of new production ranged from 0.9 trillion cubic feet (tcf) to more than 1.6 tcf in one year, and the demanded reserve backing for the new production from the pre-regulation level of 20 tcf to a commission-required 33 tcf in one year. The supply of new reserves was less than the reserve backing demanded over the years 1964-68. The demands for pre-regulation backing were 156 percent of the actual volumes of new reserves committed under contracts signed in 1963-68. The "twenty times production" reserve backing was not secured—these cumulative FPC demands were 219 percent of new contract commitments. Excess demand must have been from one-half again to one and one-fifth again as much as the actual new contract reserves committed in the late 1960s.

The shortage is probably manifest in the elimination of security of delivery for established final buyers of natural gas. If all buyers taking the last of some previous reserve commitments were to renew their purchase contracts, then almost all the reserves added to pipeline holdings over the 1964-68 period would have been necessary to provide these buyers with their usual reserve backing. The backing for these established buyers would have had to approximate 53.7 tcf of the 63.5 tcf added to pipeline holdings over these five years.[15] The remaining 9.8 tcf could have been used to increase production by 0.13 tcf per year (with the 1947 level of reserve backing) or by 0.10 tcf per year (with twenty-year reserve backing). In fact, however, production was increased by 0.7 tcf, 0.2 tcf, 0.8 tcf, 0.7 tcf, and 0.8 tcf successively over 1964-68. The backing for either new buyers or established buyers was not maintained because of the shortage. That is, if more new reserves had been supplied to the pipelines under new contracts or if there had been less additional demand from final buyers, the extent of reserve backing would not have declined below both the previous unregulated levels and the present required levels.

The FPC ceiling prices may have caused this five-year shortage, but this is not apparent from a review of inflows and outflows of gas reserve holdings. The approach to be taken in order to infer the effects of field price regulation is to find the approximate market prices and quantities for conditions similar to those when markets cleared in the 1950s and then compare these with regulated ceiling prices and quantities. It is suspected that the market-clearing prices would have been higher and would have brought forth a greater supply of reserves from new discoveries. Any remaining supply-demand difference might well have been eliminated by price-induced substitution of other fuels for gas—although it cannot be stated that this would have happened immediately.

[15] This is estimated as $[(\gamma + \epsilon)Q_t] (a + \beta T)$. If renewals are less than $(\gamma + \epsilon)$, then there is more room for expansion of deliveries to new customers.

SUPPLY-DEMAND MODELS OF GAS FIELD MARKETS

In the absence of area rates there would have been significant increases in prices in new gas contracts, and there might well have been significantly greater new reserve commitments during the 1960s. General increases in demands for energy, during a period in which the supply responses in coal and oil markets were less than might have been expected, should have led to short-term price increases for new gas reserves. The gas supply response might have been substantial–enough to fill in a major part of the excess demand shown in table 7-1; this, however, is a matter requiring estimation or forecasting under uncertainty as to the physical stock of reserves. In any event, prices probably would have cleared markets within a year or two, if there had been a continuation of the market conditions existing in the late 1950s when large demand increases from new pipelines were accommodated by price increases.

The method for showing market-clearing price and supply is first to find supply and demand schedules characterizing behavior in the 1950s and the market-clearing prices in that period. Here supply and demand models are constructed and used to determine ΔR_{tj} (reserves added in year t in producing district j) and P_{tj} (the average of the initial base prices on new contracts signed in year t in producing district j) for the years 1955 through 1960 in two markets–gas for the East Coast and Midwest (EMW) and gas for the West Coast (W). The model values for $\Delta R_{tj}{}^*, P_{tj}{}^*$ are the market-clearing values, since they solve the two-equation system; even though they differ from the actual values, these differences are small from year to year so that it would appear that markets cleared in the 1950s. The second step is to apply the model to the 1960s by inserting 1961-67 values for the outside variables into the supply and demand schedules and then solving the system for market-clearing values $\Delta R_{tj}{}^*, P_{tj}{}^*$. The model values are compared with the actual reserves and prices, and the differences are attributed to the lack of adjustment in the (regulatory ceiling) prices. The differences are seen to depend critically on lower supply at lower regulated prices.

An Introductory "Engineering" Model of Supply and Demand

There are many views in the petroleum industry on the exact nature of the process of finding and developing new gas reserves. An extreme view–one that implies the smallest possible effects from price ceilings–is an "engineering" view that gas reserves are discovered incidentally in searching for oil and then made ready for development by rapid drilling programs designed to prevent capture by adjoining developers. Reserves ΔR_{tj} added in any gas supply district are greater when total expenditures C_{tj} on drilling are greater,[16] or $\Delta R_{tj} = f_1(C_{tj})$. This view will be tested because its implications include that of no direct effect on discoveries from price ceilings. With complete information on oil exploration, resulting new gas discoveries, and development costs, this argument could be tested; here, a modified version has to be used because of very incomplete information.

The only indications of discoveries for sale are not statistics on reserves in new gas fields resulting from recent drilling operations and now offered for contract sale, but rather the amount of new gas recorded as having actually been discovered in that district and that year, whether ready for sale or not. The discovery estimates are thus for new volumes in place. Oil and gas discovery operations are continuous, and what is found in a district depends on what was found there the year before; as a result, the most concrete determinant of what is found and the amount available for use is the previous year's reserves discoveries in that district. The analysis should be modified so that

$$\Delta R_{tj} = f_1(\Delta R_{t-1,j}, C_{tj}).$$

There are no detailed statistics on costs C_{tj} of development activities in gas fields. The only data of sufficient detail show the number of gas development wells in the 1950s and 1960s, by drilling district.

[16] This last argument is the reversal of the commonplace marginal cost argument: for $C = f(Q,A)$ then $\partial C/\partial Q > 0$ for C total costs. Here, $\partial Q/\partial C > 0$ in $Q = f(C)$.

These are not wholly indicative of costs; some wells cost much more than others, because of depth and strata conditions, and wells themselves are joint factor inputs both of capital and of knowledge on surrounding geological conditions. The knowledge input is particularly important since further wells, even for development purposes alone, contribute to forming new estimates of reserves in surrounding trap or field formations. It might be expected that substituting wells $W_{t-1,j}$ for C_{tj} so that $\Delta R_{tj} = f_1(\Delta R_{t-1,j}, W_{t-1,j})$ would indicate the important factors in reserve discoveries, at least from the "engineering point of view."[17]

The demand side of gas markets has been described in a number of different ways. Here, the owners of new reserves are assumed to take and develop the discovery after bids from various pipeline buyers; the buyers set the terms of the contract in competition with each other, but they make up the side of the market that sets the terms. One of the terms is price P_{tj}, which depends on the pipelines' own costs of providing transport service from each district. The volume of discoveries is important in setting these costs—large discoveries provide the opportunity to put in large-scale gathering lines that reduce the costs of transporting a thousand cubic feet (Mcf) of gas; in simplest terms, the average new contract price in a district depends on reserves ΔR_{tj} discovered there, or $P_{tj} = f(\Delta R_{tj})$. There is another cost determinant of bid prices. The greater the distance is from the source of reserves to final home and industrial gas users, the greater the costs are to the pipelines and consequently the less the field offer price for reserves;[18] that is, $P_{tj} = f(\Delta R_{tj}, M_j)$ for mileage M of district j from the point of resale to retail distributors, where increased mileage is expected to lead to reduced bid prices. The important general demand determinant of bid prices is the final consumer, however. In years in which prices of substitutes are rising and incomes or other determinants of personal consumption are increasing, the bid prices should be higher. The first indicator of a demand increase here is an increase in the price of substitutes, as shown by the index of all fuel prices, fp_t. The price $P_{tj} = f(\Delta R_{tj}, M_j, fp_t)$, where fp_t is the index of fuel prices by consuming region, and the higher this index is, the higher the expected bid prices of the pipeline buyers of gas will be.

The engineering supply-demand relationships before field price regulation can be characterized as follows, where each variable is in logarithms:

$$R_{tj} = a_1 + b_1 \Delta R_{t-1,j} + c_1 W_{t-1,j}. \tag{1}$$

$$P_{tj} = a_2 + b_2 \Delta R_{tj} + c_2 M_j + d_2 fp_t. \tag{2}$$

The quantity of reserves discovered according to (1) equals the quantity demanded in (2) each year. This is the process of competitive market clearing, with contracts for new reserves being put in effect the year discoveries are made and the contract prices reflecting very recent changes in pipeline costs and in final consumer fuel prices. In this model, the markets clear as a result of new reserves ΔR_{tj} being made available, and then as the result of the highest bid prices of pipeline buyers being accepted by those producing companies providing the reserves.

The procedure for fitting these for the 1955-60 experience has been to complete a two-stage least squares (2SLS) regression analysis. The data series for these years are used to fit the general log linear form of the first stage equation:

$$\log \Delta R_{tj} = -\ 30.7240 + \underset{(0.1050)}{0.6547} \log(\Delta R_{t-1,j}) + \underset{(0.1479)}{0.2694} \log(W_{t-1,j})$$

$$+ \underset{(1.0398)}{0.0526} \log(M_j) + \underset{(5.3848)}{6.7685} \log(fp_t);\ R^2 = 0.5193,$$

[17] All series used for fitting functional relationships are described in detail in app. A.

[18] A diagrammatic exposition of this argument is provided in MacAvoy, *Price Formation*, chap. 3.

for sixty-six observations of eleven drilling regions for EMW. Then regression equation values $\Delta\hat{R}_{tj}$ for the given values of the independent variables in each of the sixty-six observations are used with the actual values of the exogenous variables to fit the demand least squares equation. This second stage least squares equation from the same sixty-six observations of EMW before regulation is as follows:

$$\log P_{tj} = -\ 0.9898 + \underset{(0.0160)}{0.1307} \log(\Delta\hat{R}_{tj}) - \underset{(0.1309)}{0.1260} \log(M_j) + \underset{(0.6821)}{0.8231} \log(fp_t); R^2 = 0.5498.$$

A cursory inspection shows that these fitted equations meet minimum standards for reasonable economic relations. Reserves increase the most where previous reserves and drilling activity are the greatest, at least in these years and this region. The second equation indicates that offer prices are higher where the new packages of reserves are larger, the distance to final markets is less, and the fuel price index is higher. All are in accordance with competitive relations between final demand and bid prices in the field.[19] But these are minimum standards of reasonableness. In statistical terms, the log linear form in the first equation fits to the extent that the coefficient of past reserves is statistically significant and the equation explains 52 percent of the variance in discovered reserves.[20] The demand price equation cannot be said to have statistically significant independent variables, given the presence of a fitted variable $\Delta\hat{R}_{tj}$; but the quasi-significant variable $\Delta\hat{R}_{tj}$ and the two other variables together show an R^2 of 55 percent of the variance in bid prices. Again, the interpretation of R^2 is not clear, given that the variance in $\Delta\hat{R}_{tj}$ has already been explained in the last equation. Both regressions may meet minimum requirements, in that R^2 indicates that more than half the variance in the critical ΔR_{tj} and P_{tj} variables is associated with the independent variables. This is to say that much remains unexplained.

These supply and demand functions may be only an initial description of market clearing for newly discovered reserves in the later 1950s. If this is the case, how far did actual prices and quantities depart from clearing prices and quantities in those years? Actual reserves were almost the same as those amounts that would have made supply equal to demand: the average discovery in the eleven districts over the five years was 403 billion cubic feet (bcf), while the range was from less than 10 bcf to more than 8,000 bcf; on average, more than half the market-clearing estimates of reserves were within 3 bcf of these actual volumes and more than nine-tenths of the estimates were within 9 bcf of actual reserves.[21] Actual prices, while averaging 14.7 cents per Mcf, were no more than 1 cent per Mcf greater or less than market-clearing prices in half the observations, and no more than 3 cents in almost all the observations. The supply-demand model showed no tendency for systematic departures from actual behavior (except in the South Louisiana district where the model expected less new reserves each year than the 2,000 bcf to 8,000 bcf discovered). The impression is that markets in the 1950s behaved as though they cleared each year.

The regression equations can be used in an initial attempt to simulate unregulated sales in gas field markets in the 1960s. If the conditions described by the equations for the 1950s had continued

[19] The behavior of prices and reserve volumes in competitive field markets is described in detail in MacAvoy, *Price Formation*, chap. 2-4. That description is in terms of a "reduced form" equation for prices assuming reserve supply to be dependent on geological conditions in the strictest sense, in the shortest-run time horizon. With inelastic supply, the reduced form equation becomes the demand equation given here.

[20] The coefficient b is statistically significant when the hypothesis that it has a true value of zero is rejected by $b_i/Sb_i > 2.0$ (where Sb_i is the estimated standard deviation in the distribution of computed b_i). This value of $t > 2.0$ would occur by chance in less than 5 calculations out of 100 when the zero value hypothesis held, so that it is "safe" to reject the hypothesis in favor of the alternative of b_i significantly greater than zero. The equation explains that proportion of the variance in the dependent variable indicated by R^2, defined as the ratio of variance around the regression line to total variance in the dependent variable. The values of R^2 are shown for each equation, and they are 0.52 and 0.55 respectively.

[21] The standard error of estimate around the regression equation for reserves had a log value of 1.062 so that—assuming log normality of the residuals of actual from regressed reserves—more than one-half of the actual values were within 1.062 of the regressed values.

through the 1960s, then the supply and demand curves would have produced forecast values of reserves and prices $\Delta R_{tj}{}^*$, $P_{tj}{}^*$ from introducing actual 1961-66 values of the outside variables $R_{t\text{-}1,j}$, $W_{t\text{-}1,j}$, M_j, and fp_t into the regressions. The values $R_{tj}{}^*$ and $P_{tj}{}^*$ reproduce the 1955-60 market-clearing conditions but given 1961-66 levels of well drilling, mileage, previous reserves, and final consumer demands. In that sense, the simulated $R_{tj}{}^*$ and $P_{tj}{}^*$ can be said to describe behavior in the 1960s as if there were no field price regulation.

The regulated and unregulated prices and quantities are shown in table 7-2 for the first seven years of the 1960s.[22] The two price series are not very much the same: simulated prices seem somewhat lower than the average of the area rates each year, more than 2 cents lower in 1962 and 1.5 cents lower in 1965. There would seem to have been no detectable price-reducing effect from area rate regulation—rather, area rates seem excessively high. In contrast, actual reserves were much less than reserves simulated for markets without regulation. The actual levels of reserves averaged 8 percent less than the simulated levels over the seven-year period.[23] This greater volume of discoveries at lower prices without regulation would seem to be an illusion of the model—regulation keeps prices at higher levels—except for the very strong presumption that there are results induced by the particular engineering supply conditions in the first regression equation. The reserve equation each year now forecasts more new discoveries than actually take place, which are then cleared by reduced prices in the price equation. The reserve equation forecasts are too large, probably because regulatory conditions in the 1960s changed the structural relation between past and present discoveries. In recent years, less was being found on succeeding rounds of reserve accumulation than was the case in the absence of regulation of prices in the 1950s. In this sense, the engineering model is clearly not broadly descriptive.

Table 7-2. Prices and Production of Gas for the East Coast and the Midwest, 1961-67

Year	Average price[a]	Simulated "unregulated" average price	Actual new reserves	Simulated "unregulated" reserves
	(------------ ¢/Mcf	------------)	(------------ bcf	------------)
1961	18.1	17.2	5,854	8,400
1962	19.1	16.2	5,805	5,874
1963	16.5	16.4	4,947	5,369
1964	16.8	15.2	5,558	3,869
1965	17.5	15.9	6,115	4,830
1966	17.3	16.7	4,213	5,969
1967	17.5	16.7	3,683	5,077

[a]The weighted average price $\Sigma P_{tj} \Delta R_{tj} / \Sigma \Delta R_{tj}$, where ΔR_{tj} is new reserves in drilling region j in year t.

The EMW model results can be compared with results from an analysis of production of gas under new contracts for delivery to pipelines to the West Coast. The new contract transactions in the 1955-60 period in W were used to fit the supply and demand equations as follows:

$$\log \Delta R_{tj} = \underset{}{108.9634} + \underset{(0.2466)}{0.3068} \log (\Delta R_{t\text{-}1,j}) - \underset{(0.3367)}{0.7083} \log (W_{t\text{-}1,j})$$

$$- \underset{(11.5421)}{19.0920} \log (M_j) + \underset{(7.2320)}{6.1314} \log (fp_t);\ R^2 = 0.7063.$$

[22]The years 1968 and 1969 are not shown for two reasons. First, data on actual prices and production for the eleven drilling regions have not been made available by the FPC. Also, the announced volumes of discovered reserves in year t are less in the year or two after discovery than these amounts for year t but readjusted five years after discovery so that there is a decline in discoveries for reporting reserves alone. These conditions make it unlikely that 1968 and 1969 can yet be analyzed.

[23]Both levels are based on the actual additions to reserves the previous year $\Delta R_{t\text{-}1,j}$.

$$\log P_{tj} = -19.0884 + \underset{(0.0369)}{0.0282} \log (\Delta \hat{R}_{tj}) + \underset{(1.2674)}{1.6094} \log (M_j)$$

$$+ \underset{(1.3648)}{2.2729} \log (fp_t);\ R^2 = 0.2501.$$

These equations fit the data poorly–if anything, worse than those given above describing supply and demand for EMW buyers. The reserve equation explains 70 percent of the variance in the dependent variable, but the independent variables all have statistically insignificant coefficients.[24] The second equation cannot be said to have significant coefficients because of the previous fitting of the variable ΔR_{tj}; but the sign of mileage is perverse and the values of the other coefficients do not resemble those in EMW. There are special reasons for this behavior in West Texas and New Mexico. Drilling has been following quite different patterns in one as compared with the other two districts, and the annual averages of the two patterns together with only eighteen observations over six years have had erratic effects on overall or average new reserves throughout the W area. Buying has been dominated by a single pipeline that established the policy in the early 1950s of paying higher discriminatory prices in the farther West Texas region.[25] The highly uncertain supply behavior in the first equation coupled with noncompetitive bid pricing in the second equation makes this model a poor forecast vehicle.

The models of unregulated discoveries in the EMW and the W regions in the 1950s show the great imprecision in using simple "engineering" supply-demand for assessing the effects of field price regulation. The regulated prices were similar to, if somewhat higher than, the regression-derived unregulated prices where a comparison was attempted. The regulated discoveries, however, seem to have been less than expected from markets without regulation. Both follow from the fitted supply function being an incomplete description of the actual supply function. But even with poor regression models, it seems possible to detect a systematic tendency to find less reserves in the regulated 1960s. The difference in the rate of discovery, as an estimate of excess demand, was less than excess demand shown in the first section of this review. But any final assessment of the effects of regulation requires a more detailed analysis of field market supply.

An Economic Model of Supply and Demand

The two regression equations for reserves and prices barely begin to describe the field markets for natural gas. More complete descriptions ought to be interesting in and of themselves, and they also may lead to a more revealing simulation of markets in the 1960s without regulation. The gas reserve supply equation should take account of price incentives for exploratory drilling; at the least, the changes in supply conditions after regulation shown in the engineering model should be revealed directly as disincentives from controls on prices. The demand equation should take more account of changes in final markets for gas, and there should be a direct relation established between discoveries and actual production of new reserves under new contracts, in order to assess the security of reserve backing.

The determinants of the supply of newly discovered reserves include the promised price for production of these reserves. The engineering view that $\Delta R_{tj} = f(\Delta R_{t\text{-}1,j}, W_{t\text{-}1,j})$ should hold as an initial condition. But the most widely held economic, as opposed to engineering, view is that higher prices on new contracts bring forth more gas well capacity, and the additional capacity, by drilling out more of the regions likely to contain gas, increases the amount of new reserves that are discovered.

[24]On the usual tests for the ratio of the computed coefficient to the computed standard error of the coefficient (shown below each coefficient in parentheses). The ratios do not exceed the value of 2.0 necessary to reject the hypothesis that the coefficient value is zero.

[25]See MacAvoy, *Price Formation*, chap. 5; this pattern of buyer's discrimination is explainable by entry of competing buyers only at the farther location.

It is assumed that actual new contract prices are good surrogates for those prices the drilling companies forecast for new reserves, for then the amount of drilling or wells $W_{tj} = f(P_{tj})$. Also, more gas drilling takes place where there are prospects of more profitable oil finds. This may be indicated by assuming that increased drilling occurs when there is an increase in the all-fuels price index (fp_t). Last of all, the analysis of drilling should recognize that larger amounts of information gathered in the previous year increase the relative rates of drilling for new reserves in any region. This information is obtained by increasing the amount of drilling carried out in that district in the previous year, $W_{t-1,j}$. The supply function can be taken to be

$$\Delta R_{tj} = f(\Delta R_{t-1,j}, W_{tj}),$$
$$W_{tj} = f(P_{tj}, fp_t, W_{t-1,j}),$$

for each drilling region in the EMW and W markets in the 1955-60 period.

Additional production each year depends on the amount of newly discovered reserves ΔR_{tj} and on the costs of holding these reserves. Also, it should depend on characteristics of short-term demands in final consumer markets—if, for example, prices of alternative fuels were rising while production from these alternative sources was not, then demand for more gas in the short term should increase, and these demands should be met by more production out of new reserves. In contrast, the introduction of new gas service into a final consuming region may result in gas production increases and price decreases for alternative fuels.

Although the causal relations cannot be established and measured very exactly, a first indication might be $\Delta Q_{tj} = f(\Delta R_{tj}, i_t, fp_t)$ for ΔQ production from new contracts signed in year t in district j, new reserve discoveries ΔR_{tj}, the current rate of interest i_t (as a measure of the cost of investment in reserves), and the fuel price index fp_t. The measurements for this equation are for each drilling region in the same markets and time periods as the supply of newly discovered reserves.

The demands for new reserves are assumed to conform to the market conditions posited in the two-equation engineering model. New contract prices P_{tj} bid by pipeline buyers depend on the volume of newly discovered reserves ΔR_{tj} in that drilling region; and, as an additional factor, the level of bid prices in any year depends on the sum total of all new reserves found that year (with more reserves $\sum_j \Delta R_{tj}$ reducing the bid prices). New prices are also likely to be affected by the distance M_j from the point of production to final markets and by the conditions of the final consumers' demand for gas and other sources of energy. There are likely to be two important consumers' demand conditions. First, the prices of substitute fuels should have a direct effect in that higher fuel prices usually signal increased demands for energy that increase bid prices for new gas contracts. As a result, we should observe in $P_{tj} = f(\Delta R_{tj}, \sum_j \Delta R_{tj}, M_j, fp_t)$ that a change in bid prices is in the same direction as the change in the fuel price index fp_t. Second, new contract prices for gas should be affected by the size of the final markets measured in terms of population and income. Additions to the size of final markets and to purchasing power can be termed $\Delta N \Delta Y$, for population increase ΔN and per capita income increase ΔY, and the greater the value of $\Delta N \Delta Y$, the greater the presumed offer price. The demand relation is presumed to be

$$P_{tj} = f(\Delta R_{tj}, \sum_j \Delta R_{tj}, M_j, fp_t, \Delta N \Delta Y),$$

for the same regions and time periods as for the three supply functions.

Competitive markets that clear gas reserves ΔR_{tj} and $\sum_j \Delta R_{tj}$, gas production ΔQ_{tj}, the number of development wells W_{tj}, and initial contract prices P_{tj} in a year are thus assumed to depend on the exogenous variables $\Delta R_{t-1,j}, fp_t, W_{t-1,j}, \Delta N \Delta Y, M_j$ and i_t. Data series for each of these variables have been constructed for the pre-regulatory period 1955-60 in eleven drilling regions in EMW and four drilling regions serving the W final demand territories (as described in the appendix to this paper).

These data are used to fit the supply and demand relations by two-stage least squares, with the four endogenous variables simultaneously determined. The procedure first finds "reduced form" least squares equations for each of ΔR_{tj}, ΔQ_{tj}, W_{tj}, and P_{tj} separately, given the endogenous variables, and then uses the fitted values $\Delta\hat{R}_{tj}$, $\Delta\hat{Q}_{tj}$, $\hat{W}_{tj}$, and $\hat{P}_{tj}$ in fitting the least squares supply and demand equations. The reduced form regressions for EMW for sixty-six observations in eleven drilling regions over the six-year pre-regulation period ending in 1960 are as follows:

$$\log \Delta R_{tj} = -\ 109.7031 + \underset{(0.1045)}{0.6439} \log (\Delta R_{t-1,j}) + \underset{(7.9224)}{17.1926} \log (fp_t)$$

$$+ \underset{(0.1573)}{0.3677} \log (W_{t-1,j}) + \underset{(3.6100)}{6.7137} \log (\Delta N \Delta Y) - \underset{(2.3084)}{4.8325} \log (i_t)$$

$$\text{SER} = 1.0396 \qquad R^2 = 0.5550$$

$$\log \Delta Q_{tj} = -\ 101.5247 + \underset{(0.0915)}{0.5821} \log (\Delta R_{t-1,j}) + \underset{(6.9359)}{12.6642} \log (fp_t)$$

$$+ \underset{(0.1377)}{0.6927} \log (W_{t-1,j}) + \underset{(3.1605)}{9.3609} \log (\Delta N \Delta Y) - \underset{(2.0209)}{6.0958} \log (i_t)$$

$$\text{SER} = 0.9101 \qquad R^2 = 0.6746$$

$$\log W_{tj} = -\ 54.4059 + \underset{(0.0337)}{0.1106} \log (\Delta R_{t-1,j}) + \underset{(2.5521)}{10.1146} \log (fp_t)$$

$$+ \underset{(0.050)}{0.7496} \log (W_{t-1,j}) + \underset{(1.1629)}{1.6503} \log (\Delta N \Delta Y) - \underset{(0.7436)}{0.9211} \log (i_t)$$

$$\text{SER} = 0.3349 \qquad R^2 = 0.8744$$

$$\log P_{tj} = 3.0950 + \underset{(0.0116)}{0.0800} \log (\Delta R_{t-1,j}) - \underset{(0.8806)}{0.5851} \log (fp_t)$$

$$+ \underset{(0.0175)}{0.0425} \log (W_{t-1,j}) - \underset{(0.4013)}{0.3035} \log (\Delta N \Delta Y) + \underset{(0.2566)}{0.5151} \log (i_t)$$

$$\text{SER} = 0.1156 \qquad R^2 = 0.6803$$

Little can be said of these equations since they are not direct tests of economic relations; however, they fit well in the usual statistical sense, given that many coefficients are statistically significant and the explained variation in each endogenous variable is more than 55 percent.[26] The instrument values

[26] The tests for significant coefficients indicate differences from zero for three or four coefficients in each equation. The values for R^2–indicating the percentage of dependent variable variance explained by the regression equation–are from 0.55 for new reserves to 0.87 for wells drilled.

$\Delta\hat{R}_{tj}$, $\Delta\hat{Q}_{tj}$, $\hat{W}_{tj}$, and $\hat{P}_{tj}$ vary from the observed values by negligible amounts, given that the standard error of residuals (SER) in each regression was small (as shown by the listed values of *SER*).

The values of $\Delta\hat{R}_{tj}$, $\Delta\hat{Q}_{tj}$, $\hat{W}_{tj}$, and $\hat{P}_{tj}$ have been used to find four regressions, one for the supply of new reserves, the second for the supply of wells, the third for new production, and the last for the demand for new reserves. The supply of new reserves and the number of wells equations for the 1955-60 period are as follows:

$$\log \Delta\hat{R}_{tj} = \underset{}{0.2765} + \underset{(0.0292)}{0.5894} \log (\Delta R_{t-1,j}) + \underset{(0.0489)}{0.4364} \log (\hat{W}_{tj});\ R^2 = 0.9400.$$

$$\log \hat{W}_{tj} = -32.0582 + \underset{(0.1206)}{1.0296} \log (\hat{P}_{tj}) + \underset{(0.5656)}{6.7314} \log (fp_t)$$

$$+ \underset{(0.6188)}{0.6872} \log (W_{t-1,j});\quad R^2 = 0.9848.$$

These equations show the positive cumulative effects from previous reserve findings and wells and the positive effects from both new gas contract prices and the all-fuels price index. The elasticity of reserve supply with respect to new-contract gas price is estimated to be (0.4364) (1.0296) = 0.45, so that a 10 percent price increase should have led to a general 4.5 percent increase in discovery of new reserves.

The production from the new reserves is as follows:

$$\log \Delta\hat{Q}_{tj} = 44.7285 + \underset{(0.0322)}{1.1091} \log (\Delta\hat{R}_{tj}) + \underset{(0.3514)}{1.1567} \log (i_t) - \underset{(1.9389)}{10.9032} \log (fp_t);$$

$$R^2 = 0.9528.$$

This shows a positive production-reserve relation and a positive production-interest rate relation but a negative production-fuel price relation. The elasticities of production with respect to reserves and interest seem to have been approximately one, and were quasi-statistically significant.[27] The elasticity with respect to the fuels price index was negative and larger; that is, declines in the fuels price index in the late 1950s were accompanied by more than proportionate gas production increases. This would seem to have been a result of movements in this period toward an equilibrium share for natural gas of the total energy market in regions receiving gas for the first time.

The demand relation has also been estimated in the second stage of two-stage least squares; here, the regression is as follows:

$$\log \hat{P}_{tj} = -28.8494 + \underset{(0.0055)}{0.1210} \log (\Delta\hat{R}_{tj}) - \underset{(0.0879)}{0.7138} \log (\sum_j \Delta\hat{R}_{tj})$$

$$- \underset{(0.0461)}{0.1232} \log (M_j) + \underset{(0.8726)}{6.6067} \log (fp_t) + \underset{(0.1131)}{1.3798} \log (\Delta N \Delta Y);$$

$$R^2 = 0.9205,$$

with positive elasticities for four variables and negative elasticities for two variables. The elasticity of bid prices with respect to total reserves (the reciprocal of the usual elasticity of demand) is -0.72,

[27] At the second stage of 2SLS, it is not possible to test for statistical significance; we use the terms and show the standard errors only to record the computed results as completely as in ordinary least squares (OLS).

while the elasticity with respect to the all-fuels index is +6.6 and with respect to income is +1.38. The other coefficients—for variables differentiating the drilling regions—are as expected from the economics of pipeline costs and demand. This four-equation analysis shows more than the simple engineering analysis—that the supply of newly discovered reserves and of new production depends on new contract bid prices, and these bid prices depend on the size of total discoveries, on pipeline costs, and on conditions of demand in final gas resale markets.[28]

These conditions make it likely that regulation, with ceilings on prices, has had a substantial effect on gas reserve discovery and disposition. In order to assess this possibility, the second-stage regression equations have been used, along with 1961-67 values of the independent variables, to find the values ΔR_{tj}^*, ΔQ_{tj}^*, W_{tj}^*, and P_{tj}^* that solve these four equations simultaneously (in other words, clear the gas market as if there were no price ceilings). The simulated values for EMW are compared with actual values in table 7-3.

Table 7-3. Prices, Production, and Reserves of Gas for the East Coast and Midwest, 1961-67

Year	Average price	Simulated price	Production Actual	Production Simulated	Reserves Actual	Reserves Simulated
	(------ ¢/Mcf ------)		(---------- bcf		----------)	
1961	18.1	19.2	314	400	5,854	8,189
1962	19.1	25.3	160	345	5,805	7,031
1963	16.5	26.4	461	333	4,947	6,582
1964	16.8	30.8	206	390	5,558	5,685
1965	17.5	36.8	374	418	6,115	7,134
1966	17.3	43.2	335	473	4,213	8,547
1967	17.5	42.1	582	345	3,683	7,369

The simulated reserves show the strong incentives of higher prices on discoveries when markets are allowed to clear. Simulated reserves and prices in table 7-3 are generally much greater than actual reserves and prices. The expected volumes of market-clearing reserves were 140 percent of the actual new reserves over this period. The expected prices were 14.4 cents per Mcf greater than actual prices over this period and increased over time from less than 10 cents greater up to 1963 to more than 20 cents greater in the last two years. The market-clearing process would have bid prices up greatly over the period, increasing newly discovered reserves so as to fill part of the existing supply-demand gap.

Production was greater than expected, given the amount of simulated newly discovered reserves. Simulated production from new reserves over the seven-year period averaged 386 bcf from new contracts each year, while actual production averaged 347 bcf. Although simulated reserves were greater than actual reserves, the hypothesized unregulated rate of utilization of new reserves would have been much less than the actual or regulated rate of utilization,[29] even though simulated production was greater than actual production.

One reason for the difference in rates of utilization may have been the differences between market-clearing prices and regulated prices. The average price difference results in good part from a number of large reserve dedications in South Louisiana in 1965 and 1966 that were sold for regulated prices that were half the simulated price. The price difference there was great because ceiling prices did not

[28]The four equations together make up an *equilibrium* system that "describes" well the actual prices and discoveries in the late 1960s. An indication of the closeness of description is the standard error of the regression on price, equal in log terms to 0.069, while average price was 2.6801; approximately half of the average prices were within plus or minus 0.069 of the log value of 2.6801 (given average values of the independent variables) and 90 percent of prices were within plus or minus 0.198 of the log value under these conditions. Similarly, half of the actual volumes of discovered reserves were within 0.2796 of those expected in the regression equation equilibrium system; for example, at the mean regression value of 5.9981, half the values were equal to this amount, plus or minus 0.2796.

[29]This can be seen by comparing the rates of utilization. Simulated production out of simulated reserves was 5.3 percent on average, while actual production out of actual reserves was 6.7 percent on average.

account for the advantages of location and value, and the simulated prices did show great advantages for these factors in the demand function. These sales have been followed up with extremely rapid rates of production. The higher prices in the market-clearing model would have been compatible with slower rates of production there—slower because of the lack of "emergency" needs for more gas by these particular buyers, because their final buyers would not have demanded as much.

Some of the same conditions were found for the West Coast market. The reduced form equations contained some variables with statistically significant coefficients; in particular, the wells variable lagged one period was statistically significant in the ΔR_{tj}, the ΔQ_{tj} and the W_{tj} equations, and the distance variable was statistically significant in the ΔQ_{tj} equation. The R^2 values ranged from 0.65 (for the P_{tj} equation) to 0.91 (for the W_{tj} equation). The second-stage regressions, with first-stage values of $\Delta \hat{R}_{tj}$, $\Delta \hat{Q}_{tj}$, $\hat{W}_{tj}$, and $\hat{P}_{tj}$ inserted where appropriate, were economically reasonable only in part. The supply of reserves was directly related to the previous year's reserves but inversely related to the number of exploratory wells. The number of wells was in turn inversely related to price $\hat{P}_{tj}$ so that there was a direct but small price elasticity of supply equal to the product of these two negative coefficients, or (-0.2283) (-0.9149) = 0.2088. Production in new contracts was dominated by reserve discoveries, with an elasticity of +0.9357; but the elasticity with respect to interest was negative and with respect to fuel prices was positive—the opposite of those shown in EMW. The demand price for new production was related positively to reserves $\Delta \hat{R}_{tj}$, distance M_j, income $\Delta N \Delta Y$, and fuel prices fp_t, and negatively to $\sum_j \Delta \hat{R}_{tj}$. Here the elasticity with respect to total reserves was -0.2322 (so that the reciprocal or elasticity of demand was very large).

Forecasts from this equation system are not satisfactory for a number of reasons. There are conceptual problems with the fitted equations. The model contains unlikely relations, particularly between new reserve discoveries and the number of wells drilled; it is not likely that accurate forecasts of the 1960s would follow from replicating the pattern and thus showing *increased* discoveries from *decreased* drilling. The model also suggests that significant increases in final demands in California and other western states have a very small effect on the bid prices of gas in West Texas (the elasticity was only 0.0704). This may have followed from monopsony pipeline control of prices in the 1950s, but it could not be expected to continue throughout the 1960s, given that there was substantial additional entry of pipelines on the demand side of markets there. As a result, the model based on monopsony demand should not simulate the 1960s well.

The model does not simulate well. Reserves and prices in two of the three districts in the western area are lower than simulated reserves and prices. But in the third region, more than 1,300 bcf of new reserves were discovered in each of five years, where the 1955-60 average was half this amount and the 1961-67 simulated reserves are also half this amount. There were special supply and demand reasons for the results. On the supply side, all these additional amounts were dedicated to intrastate buyers at unregulated prices from 2 cents to 6 cents per Mcf above the ceiling area rates, so that incentives available were not shown by the interstate price-reserves supply relationship. On the demand side, substantial additional intrastate demands were created by low ceiling prices, and these were centered on projects for new Texas pipelines when it was apparent that the volumes of reserves were sufficient for a large-scale pipeline system. Supply and demand were both greater than expected with the historical regression model, and prices cleared the intrastate portion of this market at higher levels than 1961-67 interstate ceiling prices.[30]

The overall impression from EMW and W together is that new reserve findings would have been more than 40 percent greater if there had been no field price regulation, and the rate of production from new reserves would have been much lower. The effects of prices on both supply and demand are

[30]This disequilibrium with two prices can occur only when all substantial sales are made intrastate, and there is still an intrastate premium. The interstate prices are nominal, for negligible sales. This seems to have occurred only in the Permian Basin in W, and thus makes the comparison of interstate-intrastate equilibrium and interstate rate ceiling disequilibrium (assuming intrastate equilibrium) impossible.

apparent. The higher market-clearing prices in the EMW buying area would have brought forth additional discovered reserves equal to roughly four-fifths of the supply-demand gap in the inventory shortage if 14.5 years of reserve backing were needed, or one-third of the gap if 20 years of reserve backing were needed. The price premium would also have reduced production from new reserve discoveries, by approximately 5 percent in each new contract year. Higher prices also should have led to some reductions in the quantities demanded of newly discovered reserves, as a result of higher ceiling prices. This last restriction is not observable, since the comparison here is between *supply* of discovered reserves at the price ceilings and *supply-demand* equilibrium without price ceilings; but it is possible that such substantial price increases would have eliminated the other one-fifth to two-thirds of the supply-demand gap, in the same way that price changes cleared markets in the 1950s.

DISTRIBUTION OF THE GAS SHORTAGE AMONG FINAL CONSUMERS

The "shortage of natural gas" has been attributed to FPC field price regulation, in the sense that the continuation of 1955-60 market processes, rather than regulatory ceiling prices, would have resulted in no excess demands for reserves. Given this attribution, the question now is who has a grievance with the commission; who experienced the shortage, and what were his economic burdens as a consequence?

The gains and losses for final consumers under conditions of reduced reserves and increased production are diverse and obscure, and evidence is mostly replaced with speculation. One view is that the shortage benefits everyone—that the increased rate of gas production at the present time is an improvement over the restricted rate of production that would have occurred without regulation. The reason given is that more gas is delivered to those favored by regulation, merely causing the reserve backing for delivery to others to be depleted earlier. All the additional new gas production occurring at present under regulation might well be going to home consumers, for whom pipeline and retail gas rates are controlled by both national and state regulation. Those without the usual reserve backing might go to close substitutes at a somewhat earlier time than they might otherwise. There is an alternative view—that increased production now has the effect of depleting the reserve stock not only of home consumers but of those long-established home consumers that had been offered twenty-year backing before field price regulation began. If this is the case, then those with the grievance are those supposed to have been protected by regulation.

The prevailing information on distribution supports the latter speculation. The information is based on dividing the history of gas sales into three periods. The first two have been described by Balestra[31] as: (1) an "innovating" period, 1950-57, for natural gas pipeline construction and initiation of service to industrial and home consumer classes, and (2) a "maturing" period, 1957-62, of extension of more service to established customers. The third phase, 1962-68, is characterized here as a "reallocation period," with shifts of service from one of the classes to the other. The allocation of quantities to each customer class can be described for each of these sub-periods; reallocation takes place when sales to one group are decreased and prices are increased (to clear final markets), while the other group gains from sales increases and price decreases.[32] The only exception is created by ceiling price regulation on gas field sales.

The relative consumption, revenue, and price for residential and industrial natural gas customers are shown in table 7-4.[33] There is little variation in proportions of total consumption of the two

[31] Pietro Balestra, *The Demand for Natural Gas in the United States: A Dynamic Approach for the Residential and Commercial Market* (Amsterdam: North-Holland Publishing Co., 1967), pp. 86-89.

[32] This description is based on W. P. Schlarb, "An Estimation of Supply Allocation among User Classes of Natural Gas Utilities" (Master's thesis, Massachusetts Institute of Technology, June 1970). Schlarb's assistance here is gratefully acknowledged.

[33] Here "price" is used crudely to denote average revenue per unit of sale.

Table 7-4. U.S. Natural Gas Sales, Total Revenue, and Average Revenue, by Customer Class

	1950	1957	1962	1968
Sales (*percent*)				
Residential	39.1	41.3	44.5	43.0
Industrial	60.9	58.7	55.5	57.0
Total	100.0	100.0	100.0	100.0
Revenue (*percent*)				
Residential	67.6	67.9	68.7	66.8
Industrial	32.4	32.1	31.3	33.2
Total	100.0	100.0	100.0	100.0
Average revenue (*therm*)				
Residential	6.11	8.33	9.50	9.39
Industrial	1.88	2.77	3.47	3.51

NOTE: Residential use includes commercial. Industrial use includes other.

SOURCE: Percentages and average revenues have been calculated from AGA, *Gas Facts*, as in W. P. Schlarb, "An Estimation of Supply Allocation among User Classes of Natural Gas Utilities" (Master's thesis, Massachusetts Institute of Technology, June 1970), table III-1.

classes between the first two periods. The fraction of gas consumed in residential use increased, and price increased greatly in both classes so that it would appear that increasing demands in both classes were met by increased production, but a larger proportion went into residential sales. In the third period, the trends were reversed and the percentage of use by residential customers declined, while the industrial percentage increased. There was a slight decline in residential average price and a very slight rise in industrial average price.

Differences in the growth patterns of the two classes are shown in table 7-5. The growth rates were much the same in the innovating period, with increased revenues keeping pace with increased consumption in the two classes. In the maturing period, revenues on industrial consumption increased faster than on home consumption because industrial prices went up faster. In the last, or reallocation, period, industrial consumption and prices both increased at a faster rate than in the home sector, resulting in an annual increase in revenues on industrial consumption 1.5 percentage points greater than on home consumption.

Those differences have had important effects on prices and profits of pipelines. Consider one or two initial indicators of general profit behavior period by period. In 1950 average revenue from industrial sales was relatively very low, less than one-third that of residential use. However, during subse-

Table 7-5. Annual Compound Rate of Increase of U.S. Natural Gas Consumption, Revenue, and Average Revenue, by Customer Class

(*percent*)

	1950-57	1957-62	1962-68
Growth of consumption			
Residential	10.9	7.5	5.5
Industrial	9.4	4.8	6.6
Growth of revenue			
Residential	15.9	9.7	5.3
Industrial	15.6	10.4	6.8
Change of average revenue			
Residential	4.5	2.7	(0.2)
Industrial	5.7	4.6	0.2

NOTE: Residential use includes commercial. Industrial use includes other. Parentheses indicate decrease.

SOURCE: Compound annual growth rate of consumption, revenue, and average revenue have been calculated from series in AGA, *Gas Facts*, as in W. P. Schlarb, "An Estimation of Supply Allocation among User Classes of Natural Gas Utilities" (Master's thesis, Massachusetts Institute of Technology, June 1970), table III-2.

quent years industrial gas prices increased at a rate almost 25 percent greater than that in the residential sector. This could be attributed to the slow response of utility regulation to increased demands in this class, because direct sales of pipelines to industrial consumers are not regulated by the FPC but only by state commissions with varying standards and effectiveness in enforcing their standards. The increased demands made the estimated marginal revenues on additional sales in the period 1950-57 greater than the average revenues in both classes of use (compare average revenue for 1950 and 1957 in table 7-4 and marginal revenues for 1950-57 in table 7-6). The marginal revenues in both classes were rising rapidly, as shown by the 4 to 5 percent annual compound rates of increase. Evidently, the demand for gas was increasing at such a rate that the pipelines had to expand output to both sectors and raise prices at the same time, by chance maintaining the initial balance of industrial and home sales.

Table 7-6. Pipeline Revenues in the U.S. Market

(*cents per therm*)

Marginal revenues	1950-57	1957-62	1962-68
Residential	9.18	11.76	9.13
Industrial	3.21	5.73	3.59

SOURCE: W. P. Schlarb, "An Estimation of Supply Allocation among User Classes of Natural Gas Utilities" (Master's thesis, Massachusetts Institute of Technology, June 1970), table III-8. The calculations are changes in revenues divided by changes in sales.

During the period from 1957 to 1962, the rates of growth were much less than earlier. Balestra, writing of this period, concludes that "in what may be termed the mature state . . . the gas industry is facing an increasingly competitive situation. Incremental demand had become a matter of "normal growth of population, household formation and the like." His econometric analysis showed "an upward trend in price elasticity"[34] in this period for new residential demand, and other generally available evidence on inter-fuel competition in industrial markets suggests increased gas price elasticity in that sector as well. In fact, prices increased at a more rapid rate and consumption at a less rapid rate in the industrial consumption class in this period. Also, the marginal revenues from additional industrial and residential sales continued to increase, so that the general impression is one of the pipelines choosing smaller volumes of higher-priced new sales in both categories. There would seem to have been no reallocation.

The third period, from 1963 to 1968, shows the effects of stabilized prices. Average revenues do not change appreciably, and are roughly the same as marginal revenues from new sales. The stabilization of prices during this period must have come about as the result of (1) FPC control of wellhead prices, and (2) interfuel competition, especially in industrial use, as Edmund Kitch suggested.[35] The ceiling field prices were passed along to residential users in near-stabilized average prices, as a matter of FPC regulation of pipeline rates. The industrial price stabilization is another matter, since there were no regulatory requirements to "pass through" area rates. It is more likely that the pipelines allocated the scarce additional gas to industrial users as long as the unregulated profit margin exceeded the regulated, and as long as the pipelines encountered at the margin more severe interfuel competition. A marked change in market growth patterns is evident in this period compared with the earlier periods, for both customer formation rate (not shown) and consumption growth (shown in table 7-5) in industrial use exceed those in residential use. Marginal revenue on industrial sales declined by more than 40 percent—without gas price controls. Prices of industrial sales had caught up to new demands, and these new demands were continuing to be filled at a *faster rate* than new home consumption demands. This can be described as reallocation from home to industrial users.

[34] Balestra, *Demand for Natural Gas*, pp. 88, 96-99.

[35] Edmund W. Kitch, "Regulation of the Field Market for Natural Gas by the Federal Power Commission," *The Journal of Law and Economics*, pp. 265-69.

These national changes in the reallocation period cover over many important changes in well-defined regional gas markets. The California markets show the period's behavior most clearly, since both industrial and home demands were growing rapidly. From the early 1950s the California market had accounted for 10 percent or more of the total U.S. natural gas utilities' consumption, and the percentage has been growing rapidly. Not only have home demands increased from the extremely rapid California population increase, but industrial demands have as well. Thermal-electric power generation companies have been a major source of industrial demand in this market. In 1968, for instance, of the total industrial gas use in the United States, about 31 percent was consumed in electric power generation, and nearly one-third of this was consumed in the Pacific region (principally in California).[36] The distinctive feature has been the lack of shortage—the western pipelines have met demands.

The relative distributions of consumption, revenues, and prices for residential and industrial use are shown in table 7-7. The proportion of gas consumption and revenues by class of customer has changed very significantly from 1950 to 1968.

Table 7-7. Natural Gas Sales, Total Revenue, and Average Revenue in the California Market by Customer Class

	1950	1957	1962	1968
Sales (*percent*)				
Residential	52.0	47.3	44.0	37.3
Industrial	48.0	52.7	56.0	62.7
Total	100.0	100.0	100.0	100.0
Revenue (*percent*)				
Residential	73.1	66.9	63.5	58.8
Industrial	26.9	33.1	36.5	41.2
Total	100.0	100.0	100.0	100.0
Average revenue (*therm*)				
Residential	5.55	7.08	8.67	8.15
Industrial	2.21	3.15	3.92	3.40

NOTE: Residential use includes commercial. Industrial use includes other.

SOURCE: Percentages and average revenues have been calculated from data in AGA, *Gas Facts*, as in W. P. Schlarb, "An Estimation of Supply Allocation among User Classes of Natural Gas Utilities" (Master's thesis, Massachusetts Institute of Technology, June 1970).

From the early 1930s to the late 1940s California utilities, either by formal regulatory controls at the state level or voluntarily, had maintained much lower prices per therm for residential use than did other sources of energy. However, as is evident in table 7-7, average revenues increased in both classes from 1950 to 1962. This came about as more demand was filled by out-of-state gas supply, higher purchase costs were incurred in the Permian and San Juan Basin areas, and higher transport costs were experienced to move the gas 1,200 miles from these sources to the California market. After 1962, the pattern of rising prices changed abruptly for both classes of use. This change was coincident with the market maturity described earlier in the analysis of U.S. sales.

Growth rates of consumption and revenue were relatively unchanged from 1950 to 1962, as shown in table 7-8. In the period 1950-57, prices or average revenues for industrial use increased at a rate appreciably greater than that for residential use. In the period 1957-62, prices increased at approximately the same rate in both classes. In the final period, 1962-68, *prices* for industrial use *declined* at more than twice the rate of that for residential use. In this last period, in contrast to the price decline there were significant relative *consumption increases* for industrial use. The pipelines preferred to provide more gas—at twice the rate of annual increase in residential consumption—to industrial users, even though industrial prices were declining. To be sure, there were strikingly different conditions of industrial demand in this last period. In 1960 the FPC authorized Pacific Gas Transmission to

[36] AGA, *Gas Facts* (1969), table 124, p. 132.

Table 7-8. Annual Compound Rate of Increase of Natural Gas Consumption, Revenue, and Average Revenue, in the California Market, by Customer Class

(*percent*)

	1950-57	1957-62	1962-68
Growth of consumption			
Residential	6.9	6.2	4.0
Industrial	9.8	9.2	8.9
Growth of revenue			
Residential	10.6	10.8	2.9
Industrial	15.4	14.1	6.4
Change of average revenue			
Residential	3.5	4.1	(1.1)
Industrial	5.2	4.5	(2.4)

NOTE: Residential use includes commercial. Industrial use includes other. Parentheses indicate decrease.

SOURCE: Compound annual growth rate of consumption, revenue, and average revenue have been calculated from data in AGA, *Gas Facts*, as in W. P. Schlarb, "An Estimation of Supply Allocation among User Classes of Natural Gas Utilities" (Master's thesis, Massachusetts Institute of Technology, June 1970).

construct a transmission system to take gas from the province of Alberta in Canada to California.[37] Pacific Gas and Electric controlled 50.5 percent of the Pacific Gas Transmission system and dictated the transmission company policy; thus sales were allocated to the industrial sector as boiler fuel for its plants. Demands were changed further when in January 1964 the Board of Supervisors of Los Angeles County enacted Rule 62.1 which extended the existing Rule 62 ban on the burning of high-sulfur fuels to a year-round basis.[38] This meant that virtually all demand for industrial fuel in the Los Angeles basin was demand for natural gas. These actions greatly increased demands, but the allocations for these purposes exceeded the usual response by taking new sales in this class to the point of initiating price reductions. This is classic reallocation, because it required movement down the demand schedule of the industrial class while maintaining the earlier location on the demand schedule of the residential class.[39] The industrial user received proportionately more of all new reserves.

At this point, there is room for more speculation. The pipelines in a period of short supply additions seem to have been allocating relatively more of this new gas to industrial users, probably as a result of loose or nonexistent regulation of industrial resale, while there was strong regulation on home resale. The regulated rates for deliveries to retail public utility companies increased only when field gas prices or delivery costs changed, and ceiling prices in the field greatly limited these increases. Price could not "ration" new home users. In contrast, prices to industrial consumers depended only on conditions in markets for energy for boiler fuel or industrial use—conditions that continued to favor natural gas, particularly when new requirements for pollution-free sources of energy were made by local governments. Here, prices could ration use, and pipelines added to these sales to the point where marginal revenues were close to those on regulated residential resale. This induced priority sales to industrial users.

Regulation in natural gas fields has produced a more rapid rate of production from new reserves in order to increase the rate of sales to industrial users at unregulated prices. The inventory analysis in the first section showed that new demands for reserves by both sectors together exceeded new supplies by more than 50 percent and perhaps as much as 120 percent. There was some indication in the second section that the excess demand could be cleared by 40 percent more supply and by inducing buyers to use alternative fuels—both achieved by 10-cent to 20-cent price increases.

[37] 24 FPC 134 (1960).

[38] 31 FPC 434 (1964).

[39] The price reduction for home users might suggest a movement down that schedule as well. This reduction, as required by regulation, probably had no such effect since new supplies were allocated at less than the growth rate of households. The most likely occurrence was that of marginal rationing to households.

The last section indicated that regulation has had the effect of taking the short supplies and reallocating them from home consumers under the aegis of the FPC to the industrial users buying in unregulated, more or less competitive energy markets. The home consumer might well look to the regulator operating in his name and ask whether some other rationale for field price regulation could be found.

APPENDIX: DATA SOURCES

The data used in the regressions and simulations were all obtained from publicly available sources. These consisted of the official publications of various agencies of the United States government and publications of professional associations related to petroleum activities. A brief description of the variables and their respective sources of measurement follows.

Reserves ΔR_{tj}: The figures used represent the additional reserves discovered in a given oil-producing district in a given year. The series is taken from *Reserves of Crude Oil, Natural Gas Liquids and Natural Gas in the United States and Canada as of December 31st, 1968,* vol. 23 (1969), published jointly by the American Gas Association, the American Petroleum Institute, and the Canadian Petroleum Association, table XVII, "Summary of Estimated Annual Discoveries of Natural Gas Reserves." The data file contains the values for the 1954-68 period, expressed in bcf.

Production ΔQ_{tj}: The figures in this series refer to the deliveries by natural gas producers to interstate pipeline companies made in 1967, in the first year of a basic contract. The figures were obtained from Form 2 reports submitted by purchasing pipeline companies to the FPC, either in the succeeding year of the contract or two years thereafter.

It should be noted that these figures include additions to and modifications of the original contract. They exclude intrastate sales. The contracts have a nominal life of twenty years, although they remain in force until such time as the FPC issues a certificate permitting the discontinuation of supply from a given contract. The production is expressed in bcf.

Prices P_{tj}: The price series used relates to the production series described above. It is the average price at which the production of the contracts in a given year in a given district was sold to the interstate pipeline companies. The data source is the independent producer rate schedule filings for certificate applications.

Development drilling W_{tj}: Figures for the period 1954-66 were obtained from the "Review Forecast" issue of the *Oil and Gas Journal*, published annually the last week of January. The numbers refer to the Field Well Completions–Gas Producers, and include the condensate producers.

Distance M_j: The variable expresses the pipeline distance between the center of a drilling region and the center of the area of consumption. Los Angeles is the consumption center for gas produced in the Texas Railroad Commission (TRRC) Districts 7e, 8, 8a, and 9, New Mexico Northwest (the San Juan Basin), and New Mexico Southeast (the Permian Basin). Cleveland is the consumption center for the TRRC Districts 1, 2, 3, 4, 5, 6, 7b, and 10, Louisiana Northern, Louisiana Southern and Offshore, Oklahoma and Kansas.

Fuel Price Index fp_t: The wholesale price index for fuel, power, and lighting materials was obtained from U.S. Bureau of Labor Statistics, *Wholesale Prices and Price Indexes*, which is published annually. The average annual figures for the years 1955-68 were used.

Population and Income Increase $\Delta N \Delta Y$: The increases in population that are potential new users of gas are defined as $\Delta N = N_t - (1-r)N_{t-1}$, where N_t is U.S. population in year t, and r is the rate of depreciation of past population. The increase in per capita incomes for potential gas consumption is similarly $\Delta Y = Y_t - (1-r)Y_{t-1}$, where again r is a rate of depreciation. In this context, r can be considered the rate of change of tastes of past users N_{t-1} of other fuels, with past per capita incomes Y_{t-1}. The rate r is taken as 0.11, from the results given in P. Balestra and M. Nerlove, "Pooling Cross Section and Time Series Data in the Estimation of a Dynamic Model: The Demand for Natural Gas," *Econometrica*, vol. 34, no. 3 (July, 1966), page 585. The values of N_t are from the *Statistical Abstract of the United States*, for particular regions; the values of Y_t are from the *Survey of Current Business* for the same regions. For region W, the following states were aggregated: Washington, Oregon, California, New Mexico, Arizona. For region EMW, the totals were from Maine, New Hampshire, Vermont, Massachusetts, Rhode Island, Connecticut, New York, New Jersey, Pennsylvania, Ohio, Indiana, Illinois, Iowa, Kansas, Missouri, Oklahoma, and Nebraska.

The Rate of Interest i_t: The arguments for use of one or the other interest rate as the economic indicator are extensive and controversial and cannot be dealt with here. On institutional grounds, Moody's composite average of yields on public utility bonds was used; this rate indicated the costs of a major source of public utility funds, including pipeline company funds. The rate on equity capital could not have been independent of the composite bond yield, although it may not have been determined by the yield.

8 PRICE, REGULATION, AND THE SUPPLY OF NATURAL GAS IN THE UNITED STATES

*Edward W. Erickson and Robert M. Spann**

The quantity of natural gas reserves discovered in any year depends on the price of natural gas, the price of oil, the relationship between the search for oil and the search for gas, and the effects of various public policies that affect not only prices but also expectations and the conditions of production for both oil and gas. The relative responsiveness of either gas or oil discoveries to changes in their own price, other relevant influences unchanged, is indicated by the own price elasticity of supply of each output of the exploration process. The effect of the price of oil on the quantity of gas discovered is indicated by the cross elasticity of supply of gas with respect to the price of oil. The reverse effect is the cross elasticity of supply of oil with respect to the price of gas. The cross elasticities of supply are the net effects of an interdependent set of shifts in the supply curve of either output as a result of changes in the price of the other.

Changes in the price of either oil or gas are not the only things that shift the supply curve of discoveries of the alternative output. Technological advances shift each curve to the right. The imposition of severe market-demand prorationing of oil production shifted the supply curve of oil to the left. The long-run exhaustion of attractive prospective drilling sites may shift the curves to the left. There is also a strong random effect in new discoveries. The same volume and composition of exploratory effort in the same area will not produce the same results each time. There are also geographic effects among regions. A given wellhead price of gas or oil will call forth a different volume, composition, and result of exploratory effort in Appalachia than in the Gulf Coast.

All of these diverse influences can be taken into account in an appropriately specified model of the supply of petroleum discoveries. The model can then be statistically estimated. One econometric model for estimating the effects of economic incentives on petroleum discoveries is the Fisher-Erickson model for crude oil discoveries.[1] This model relies on an interdependent set of hypotheses about the economic behavior of firms engaged in the search for crude oil to estimate their own price elasticity of supply of crude oil. In the model, cross-section and time-series data are combined to estimate three equations of the form

*Edward Erickson is professor of Economics, Department of Economics, North Carolina State University; and Robert Spann is a graduate student in economics at the university.

The authors wish to state that the results reported here have benefited from criticisms of an earlier draft by Franklin M. Fisher, Roland S. Homet, Paul R. Johnson, Richard B. Mancke, Dudley H. Chapman, Wade Sewell, Bruce Netschert, Alfred E. Kahn, and Richard L. Gordon. All their suggestions are not reflected in this particular paper, and they are not responsible for any errors.

[1] Franklin M. Fisher, *Supply and Costs in the U.S. Petroleum Industry* (The Johns Hopkins Press, 1964) and Edward W. Erickson, "Economic Incentives, Industrial Structure and the Supply of Crude Oil Discoveries in the U.S., 1946–58/59" (Ph.D. dissertation, Vanderbilt University, 1968). For a discussion of this model and an example of its application to a policy problem, see James C. Burrows and Thomas A. Domencich, *An Analysis of the United States Oil Import Quota*, a Charles River Associates Research Study (Lexington, Mass.: Heath Lexington, 1970).

$$Y_{ijt} = f(P_{jt}, X_{ijt}, Z_{jt}), \qquad (1)$$

where j and t are cross-section and time subscripts, P is the deflated wellhead price of crude oil, the Z's are cross-section dummies, and the X_i's are various structural variables. The Y_j's are alternatively, (1) wildcat wells drilled, (2) the ratio of successful to total wildcats, and (3) average discovery size per successful wildcat.

Total discoveries in any year are the product of the three Y_j's for that year. Each of these components of total discoveries reacts differently to changes in price. The sum of their separate effects determines the price elasticity of supply of crude oil discoveries. The internal consistency of the estimate is verified by the pattern of signs and magnitudes on the coeffieients of the X_{ijt}'s and P_{jt}'s in the individual equations for drilling, the success ratio, and average discovery size.

In this paper, the model is expanded to include the search for gas, and the estimations are extended to include the years 1960-67. The specific problems and results of the respecification and extension are discussed in detail below. But it is appropriate to discuss here one general assumption under which the estimations are made. This assumption is that the field markets for natural gas are characterized by competitive structures. All firms are price takers. In competitive markets, the general level and pre-regulation trend of prices are set by arms-length bargaining between buyers and sellers. There is no collusion or conspiracy among sellers. Firms equate marginal cost with the ruling market price in either regulated or unregulated markets.

It is important that this general assumption be clearly stated at this stage in the analysis because so much has been said on both sides of the issue, and it has been a prominent feature of area rate proceedings. It is my belief that the evidence supports the position that field markets for oil and gas are effectively competitive.[2] On the supply side, the competitive model is the appropriate analytical framework.

Policy makers require technical analysis in order to inform themselves about the state of the world. Such analysis is crucial in defining the possibilities space for policy options. The fundamental question is always, "If this policy is pursued, what results are achieved at what cost?" Policy makers may justify their decisions to the public in terms of extraneous considerations that are not central to the issue being resolved. Peripheral, irrelevant, or unnecessary features are introduced into the underlying analysis only at the risk of falsely truncating the possibilities space and foreclosing what might otherwise be an efficacious policy option. Further quibbling about the competitiveness of field markets for natural gas is such a feature in the analysis necessary to inform decision makers about the state of the world relevant to public policy regarding the price of natural gas.

Policy makers are frequently involved in accommodating conflicting interests in terms of some conception of an overall goal. In economic terms, the resulting accommodation often results in a change in the level and distribution of real income. In the long run, the viability of the accommodation is determined by the political constituents of the policy makers. If it is not viable, the regime is changed, the policy makers replaced, and the policy altered. The question that policy makers must continually ask themselves, and that is ultimately answered in the political process, is whether or not the changes in the level and distribution of real income attendant upon some policy implementation are worthwhile in terms of achievement of the overall goal.

Many economists have a simple view of the overall goal and the means for achieving it that abstracts from the difficult detailed decisions that administrative policy makers and their political patrons must face. This particular economic view is that the overall goal should be to maximize national income

[2] For a review of this evidence, see James W. McKie, "Market Structure and Uncertainty in Oil and Gas Exploration," *Quarterly Journal of Economics*, vol. 76 (February 1962), pp. 98-121; Paul W. MacAvoy, *Price Formation in Natural Gas Fields* (Yale University Press, 1962); Edmund W. Kitch, "Regulation and the Field Market for Natural Gas," *Journal of Law and Economics*, vol. 11 (October 1968), pp. 243-80; and Clark A. Hawkins, *The Field Price Regulation of Natural Gas* (Florida State University, 1969).

through an efficient allocation of resources with prices equal to marginal costs and income redistribution accomplished through means that do not directly affect relative prices.[3]

In an important sense, the price of natural gas is not equal to its marginal cost. The regulated price is equal to the marginal cost of gas, but the regulated price is below the demand prices of some current gas consumers for additional gas service and also below the demand prices of some potential gas users who are now excluded from the market because of ad hoc non-price rationing, or who will be excluded from the market by end-use controls. Some consumers would pay higher prices for beginning or additional gas service than the current regulated price, which is equal to marginal cost.[4] Thus, there is an inefficient allocation of resources. Some users, or potential users, are willing to pay more for an increase in output than the real cost of generating that increase.

The following section of this paper applies the theory of joint production and supply to the problem of oil and gas discoveries. The implications of this theory are then used to specify and test a model that estimates the own and cross elasticities of supply for oil and gas. Alternative assumptions about the magnitude of the shortage in the natural gas market and the price of oil are then coupled with a simple model of the relation between gas discoveries, reserves, and production to determine a range for the magnitude of price increase necessary to eliminate the shortage.

JOINT COSTS IN OIL AND GAS EXPENDITURES

Assume an industry has a production function of the following nature:

$$X = h(q_1, q_2) \tag{2}$$

where q_1 and q_2 equal two outputs and X equals a common input. For any given X, there is a product transformation curve like that in figure 8-1.

In terms of oil and gas exploration, shifts in exploratory effort from prospects with high expected oil/gas discovery ratios to prospects with lower ratios are movements along a product transformation curve like that in figure 8-1. The assumption underlying the definition of the product transformation curve is that the level of exploratory inputs is held constant. The product transformation curve refers to planned output. The composition (and amount) of realized output may differ from planned output. If it is impossible to produce either output to the total exclusion of the other, the product transformation curve terminates before reaching either output axis.

The product transformation function is concave to the origin because the probability of finding gas rather than oil, and vice versa, is higher on some prospects than on others. As inputs are allocated away from the search for oil to the search for gas, higher probability oil prospects are given up for lower probability gas prospects. Thus, in an expected sense, larger and larger quantities of oil discov-

[3]For example, the marginal cost (including rents to host governments) of world crude oil production delivered in the United States is about $2.00 per barrel. The U.S. price of crude oil substantially exceeds $3.00 per barrel. Foreign sources of supply are eager to sell their oil in the United States at a delivered price of $2.00, and there is more U.S. oil available than demanded at the current domestic price. This discrepancy between quantity supplied and quantity demanded is maintained through import controls on foreign oil and restrictions on domestic production. In the absence of some compelling national security argument, an appropriate policy that promotes the efficiency of resource allocation and the maximization of national income is one that moves the U.S. price of oil in the direction of the world price of oil. The executive agencies responsible for national security–State, Defense, and the Office of Economic Preparedness–all believe that there is room for relaxation of the current level of oil import controls rather than to maintain the controls at their current level and support domestic increases in the price of oil. (See Cabinet Task Force on Oil Import Control, *The Oil Import Question: A Report on the Relationship of Oil Imports to the National Security* [February 1970].) Economists would then conclude that the present level of oil import controls are not worthwhile in terms of their contribution to the overall goal, even though policy makers, perhaps reflecting a more limited constituency, might not reach the same conclusion.

[4]In a competitive industry, setting an effective price ceiling insures that marginal costs will equal the ceiling price. Firms will commit resources to the industry until the cost of the last units of resource, the marginal cost of output, just equals the return to be gained from their employment in the industry, the ceiling price.

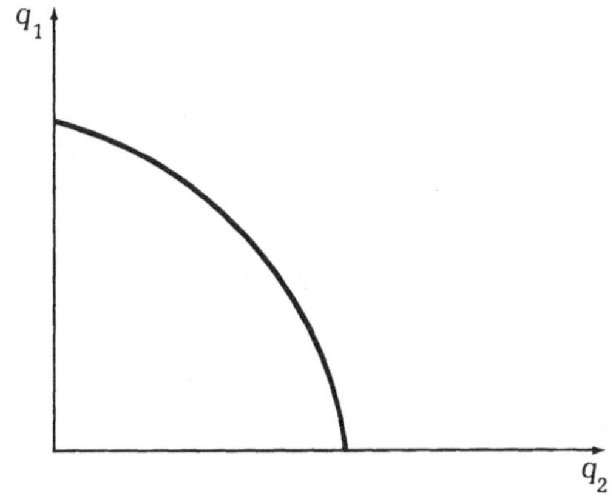

FIGURE 8-1.

eries are given up for each equal increment in gas discoveries as the search for oil and gas becomes more gas intensive.

Alternatively, holding the probability of discovering either oil or gas constant, larger prospects are drilled before smaller prospects at any level of exploratory effort. Intensifying the search for one output again means giving up larger and larger amounts of the alternative output per unit discovered of the favored output, and the product transformation function is concave to the origin.[5] The shape of this function means that the marginal opportunity cost of additional units of discoveries of either output increases as exploration becomes more intensive with respect to that output.

The profits, π, of firms engaged in the search for oil and gas are defined as

$$\pi = R - C \tag{3}$$

where

$$R = p_1 q_1 + p_2 q_2, \tag{4}$$

and

$$C = r h(q_1, q_2), \tag{5}$$

and the firm is a price taker in the output and factor markets so that p_1 and p_2, the product prices, and r, unit factor costs, are independent of q_1 and q_2.[6]

Equations (3), (4), and (5) can be combined so that

$$\pi = p_1 q_1 + p_2 q_2 - r h(q_1, q_2). \tag{6}$$

The conditions for profit maximization can be obtained by differentiating (6) with respect to q_1 and q_2 and setting the partials equal to zero. Necessary conditions for a regular maximum are

[5]The product transformation function may also be concave to the origin because some inputs are relatively more efficient in the search for gas than in the search for oil, or vice versa. For example, those prospects in traditional oil areas that can be easily located with seismic or other geophysical techniques have already largely been drilled up. Thus, factor proportions may not be the same in the production of each output. This means that diminishing returns to some factors are more quickly encountered as exploratory effort becomes either more gas or more oil intensive than would otherwise be the case.

[6]In the market for gas, unit prices for large discoveries are usually higher than those for small discoveries of the same quality and location. This is because unit costs of providing pipeline service are lower for large deposits than for small (see MacAvoy, *Price Formation*). The price-taker assumption in output markets simply means that firms face a market-determined price for gas discoveries of any size, location, and quality.

$$\frac{\partial \pi}{\partial q_1} = 0 = p_1 - r h_1 \tag{7}$$

and

$$\frac{\partial \pi}{\partial q_2} = 0 = p_2 - r h_2 \tag{8}$$

where h_i indicates partial differentiation with respect to the ith argument of h. The conditions specified by (7) and (8) can be summarized as

$$\frac{p_1}{p_2} = \frac{h_1}{h_2}. \tag{9}$$

Equation (9) simply states that maximum profits require that firms allocate inputs so that the ratio of marginal opportunity costs of using factors in the search for oil or gas (the slope of the product transformation curve) equal the ratio of product prices.[7]

The marginal cost of the first output is rh_1 and the marginal cost of the second output is rh_2. The second-order conditions for profit maximization require that the Hessian be negative definite, that is,

$$-rh_{11} < 0 \tag{10}$$

and

$$\begin{vmatrix} -rh_{11} & -rh_{12} \\ -rh_{21} & -rh_{22} \end{vmatrix} > 0. \tag{11}$$

The conditions in (10) and (11) are equivalent to requiring that h_{11} be greater than zero, or that the marginal costs of increasing the output of q_1 be increasing. In addition, since h_{12} is equal to h_{21},[8] we have

$$r^2 \, [h_{11} h_{22} - (h_{12})^2] > 0. \tag{12}$$

The condition in (12) states that any increase in the marginal cost for q_1 of an increase in the output of q_1 is greater in absolute value than any shift in the marginal cost of q_2 associated with the increase in the output of q_1.

To determine the supply curves of the products, we differentiate (7) and (8) totally so that

$$dp_1 = rh_{11}\, dq_1 + rh_{12}\, dq_2 \tag{13}$$

[7]From (2), we have

$$dX = 0 = h_1 dq_1 + h_2 dq_2$$

and

$$\frac{dq_1}{dq_2} = \frac{-h_2}{h_1}.$$

[8]This is a property of continuous functions. The order of partial differentiation does not matter.

and

$$dp_2 = rh_{21}\, dq_1 + rh_{22}\, dq_2. \tag{14}$$

A heuristic explanation of the relationship specified in (13) and (14) is as follows. A change in price elicits a change in the quantity supplied of either product. If price increases, quantity supplied increases until the increase in price is equaled by an increase in marginal costs. Under conditions of joint production, the increase in marginal costs is composed of two parts. The first part is a movement along the marginal cost curve of the output in question. The marginal cost curve is defined by the technology of production, factor prices, and the volume of output of the joint product. By the very nature of joint production, a change in the volume of production of one output changes the marginal costs of producing the second. In this case, an increase in output of the first product in response to an increase in the price of that product shifts the marginal cost of the second output and, price of the second output unchanged, leads to a change in the volume of production of the second output.

The volume of production of the second output is a determinant of the marginal cost of the first output. Thus, an increase in the volume of production of the second output as a result of a change in the price of the first output leads in turn to a shift in the marginal cost curve of the first output. The second part of the change in the marginal costs of the first product is the result of this induced shift in the marginal costs curve of the first output. Under competition, these marginal cost changes exhaust the price change.

We designate the movement along the original marginal cost curve as the "own" effect and the shift in the marginal cost curve as the "cross" effect. In (13) and (14), $rh_{11}\, dq_1$ and $rh_{22}\, dq_2$ are own effects while $rh_{12}\, dq_2$ and $rh_{21}\, dq_1$ are cross effects. These effects are illustrated in figure 8-2. The assumption underlying figure 8-2 is that an increase in the output of either product reduces the marginal cost of the other product. An increase, dp_1, in the price of the first product from p_1 to p_1' results in a movement along MC_1 from e to a. As the output of q_1 increases, the marginal cost of q_2 shifts, such as from MC_2 to MC_2'. At lower marginal costs, more of the second product is supplied at a price of p_2. Output increases from q_2 to q_2'. This increase results in turn in a shift in the marginal cost of q_1 from MC_1 to MC_1'. At output $\bar{q}_1$, the total increase in the marginal costs of the first output is the sum of the positive change in marginal costs, ba, caused by moving along MC_1 plus the negative change, ac, caused by the shift from MC_1 to MC_1'. The total change in marginal costs at output $\bar{q}_1$ is bc and is the sum of the own effect, ba, plus the cross effect, ac. Since the total change in mar-

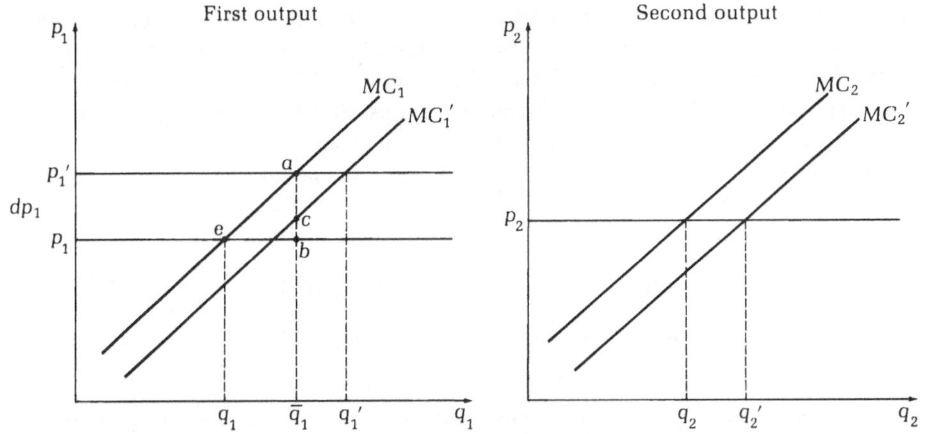

FIGURE 8-2.

ginal costs at $\bar{q}_1$ does not exhaust dp_1, output continues to expand to q_1' where the sum of the own and cross effects equals the price change.[9]

The total output response of the first product to an increase in its own price is the sum of these two effects–a change in output resulting from a movement along a marginal cost curve and a change in output resulting from a shift in the marginal cost curve. In the example presented in figure 8-2, output of the first product expands from q_1 to q_1', or until MC_1' equals the new price, p_1'. The total change in output, dq_1, equals q_1' minus q_1. The own effect on output is $q_1\bar{q}_1$ and the cross effect is $\bar{q}_1 q_1'$. The solutions for dq_1 and dq_2 are as follows:

$$dq_1 = \frac{r\,dp_1\,h_{22}}{r^2\,[h_{11}\,h_{22} - (h_{12})^2]} - \frac{r\,dp_2\,h_{12}}{r^2\,[h_{11}\,h_{22} - (h_{12})^2]} \tag{15}$$

and

$$dq_2 = \frac{r\,dp_2\,h_{11}}{r^2\,[h_{11}\,h_{22} - (h_{12})^2]} - \frac{r\,dp_1\,h_{12}}{r^2\,[h_{11}\,h_{22} - (h_{12})^2]}. \tag{16}$$

The reciprocal of the slope of the own supply curve for either product is

$$\left.\frac{dq_1}{dp_1}\right|_{dp_2 = 0} = \frac{rh_{22}}{r^2\,[h_{11}\,h_{22} - (h_{12})^2]}. \tag{17}$$

The value of (17) is necessarily greater than zero as a result of the second-order conditions stated above. From (13) through (16), it follows that the reciprocals of the slopes of the cross supply curves are

$$\frac{dq_1}{dp_2} = \frac{dq_2}{dp_1} = \frac{-r\,h_{12}}{r^2\,[h_{11}\,h_{22} - (h_{12})^2]}. \tag{18}$$

The cross effects of price changes are symmetrical and are determined by factor costs and the ratio of (a) the shift in the marginal cost of the second product relative to a price change in the first, and (b) a function of the difference between own effects and cross effects. The ratios presented in (18) are not elasticities but reciprocals of slopes of cross supply functions. Elasticities are *weighted* reciprocals of slopes where the price and quantity weights can be combined to provide indexes of market sizes. The magnitudes of the cross elasticities of supply depend on relative market sizes. This will be discussed in detail below. The signs of the cross elasticities of supply depend on the sign of h_{12}.

$$\left.\frac{dq_i}{dp_j}\right|_{(i \neq j)} \text{ is } \begin{matrix} > 0 \text{ iff } h_{ij} < 0 \\ < 0 \text{ iff } h_{ij} > 0 \end{matrix}.$$

[9]This process does not result in output expanding to infinity or declining to zero (depending on the direction of the price change of the first product and the effect of changes in the output of one product and the marginal cost of the other) because of the condition stated in (12) that own effects dominate cross effects.

The sign of h_{12} describes the direction of the shift in the marginal cost of the first output, q_1, due to an incremental change in the second output, q_2. If h_{12} is negative, then MC_1 decreases as q_2 increases. Increases in p_2 will increase the supply of q_1 and vice versa if h_{12} is of the opposite sign. We can now define[10] two linear supply curves such as

$$q_1 = \alpha p_1 + \beta p_2 \tag{19}$$

and

$$q_2 = \beta p_1 + \psi p_2 \tag{20}$$

where α and ψ are greater than zero and β may be positive, negative, or zero.

Equations (19) and (20) are, of course, the reduced form of the marginal cost functions, since price equals marginal cost under competitive conditions.

EFFECTS OF REGULATION AND OTHER PUBLIC POLICY

Under conditions of competitive equilibrium with no import controls or market-demand prorationing in crude oil markets and no field price regulation in natural gas markets, p_1, p_2, q_1, and q_2 are determined by the simultaneous solution of the demand and supply equations.[11]

If we designate p_1 and q_1 as the price and quantity of oil and p_2 and q_2 as the price and quantity of gas, we can examine the effects of alternative public policies on oil and gas supply. Suppose, for example, that the price of oil decreases due to increases in imports. If β is greater than zero,[12] decreases in oil production and discoveries cause increases in the marginal cost of gas discoveries. The supply curve of gas discoveries shifts to the left and gas prices rise and quantity falls. This shift in turn causes a leftward shift in the oil supply curve.

In figure 8-3, the gas and oil markets are initially in equilibrium with curves D and S representing domestic demand and supply in each market. If the world supply of oil is S_w, increased oil imports will decrease the domestic price and reduce the equilibrium volume of domestic discoveries.[13] The re-

[10]This definition depends on the assumption that

$$\frac{\delta q_j}{\delta p_j} \text{ (all } i, j)$$

are constants. This condition requires that the slopes of the cross supply functions be independent of the levels of oil production and gas price, etc.

[11]The supply functions are given in (19) and (20). If we temporarily suspend disbelief and assume no cross effects in demand so that

$$q_1 = a - cp_1 \text{ and } q_2 = d - ep_2, \text{ then}$$

$$p_1 = \frac{a(\psi - e) - \beta d}{(\alpha + c)(e + \psi) - \beta^2}$$

and

$$p_2 = \frac{d(\alpha - c) - \beta\delta}{(\alpha + c)(e + \psi) - \beta^2}.$$

[12]This occurs if h_{12} is less than zero so that increases in oil discoveries reduce the marginal costs of gas discoveries and vice versa.

[13]This whole example assumes no artificial restraints in U.S. oil exploration and production such as those imposed by market demand prorationing. If market demand prorationing and its attendant practices would disappear at lower oil prices, the existence of such an inhibiting factor introduces an opposite and offsetting shift in the supply of oil. This is clearly a real consideration in evaluating public policy but is ignored for simplicity's sake in this example.

duction in the volume of oil discoveries (if $\beta > 0$) causes the supply curve in gas to shift to S' because $\frac{\partial q_2}{\partial p_1}$ is greater than zero and p_1 acts as a shifter in the supply curve of gas. The resulting reduction in gas discoveries in turn raises marginal costs in the oil industry and shifts the oil supply curve from S to S'. The final result of our hypothetical example is a leftward shift in the gas supply curve, an increase in gas prices, a reduction in the volume of gas drilling and quantity of gas discoveries, and a leftward shift in the supply of oil.[14]

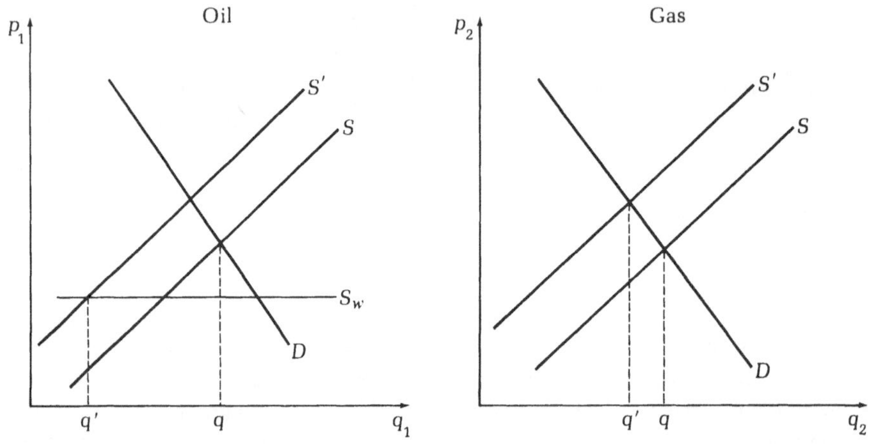

FIGURE 8-3.

The symmetrical case is that where β is less than zero.[15] In this situation, the direction of shifts in the marginal costs of one output is positively related to the volume of the other output. To use our same example, an increase in the volume of oil imports reduces oil price and the volume of oil discoveries.[16] This lowers the marginal cost in gas discoveries ($h_{12} > 0$ and $\beta < 0$) and shifts the supply of gas discoveries from S to S' in figure 8-4. The shift in the supply of gas discoveries in turn raises the marginal costs in oil discoveries and shifts the supply curve of oil discoveries from S to S'. In the case of β less than zero, increased oil imports induce a leftward shift in the supply curve of oil discoveries (just as in the case of $\beta > 0$), but there are more gas discoveries and a lower gas price.

The analysis is reversible. Suppose that the price of gas is fixed below the equilibrium level and that β is greater than zero ($h_{12} < 0$). Because decreases in the volume of natural gas discoveries shift the marginal cost of oil discoveries upward, the supply curve of oil discoveries shifts to the left. This results in a decrease in the volume of oil discoveries at all prices and in turn shifts the supply of gas discoveries to the left.[17] The ultimate effect is that we get a larger shortage of gas discoveries than we would have expected if we had ignored the joint cost features of the search for oil and gas. Interaction between the markets for gas and oil discoveries aggravates the shortage of natural gas caused by an artificially low price regardless of the technical relations (that is, the signs of h_{12} and β) of the conditions of joint production.

[14] For a similar discussion of this problem see the submission of Richard J. Gonzalez to the Cabinet Task Force on Oil Import Control. The Gonzalez analysis is based on the implicit assumptions that there is no directional drilling for gas or oil and that average associations equal marginal associations. These problems are avoided here.

[15] This occurs if h_{12} is greater than zero so that increases in oil discoveries increase the marginal costs of gas discoveries and vice versa.

[16] Again, the simplifying assumption to ignore effects from the dissolution or reinforcement of market demand prorationing is in force.

[17] This example is essentially fig. 8-3 with the "oil" and "gas" notation reversed.

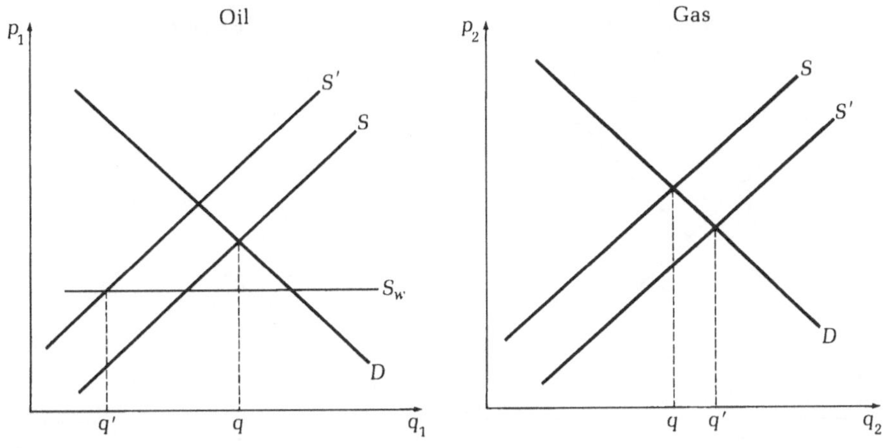

FIGURE 8-4.

STATISTICAL ESTIMATION OF THE PRICE ELASTICITY OF SUPPLY OF NATURAL GAS

The Fisher-Erickson model examines only the supply side of the market for gas and oil discoveries. The assumption at work here is that the demands for natural gas and oil have tended to be much more volatile in terms of being affected by influences other than the price of either output than have gas or oil supplies. In particular, the growth of income and population and the effect of this growth on the use of motor vehicles, electric power, and space heating and cooling have shifted the demands for these fuels rapidly to the right. This implies that the records of market equilibrium prices and quantities approximately trace out and identify a supply curve.[18] This assumption, particularly for more recent years, is more true for natural gas than for crude oil.

Two features of field markets for natural gas reserves have tended to cause observed prices and quantities to be off the demand curve. The first of these features is the monopsony power that pipelines probably had in gas field markets in the decade or so after World War II.[19] It was not until sometime in the 1950s that the interstate pipeline network lacing the major gas producing areas became sufficiently dense that the producers of any large discovery could bargain effectively with several alternative pipelines for service. The second feature that has caused market prices and quantities to be off the demand curve is regulation itself.

Because monopsonistic buyers perceive the effects of their own demands on the price at which they can buy, they recognize a marginal outlay curve such as *MO* in figure 8-5. Although monopsonists offer prices and demand quantities (such as p_o, q_o) that pick supplies off the supply curve, the observed prices and quantities are not on the demand curve. Regulation has the same effect. Imposition of a ceiling price results in observed prices and quantities (such as p_1, q_1) that are on the supply curve but not on the demand curve. For these reasons, estimation of the supply curve alone may be less troublesome than simultaneous estimation of both supply and demand.

The basic relationships estimated in the model are the price elasticities of wildcat well drilling, the success ratio, and average discovery size. Wildcat well drilling is a measure of the amount of exploratory effort undertaken in any period. The success ratio and average discovery size are measures of the results of this activity. Total discoveries in any period are the product of the number of wildcat wells times the fraction of the wildcat wells that are successful times the average discovery per wildcat well.

[18]The possible sources, directions, and magnitudes of bias in the estimation of the gas and oil price elasticities, which result from, among other things, ignoring the demand side of the markets, will be examined in a subsequent paper.

[19]MacAvoy, *Price Formation.*

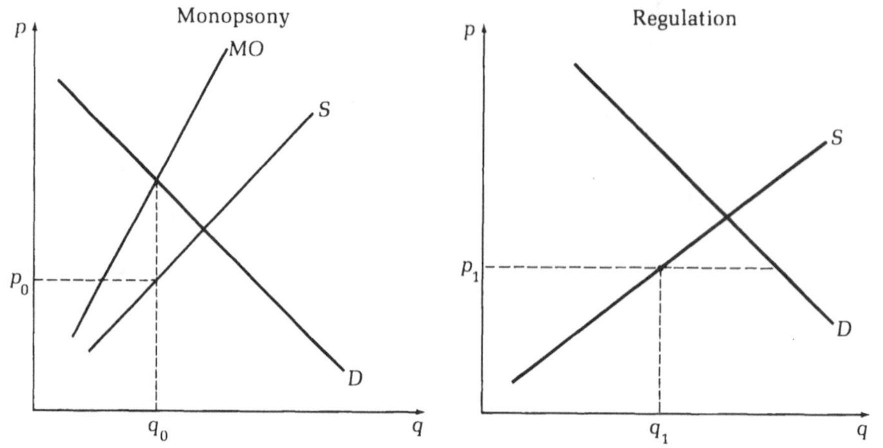

FIGURE 8-5.

When drilling, the success ratio, average discovery size, and prices are all measured in logarithms. The price elasticity of total discoveries is the sum of the price elasticities of each of the component elements whose product is total discoveries.

Each of the components of total discoveries—wildcat wells, the success ratio, and average discovery size—responds separately to economic incentives. Wildcat drilling is a positive function of the prices of both gas and oil. When prices rise, more wells are drilled and vice versa. But the average characteristics of the prospects accepted change. At higher prices, it is worthwhile to drill poorer prospects. The success ratio and average discovery size are measures of the characteristics of prospects accepted. The success ratio is not a probability coefficient given by nature. It reflects the distribution of prospects accepted, not the total distribution of prospects. When prices rise, it is worthwhile to accept prospects with lower chances of success. Therefore, we should expect to see the price coefficient of the success ratio be less than zero. A similar phenomenon occurs with average discovery size. One of the principal determinants of the ex post profitability of discoveries is their size.[20] If firms can at least roughly rank drilling prospects ex ante according to their expected size, then higher prices will justify moving through the distribution of prospects and accepting some prospects that would have been too small to drill at lower prices. Thus, the price coefficient of average discovery size should also be negative.

Over time there tends to accumulate an inventory of relatively small, relatively certain prospects. This inventory represents an aspect of the intensive margin in exploration. These prospects are just too small, even though they are relatively certain, to be drilled at the level of prices prevailing during their accumulation. If prices rise, firms drill into this inventory. The inventory is replaced as information from dry holes, and large, risky discoveries at the extensive margin builds up. The existence of this inventory implies that there should be a definite pattern of signs on the coefficients of the average characteristics of prospects accepted in the previous year when they are included in the estimations. This is the case.

For example, if there exists an inventory of relatively small, relatively certain prospects, then if the success ratio goes up, this implies that the inventory is being used up. This means that in the following year there are fewer small prospects to drill than would otherwise have been the case. Thus, when the success ratio for the previous year is included in the equation for average discovery size in the current year, the coefficient should have a positive sign—a high success ratio last year is associated with larger average discovery sizes this year. This is what we observe. By the same reasoning, because a higher suc-

[20]See, for example, the summarization of the results of exploratory activity, and especially the classification of past discoveries by field sizes, in the June issues of the *Bulletin of the American Association of Petroleum Geologists*.

Table 8-1. Summary of Variables and Notation

Symbol	Explanation
j	Subscript denoting petroleum administration for defense district.
t	Subscript denoting year.
W	Number of new field plus new pool wildcats drilled.
F	Success ratio (ratio of productive to total new field plus new pool wildcats).
S	Average size of oil discoveries per productive new field plus new pool wildcat (thousands of barrels), 1965 NPC data.
S'	Average size of oil discoveries per productive new field plus new pool wildcat (thousands of barrels), yearly API data.
N	Average size of gas discoveries per productive new field plus new pool wildcat (thousands of cubic feet), 1965 NPC data.
N'	Average size of gas discoveries per productive new field plus new pool wildcat (thousands of cubic feet), yearly API data.
D	Average depth of wildcats (feet).
P	Deflated price per barrel of crude oil at the wellhead (constant 1947-49 dollars).
G	Deflated wellhead price per thousand cubic feet of gas by year of basic contract (constant 1947-49 dollars).
M	Wildcats drilled by major companies in a given district in a given year as a percent of total U.S. wildcats drilled by all companies in that year.
R	Wildcats drilled by major companies in a given district in a given year as a percent of total U.S. wildcats drilled by those companies in that year.
X^1, X^2	Texas shutdown days, X_t equals Texas shutdown days. $X^1_{jt} = X_t$ if $j = 3$, 0 otherwise; $X^2_{jt} = 0$ if $j = 3$, X_t otherwise
$Z_1, \ldots, Z_4$	Dummy variables distinguishing among petroleum districts.

cess ratio in the previous year means that part of the inventory was drilled up and there are fewer prospects available in the current year than would otherwise be the case, we should observe a negative relation between the lagged success ratio and current drilling.[21] This is also what we observe.

The basic data and notation are the same as previous formulations of the model with two exceptions. The first exception is that there is now available a series on gas prices for gas delivered in this year under contracts dated in previous years with contract year specified.[22] For our purposes, this is clearly superior to the totally unusable weighted average wellhead price of gas delivered under contracts dated in all previous years. The second exception is that in the estimations for the period 1959-67 the basis for average discovery size is the additions to reserves through discoveries in the new fields and new pools series of the American Petroleum Institute (API). The variables and notation are shown in table 8-1.

When the price of gas and an equation for average gas discovery size are included in the model, the principal equations[23] for the period 1946-58/59 are as follows:

[21] For a complete discussion of these relations, see Fisher, *Supply and Costs*, and Erickson, "Economic Incentives."

[22] Table F, "Sales by Year of Basic Contract, Producers Whose Combined Sales Were 2 Million Mcf or More, 1968," in FPC, *Sales by Producers of Natural Gas to Interstate Pipeline Companies, 1968* (1969).

[23] These equations are reported with the specification last used in Erickson, "Economic Incentives."

Equation (21), Wildcatting

$$\log W_{jt} = -8.21 + \underset{(0.464)}{1.48} \log P_{jt} + \underset{(0.416)}{0.352} \log G_{jt}$$

$$- \underset{(0.263)}{0.777} \log F_{jt-1} + \underset{(0.0129)}{0.0320} R_{jt}$$

$$+ \underset{(0.598)}{1.76} \log D_{jt-1} + \underset{(0.00124)}{0.00176} X_{jt}^1$$

$$+ \underset{(0.00089)}{0.00129} X_{jt}^2 - \underset{(0.686)}{2.30} Z_1 + \underset{(0.508)}{1.55} Z_2 - \underset{(0.525)}{0.416} Z_4$$

$R^2 = 0.972$ d.f. = 33

Equation (22), The Success Ratio

$$\log F_{jt} = 2.93 - \underset{(0.307)}{0.134} \log P_{jt} + \underset{(0.284)}{0.0109} \log G_{jt}$$

$$- \underset{(0.0150)}{0.00807} M_{jt} + \underset{(0.000799)}{0.000399} X_{jt}^1 - \underset{(0.000574)}{0.000888} X_{jt}^2$$

$$+ \underset{(0.356)}{0.136} Z_1 - \underset{(0.201)}{0.126} Z_2 - \underset{(0.197)}{0.184} Z_4$$

$R^2 = 0.577$ d.f. = 35

Equation (23), Average Oil Discovery Size

$$\log S_{jt} = 4.64 - \underset{(0.852)}{0.415} \log P_{jt} + \underset{(0.797)}{0.713} \log G_{jt}$$

$$+ \underset{(0.540)}{0.773} \log F_{jt-1} - \underset{(0.00240)}{0.00779} X_{jt}^1$$

$$- \underset{(0.00196)}{0.00308} X_{jt}^2 - \underset{(0.729)}{4.30} Z_1 - \underset{(0.442)}{1.19} Z_2 + \underset{(0.493)}{0.0371} Z_4$$

$R^2 = 0.893$ d.f. = 31

Equation (24), Average Gas Discovery Size

$$\log N_{jt} = 10.4 - \underset{(1.05)}{1.50} \log P_{jt} + \underset{(0.984)}{0.329} \log G_{jt}$$

$$- \underset{(0.667)}{0.241} \log F_{jt-1} - \underset{(0.00297)}{0.00166} X_{jt}^1$$

$$- \underset{(0.00241)}{0.0000184}\, X_{jt}^2 \; - \; \underset{(0.900)}{0.00424}\, Z_1 \; - \; \underset{(0.546)}{1.74}\, Z_2 \; - \; \underset{(0.609)}{0.994}\, Z_4$$

$R^2 = 0.599$ d.f. $= 31$

These results, especially with regard to oil price, are basically similar to those reported and discussed in detail in the original formulations and estimations of the model. The observed patterns of signs on log $F_{jt\text{-}1}$ in equations (21) and (23) are consistent with effects of the inventory of small prospects. The variable relating to how majors allocate their effort among districts, R_{jt}, has the same sign and plays the same role as before in equation (1) for drilling. But M_{jt} becomes insignificant and changes sign in (22) for the success ratio. The success ratio equation has consistently given the least satisfactory results.

The variables for Texas shutdown days, X_{jt}^1 and X_{jt}^2, indicate that increasingly severe market demand prorationing shifts the supply of crude oil discoveries to the left. This effect is picked up by the significant negative signs of the coefficients of X_{jt}^1 and X_{jt}^2 in equation (23), the equation for crude oil average discovery size. The negative coefficient on X_{jt}^1 arises because market demand prorationing restricts production from large discoveries proportionately more than from small discoveries and thus makes the commitment of resources to the search for large discoveries relatively less attractive at any level of prices. The negative coefficient on X_{jt}^2 arises because increasingly severe production restrictions in District III signal the intentions of the state-operated cartel to maintain the going price. This reduces uncertainty in price expectations and encourages drilling a larger number of smaller prospects in the other districts than would otherwise be the case.

The own price elasticity of supply for crude oil discoveries is +0.93. This elasticity is the sum of the elasticities of +1.48 for wildcat drilling, -0.13 for the success ratio, and -0.42 for average crude oil discovery size. This compares to an own price elasticity of +0.91 for crude oil discoveries in the earlier Erickson estimations.[24] The own price elasticity of gas discoveries is +0.69. This elasticity is the sum of the elasticities of +0.35 for wildcat drilling, +0.01 for the success ratio, and +0.33 for average gas discovery size. Some details of the data and estimations that are involved in the inclusion of gas prices and discoveries are discussed next.

If total discoveries are equal to the number of wells drilled times the fraction of those wells that are successful times the average discovery size per successful well, then it might be contended that an estimation of the own price elasticity of gas discoveries requires a differentiation between gas and oil wells in the equations for drilling and the success ratio, as well as in the equation for average gas discovery size. We do not do this for two reasons. One reason is practical and the other is conceptual, but they are connected. It is neither possible nor, for the analysis reported here, advisable to make such a distinction.[25]

There is no way to tell from the data what fraction of wildcat wells are considered ex ante to be gas wells as opposed to oil wells. These data are available ex post for all exploratory wells, but all exploratory wells are not wildcat wells and wildcats are the source of major new additions to reserves through exploration. Language follows life.[26] Development is not exploration, and extensions are not wildcats. It is reasonable to suppose that these exploratory wells are drawn from a different distribution of prospects than are wildcat wells. Using the ex post distribution of oil and gas wells among all successful exploratory wells as a yardstick for the ex ante identification of oil and gas wells among wildcats

[24] These earlier estimations did not include the price of gas. The inclusion of the price of gas should not materially affect the estimation of the own price elasticity of crude oil as a result of the covariance between gas and oil prices. This is commented on below and discussed in detail in yet to be published work.

[25] Development and extension drilling that is induced solely by regulatory or market changes in the incentives to produce gas are very important in the analysis of short-run supply response. These responses are not discussed here.

[26] For another demonstration of this principle in a different economic context, see J. W. McKie, *Industry Classification and Sector Measures of Industrial Production*, Working Paper no. 20, U.S. Department of Commerce, Bureau of the Census (1965).

(including successes and dry holes) would introduce biases into the estimations. The directions and net effects of these biases would probably be indeterminant, and in any event we do not want to introduce biases that result from drawing observations from separate and distinct distributions.[27]

The conceptual reason for not introducing a distinction between oil and gas into the drilling and success ratio equations is that such a procedure would entail a tacit assumption that there are limited joint costs in exploration. If all wells could be perfectly identified ex ante as either oil or gas wells, then there would be significantly less reason to consider the joint cost aspects of the problem. The ex ante identifiability of gas wells has been a feature in the area rate proceedings. In this context, it is referred to as "directionality."[28] We do not believe the narrow definition of directionality applies—that is, that the search for gas can be completely separated from the search for oil. Neither, apparently, do others.[29]

In equations (21) through (24), and in all estimations to follow, Petroleum Administration for Defense (PAD) District V is omitted. This is because the overwhelming bulk of activity in District V is in California. California is outside the jurisdiction of the Federal Power Commission (FPC), because all sales of gas discovered and produced in California are intrastate sales. As a result, California is not included in the gas price series that reports prices by year of contract, and the only available prices are unusable. In addition, prices are not available for some years for Districts I and IV. These years are also omitted. This reduces the nominal degrees of freedom from $j \times t$ minus the number of estimated parameters to the values reported for each equation.

A higher price for natural gas has effects that increase the volume of proved reserves declared for some discoveries. A higher price will make it possible to dedicate a larger volume of reserves out of a typical just-discovered reservoir. This leads to larger volumes of reserves being reported as discoveries at higher prices for any volume of exploratory activity. The major reserve estimating and reporting bodies take explicit account of existing, or ruling, prices, costs, and technology.[30] That the effect of higher prices is to cause the same discovery to yield higher proved reserves cannot be observed, but it is apt to be relatively constant over time. Also, it is a real effect of price on the availability of gas reserves. At the exploration margin, the price of natural gas has a positive effect on the volume and composition of exploratory activity. By including the price of natural gas in the equations for exploratory activity, in the context of joint costs for oil and gas discoveries, we believe that we can obtain an acceptable estimation of the own price elasticity of gas discoveries for at least one period of interest.

INTERPRETATION OF THE RESULTS FOR 1946-58/59

The pattern of signs and magnitudes for some of the coefficients in equations (21) through (24) may at first seem puzzling. This is especially true with regard to the own and cross elasticities of average oil and gas discovery sizes. Why should the own price elasticity of average oil discovery size be

[27] See Franklin M. Fisher, *A Priori Information and Time Series Analysis* (Amsterdam: North Holland Publishing Company, 1966).

[28] See, for example, the testimony and exhibits of C. M. Allen in FPC, *In the Matter of Area Rate Proceedings*, Docket nos. AR 64-1 and AR 64-2.

[29] For example, see the submission of R. J. Gonzalez to the Cabinet Task Force and the separate report prepared jointly by the chairman of the FPC and the secretaries of the interior and commerce. Gonzalez represents the industry view. The thrust of his statement was a criticism of the Charles River Associates for not emphasizing in their report to the Office of Science and Technology the possible effects of decreases in oil prices on gas supplies. This criticism was also voiced to the staff of the task force in conversations with other oil industry representatives. By not making a direct distinction between gas and oil wells, we are in fact reflecting the judgment of the industry, as revealed by its position on public policy issues, that narrow directionality is dominated by joint costs. However, by including gas prices in the equations of our model, we are accepting a broad definition of directionality to the effect that gas discoveries respond to their own incentives.

[30] See, for example, American Gas Association, American Petroleum Institute, and Canadian Petroleum Institute, *Reserves of Crude Oil, Natural Gas Liquids, and Natural Gas in the United States and Canada as of December 31, 1967*, vol. 22 (May 1968), p. 15; and National Petroleum Council, *Proved Reserves and Productive Capacity* (1961), pp. 7-8, and (1965), p. 7.

negative while the own price elasticity of average gas discovery size is positive? Can the positive cross elasticity of gas price on average oil discovery size be reconciled with the negative cross elasticity of oil price on average gas discovery size? To a large extent, this pattern can be explained. The explanation depends on a clear understanding of the interaction between price increases and selection from the prospect inventory. The observed pattern may emerge from the data because of differences in the characteristics of the gas and oil prospect inventories in this period, and because the change in incentives for a given percentage increase in the average price of oil in a PAD district is more uniform across the distribution of oil prospects than the change in incentives across the distribution of gas prospects for a given percentage increase in the price of gas.

Consider the distribution by size of undrilled prospects in crude oil. For simplicity, assume that there are only two sizes of prospects–large and small. The early preliminary geophysical, geological, and core-drilling work may be essentially the same for identifying small prospects as for large.[31] But the distribution of prospects available in nature, as evidenced by the historical record of discovery sizes, is heavily weighted toward small prospects. The additional resources necessary to find and identify an additional large prospect are much larger per additional prospect for large prospects than for small prospects. These factors suggest marginal cost curves for prospect types of the relative slopes and positions of those presented in figure 8-6. The subscripts L and S refer to "large" and "small."

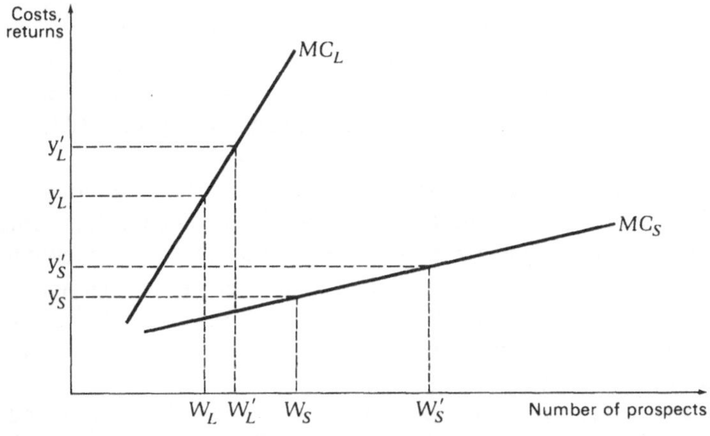

FIGURE 8-6.

If a barrel of oil (adjusted for quality and location) has the same market value regardless of whether it comes from a small or a large prospect, and if large prospects are some typical size multiple of small prospects, then the returns from discovering a large prospect are greater than those from discovering small prospects in the same proportion as the relative prospect sizes.[32] For simplicity, we assume that

[31] This is an overstatement. For the period in question, many wells were drilled with little or no formal investment in search and identification. (See McKie, *Industry Classification*, and the June issues of the *Bulletin of the American Association of Petroleum Geologists*.) If this was the case and many small, shallow wells were drilled on "oil sense" alone, the argument that follows is strengthened.

[32] This example abstracts from such considerations as differential effects of prorationing that tend to reduce the returns of large prospects relative to small prospects. It also abstracts from differences in gathering and production costs, which also vary with prospect size and may cause returns to vary in a larger proportion than prospect size. It also abstracts from the tendency for large prospects to be drilled at higher levels of uncertainty than small prospects so that *expected* returns may diverge in a smaller proportion than prospect size. All these complicating factors could be introduced without altering the usefulness or explanatory power of the example, which, as will be illustrated below, depends on the relative slopes of the marginal cost curves and the proportional changes in net returns across prospect sizes. These are empirical questions and although they cannot be observed directly, their net effects are contained in the coefficients in equations (21) through (24).

large prospects are uniformly twice as big as small prospects. This means that y_L, the returns from a large prospect, are twice as large as y_S, the returns from a small prospect. Firms will pursue both large and small prospects until the marginal costs of identifying and drilling a prospect of either type are just equal to the returns from doing so.[33] The number of small prospects drilled is W_S and the number of large prospects drilled is W_L. If Q_L is the number of barrels in a large discovery and Q_S is the number of barrels in a small discovery ($Q_S = 0.5\ Q_L$), then average discovery size S is equal to

$$S = \frac{Q_L W_L + Q_S W_S}{W_L + W_S}. \tag{25}$$

If the average price of oil increases, the returns to discovering both large and small prospects increase to y'_L and y'_S. But a given percentage increase in the price of oil has different absolute effects on the returns to each type of prospect. Under our assumption that large prospects are uniformly twice as big as small prospects, $y'_L - y_L$ is twice as big as $y'_S - y_S$. In response to the increase in returns, firms drill more of each type of prospect and W_L and W_S increase to W'_L and W'_S. Average discovery size may increase, remain constant, or decrease. If the distribution of wells drilled among large and small prospects changes, however, average discovery size will change. If the ratio of small to large prospects accepted increases so that W'_S / W'_L is greater than W_S / W_L, average discovery size will decrease. The condition for this to happen is that the ratio of the slope of MC_L relative to the slope of MC_S be greater than the ratio of Q_L to Q_S. It is the effect of the operation of this condition that we observe when the estimated own price elasticity of average crude oil discovery size is negative.[34] The own price elasticity of average crude oil discovery size estimated in equation (23) is -0.42.

The same phenomenon is at work in the gas market, but in a different way. As large oil prospects were preliminarily identified during the postwar years, resources were committed to confirming their size and other characteristics. These prospects were drilled as soon as enough information was accumulated to justify drilling them at the going level of prices, or as soon as prices moved up sufficiently to justify drilling them at their given level of expected size and uncertainty. In a state of the world in which even large gas discoveries were unmarketable because of lack of pipeline connections, one of the factors that tended to reduce the attractiveness of a prospect (other things equal) was a relatively high probability that it would be gas rather than oil.[35]

As a result, when a prospect was identified as apt to be gas rather than oil, it tended to be shelved in the inventory of undrilled prospects regardless of its size. The consequences of this market condition were that the inventory of undrilled gas prospects had different characteristics than the inventory of undrilled oil prospects when the network of interstate pipelines began to become dense in the gas producing provinces. The inventory represented an accumulation of many years, and only large gas prospects were kept in the inventory. Small gas prospects tended to be generated only by current activity. The ratio of the number of large gas prospects to small gas prospects was probably higher than the comparable ratio in crude oil. Also before being placed in the undrilled inventory, large gas prospects probably tended to have had some post preliminary geophysical and geological work done on them in the hopes that they were oil prospects. Thus, the additional resources necessary to prove

[33] Separately identifying y_L and y_S in no way implies that there are separate markets for oil from large and small discoveries. The marginal costs per barrel are identical for large and small discoveries and equal to the market price per barrel.

[34] There is evidence to suggest that MC_S is very flat indeed relative to the slope of MC_L. See Erickson, *Economic Incentives*, pp. 56-66; Stephen L. McDonald, "Percentage Depletion in Federal Taxation of Income from Oil and Gas," *1962 Proceedings*, National Tax Association; and the forthcoming work of Professor R. Mancke of the University of Chicago School of Business in the *Antitrust Bulletin.*

[35] During personal interviews with the authors, executives of exploration firms have emphasized that in these years small oil discoveries were more valuable than even large gas discoveries because, "If worse comes to worst, you can always truck the oil out."

up a large gas prospect for drilling were probably less, relative to a small prospect, than in crude oil. These conditions show up in the relative slopes and positions of the illustrative MC_L and MC_S in the gas discoveries market in figure 8-7. The configuration of MC_L and MC_S in figure 8-7 was local in time to the postwar years and probably does not exist now. The cost conditions of identifying gas prospects by size are probably now similar to those in figure 8-6.

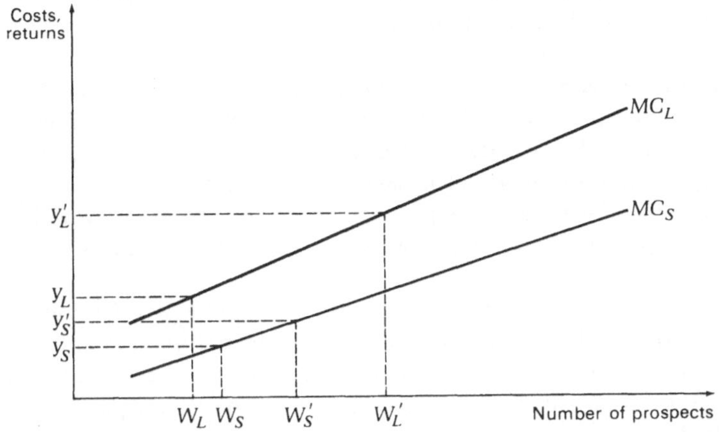

FIGURE 8-7.

There is, however, another special feature of the gas market. This feature is permanent and is as important as the transitory inventory configuration in determining the observed pattern of coefficients. Large gas discoveries are worth more than small gas discoveries by a factor that is more than proportional to the difference in their sizes. This is because of the tremendous economies of scale in long-distance gas pipelines and the deliverability requirements in FPC regulations.[36] Thus, a given percentage increase in the price of gas for interstate shipment will have a disproportionate effect on the relative attractiveness of large prospects.

Assume gas markets are in equilibrium at y_S, y_L, W_L and W_S. We again assume that large prospects are uniformly twice as big as small prospects ($Q_S = 0.5\ Q_L$). For convenience, we also assume that gas production is totally used in local energy markets or carbon black operations. The marginal cost of discovering another cubic foot of gas just equals the price of gas net of production costs. Enter long-distance pipelines that are willing to pay a premium for large gas discoveries. Under this condition y_S'/y_S no longer equals y_L'/y_L. The premium for large discoveries makes y_L'/y_L greater than y_S'/y_S. In order to equalize prices and marginal costs across prospect sizes, relatively more resources have to be committed to the search for large prospects. For the conditions illustrated in figure 8-7, the result is clearly that W_L'/W_L is greater than W_S'/W_S. This means that average discovery size increases. The own price elasticity of average gas discovery size estimated in equation (24) is +0.33.

Thus, it is possible for an increase in the average price of gas to be associated with an increase in average gas discovery size. This effect could persist for some time as MC_L shifted to the left and rotated. The leftward shift and upward rotation of MC_L would be the result of pipelines pushing into new areas and causing the inventory of large gas prospects to be drilled up. After this occurred, the search for and identification of large gas prospects would take place under conditions similar to those in crude oil and illustrated in figure 8-6.

The negative cross elasticity of oil price on average gas discovery size follows immediately from the direct effect just discussed. Although exploring firms are able to roughly rank prospects ex ante in terms of the probability of gas or oil, this ranking is likely to be much less precise than their ranking of prospects by size—regardless of gas versus oil potential. In addition, even if the ranking by proba-

[36] MacAvoy, *Price Formation.*

bility of gas versus oil is precise, it is still probabilistic. Suppose there is a class of small prospect with characteristics that indicate that the probability of oil rather than gas is +0.6. If the price of oil goes up and enough members of this class are successfully drilled so that 100 discoveries are made, then about 40 new small gas discoveries will be made. If the elasticity of wildcat drilling with respect to the price of oil is considerably larger than the same elasticity with respect to gas, then we would expect to see a relatively large negative elasticity of average gas discovery size with respect to oil price. The price elasticity of wildcat drilling with respect to oil price is +1.48. With respect to gas price, the elasticity of wildcat drilling is +0.35. The cross elasticity of average gas discovery size with respect to oil price estimated in equation (24) is - 1.50.

The symmetrical argument holds with regard to the cross elasticity of average oil discovery size with respect to gas price. If the own price elasticity of average gas discovery size is positive, and if the ex ante separation of prospects into oil versus gas is at best probabilistic, then an increase in the price of gas will result in some large structures being drilled, some of which turn out to be oil discoveries. Holding other things constant, the effect is for average oil discovery size to increase with increases in the price of gas. The cross elasticity of average oil discovery size with respect to gas price estimated in equation (23) is +0.71.

The price elasticity of the success ratio is not very large with respect to either the price of oil or the price of gas. However, the sign of the elasticity of the success ratio with respect to the price of oil fits into the predicted pattern. The oil price elasticity coefficient of the success ratio estimated in equation (23) is - 0.13. The gas price elasticity of the success ratio estimated in equation (23) is +0.01 but is dwarfed by its standard error.

The overall cross elasticity of gas supply with respect to oil price is the sum of the direct elasticities of wildcat drilling, the success ratio, and the cross elasticity of average gas discovery size. These elasticities are respectively +1.48, - 0.13 and - 1.50. Their sum is - 0.15. This can be interpreted to mean that in the period in question a 1 percent increase in oil price was associated with a - 0.15 percent decrease in gas discoveries.

Similarly, the overall cross elasticity of oil supply with respect to gas price is the sum of the direct elasticities of wildcat drilling, the success ratio, and the cross elasticity of average oil discovery size. These elasticities are respectively +0.35, +0.01 and +0.71. Their sum is +1.07. This can be interpreted to mean that in the period in question a 1 percent increase in gas price was associated with a +1.07 percent increase in oil discoveries.

The difference in sign and magnitude of these cross elasticity estimates requires comment. From the treatment of production theory in the joint cost section above we know that

$$\frac{\partial q_1}{\partial p_2} = \frac{\partial q_2}{\partial p_1}.$$

These slopes of cross supply curves can be converted to elasticities.[37] From this conversion it appears that the cross elasticities of supply must have the same sign and be proportional to relative market

[37]Multiplying each side by

$$\frac{p_2 p_1}{q_1 q_2}$$

and simplifying we obtain

$$\frac{p_1}{q_2}\eta_{12} = \frac{p_2}{q_1}\eta_{21}, \text{ or } \frac{p_1 q_1}{p_2 q_2}\eta_{12} = \eta_{21}.$$

If the subscript 1 designates oil and the subscript 2 designates gas, and if the oil market is bigger than the gas market ($p_1 q_1 > p_2 q_2$), then η_{12} is less than η_{21}. The elasticity of oil supply with respect to gas is η_{12} and the elasticity of gas supply with respect to oil price is η_{21}.

sizes. Thus, if either $\partial q_1/\partial p_2$ or $\partial q_2/\partial p_1$ is positive, it appears that both elasticities should be positive and that the cross elasticity of oil price on gas discoveries should be considerably larger than the cross elasticity of gas price on oil discoveries. This is not what we observe.

This is easily explained. The standard joint cost case presented in the second section of the paper is not complicated by the existence of inventories of undrilled prospects. But even if it were, the equality of sign and determination of relative magnitude of the cross elasticities of supply of discoveries would be as just described—providing the effect of the inventories in each market were the same. As discussed at length above and illustrated by the estimations presented in equations (21) through (24), this is not the case. If the marginal costs and incentives across prospect sizes were like those illustrated in figure 8-6 in both gas and oil discoveries in the period for the estimations, the sign and relative magnitude conditions on the cross elasticities would hold. Under the disparate inventory conditions that existed for most of the period, the only place that we expect anything resembling the easily specified formal relationships between the cross elasticities to hold is with regard to the direct own elasticities of wildcat drilling. The oil price elasticity of wildcat drilling is +1.48 and the gas price elasticity is +0.35.

Too much emphasis must not be placed on the particular values of the cross elasticities that are derived from equations (21) through (24). They are a product of a special era. There is every reason to believe that they have subsequently changed. It is not too early in this paper to admit that we currently have much less success with the estimations for the 1960s. Rather than use values that are relevant to a different process than is now likely to exist, it may be more appropriate to rely on informed judgment.[38] There can be little doubt that both cross elasticities are now positive, but their absolute and relative magnitudes are open to speculation.

ESTIMATED SUPPLY RESPONSE, 1959-67

There are several important reasons to consider separately the data for the period 1959-67. It is impossible to pinpoint a single year and say that it was the year in which the long-distance pipeline network integrated the whole Gulf Coast and Southwest into a single market, but this process was certainly nearly complete by 1959. Horizontal integration of the gas market implies that the special effects of the inventory of gas prospects discussed above were exhausted. In addition, 1959 can be pinpointed as the year in which important changes in regulatory practices affecting both the oil and gas markets were made. Mandatory import quotas were imposed on the U.S. crude oil market in March 1959. This resulted in the real price for U.S. crude oil being dramatically stabilized (see table 8-2) and price expectations were revised accordingly.

At the same time, the FPC unilaterally altered the rules under which price took place in natural gas contracts. Again, the relation between the observed current wellhead price of output and future

Table 8-2. Crude Oil Price Variation, 1951-54 and 1960-63

PAD District	Variable	1951-54 Mean	1951-54 Standard deviation	1951-54 CV^a	1960-63 Mean	1960-63 Standard deviation	1960-63 CV^a	$\frac{CV_{60-63}}{CV_{51-54}}$
I	$\log P_{jt}$	0.890	0.107	0.120	0.939	0.0193	0.0206	0.172
II	$\log P_{jt}$	0.509	0.0549	0.108	0.517	0.02299	0.00578	0.0535
III	$\log P_{jt}$	0.501	0.0632	0.126	0.548	0.00472	0.00861	0.0683
IV	$\log P_{jt}$	0.384	0.0649	0.169	0.399	0.00548	0.0137	0.0811
V	$\log P_{jt}$	0.381	0.0853	0.224	0.356	0.0144	0.0404	0.180
VI	$\log P_{jt}$	0.533	0.204	0.383	0.552	0.212	0.384	1.00

[a]CV is the coefficient of variation and is defined as the mean divided by the standard deviation.

[38]See, for example, app. D of Cabinet Task Force, *The Oil Import Question*.

profits was altered. An important assumption that underlies the estimations for the 1946-58/59 period is that the elasticity of expectations with regard to the relationship between current and future prices was +1.0–a given percentage increase or decrease in current prices reflected the same percentage increase or decrease in future prices. For natural gas, the elasticity of expectations was probably subject to wide swings during the 1959-67 period. The reform of price escalation clauses probably reduced it to zero. The industry reaction to the Permian Basin decision indicates that, for a period of a year or so, the elasticity of expectations was negative. Now, with substantial regulatory reform in the offing, the elasticity of expectations may be greater than +1.0. With these kinds of swings in such a crucial relationship, it is not surprising that the unconstrained estimations for 1959-67 are disappointing. Moreover, the structure of the market changed from one of monopsony to one of superfluous regulation.[39] For these reasons, it is necessary to regard the data for 1959-67 as a separate set of observations with separate processes at work than those observed in 1946-58/59.

We argued earlier that the form of regulation imposed on natural gas markets by the FPC makes the appropriate model for measuring supply response to price incentives a nonsimultaneous system such as the Fisher-Erickson model. This is because the observed prices and quantities are on the supply curve but off the demand curve.[40] This may be true, and there may be insufficient variation in the price data to support any kind of estimation of the relevant elasticities. Alternatively, the supposition that a nonsimultaneous model is the appropriate vehicle may be wrong and the estimations may suffer from severe (but correctable) identification problems. In any event, the estimations reported in equations (26) through (29) are very unsatisfactory.

Equation (26), Wildcatting

$$\log W_{jt} = \underset{}{9.50} + \underset{(0.761)}{3.03 \log P_{jt}} - \underset{(0.396)}{0.832 \log G_{jt}}$$

$$- \underset{(0.0912)}{0.00550 \log F_{jt-1}} - \underset{(0.291)}{0.127 \log D_{jt-1}}$$

$$+ \underset{(0.00662)}{0.00326\, R_{jt}} - \underset{(0.00211)}{0.00000409\, X^1_{jt}} + \underset{(0.00132)}{0.000508\, X^2_{jt}}$$

$$- \underset{(0.939)}{4.54\, Z_1} - \underset{(0.889)}{0.346\, Z_2} - \underset{(0.894)}{1.25\, Z_4}$$

$R^2 = 0.997$ d.f. $= 25$

Equation (27), The Success Ratio

$$\log F_{jt} = -0.153 + \underset{(1.19)}{6.00 \log P_{jt}} + \underset{(0.572)}{0.100 \log G_{jt}}$$

$$+ \underset{(0.0101)}{0.00237\, M_{jt}} - \underset{(0.00285)}{0.00237\, X^1_{jt}} - \underset{(0.00194)}{0.00300\, X^2_{jt}}$$

$$- \underset{(1.08)}{1.66\, Z_1} + \underset{(0.924)}{0.124\, Z_2} + \underset{(0.873)}{0.459\, Z_4}$$

[39] Some of the possible effects of regulation on both supply response itself and the problem of estimating supply response are discussed by Clark Hawkins in his paper (chap. 6).

[40] See pp. 201-2 above and figure 8-5.

$R^2 = 0.888$ d.f. = 27

Equation (28), Average Oil Discovery Size

$$\log S'_{jt} = \underset{}{10.5} + \underset{(4.60)}{3.53 \log P_{jt}} - \underset{(2.24)}{3.37 \log G_{jt}}$$

$$+\underset{(0.520)}{0.631 \log F_{jt\text{-}1}} - \underset{(0.0109)}{0.00149\, X^1_{jt}}$$

$$+\underset{(0.00710)}{0.00000382\, X^2_{jt}} - \underset{(3.48)}{5.88\, Z_1} - \underset{(2.99)}{2.14\, Z_2} - \underset{(3.09)}{0.680\, Z_4}$$

$R^2 = 0.937$ d.f. = 27

Equation (29), Average Gas Discovery Size

$$\log N'_{jt} = 10.6 - \underset{(3.14)}{3.15 \log P_{jt}} - \underset{(1.53)}{0.241 \log G_{jt}}$$

$$+\underset{(0.354)}{0.125 \log F_{jt\text{-}1}} + \underset{(0.00741)}{0.00128\, X^1_{jt}} - \underset{(0.00484)}{0.00351\, X^2_{jt}}$$

$$+\underset{(2.37)}{1.46\, Z_1} - \underset{(2.04)}{0.598\, Z_2} - \underset{(2.11)}{0.637\, Z_4}$$

$R^2 = 0.769$ d.f. = 27

For our purposes here, the most striking thing about equations (26) through (29) is the negative sign on the coefficient of the log of gas price in equation (26) for wildcatting. According to equation (26), a 1 percent increase in the price of natural gas leads to a -0.83 percent *decrease* in the amount of wildcat drilling undertaken. We have no reason to believe that this is the true direction of the relationship, that an increase in the incentive to drill wells leads perversely to a decrease in the actual number of wells drilled. Thus, we must conclude that equation (26) does not capture the real relation between gas prices and wildcatting activity.

The coefficient of the log of gas price in equation (27) for the success ratio is +0.10. The own price elasticity of average gas discovery size estimated in equation (29) is -0.24. Combined with the negative elasticity of wildcat drilling with respect to gas price, these latter two elasticities yield an overall own elasticity of total gas discoveries with respect to gas prices of -0.97. We reject this result.

There are several possible explanations for the unsatisfactory outcome of the 1959-67 estimations. One possible explanation is that the appropriate model actually is a simultaneous model that includes both demand and supply. This is an alternative that is open to testing. Estimation of an appropriately specified simultaneous model, however, is apt to become snagged on a problem that we believe is a major contributing factor to the unsatisfactory results attained with our model. This problem is the high covariance between gas and oil prices. The simple correlation coefficient between the price of gas and the price of oil for the period 1959-67 is ±0.959. For the period 1946-58, it is +0.608. Somewhere between these values, a critical point is passed, and the statistical properties of the data cause the estimations obtained from the model, as originally specified, to lose meaning.[41] Although satisfactory overall results were obtained for the earlier period, equations (26) through (29) do not divulge the price elasticity of natural gas discoveries for the period immediately at hand.

We desire to know the elasticity of supply of natural gas discoveries because there is an increasing sense of crisis with regard to natural gas supply. It is clearly open to doubt that the way to meet this crisis is to pursue more vigorously the same set of policies and responses that have produced it. If we change policies, however, we wish to know the likely consequences of our acts. Together with the elasticity of demand for natural gas, the relationships that determine the ratio of production to reserves, and the net time trends of the important shifters in the supply and demand functions, the elasticity of supply of natural gas discoveries determines what these consequences are. We have failed to obtain a reliable point estimate of the elasticity of supply of natural gas discoveries that is relevant to the contemporary problem. But the alternative of using a plausible range of values for the key variables in order to determine the effects of one possible policy change is open to us. In the next section, this device is employed to explore the question of how large an increase in the field price of natural gas would be necessary to eliminate the current regulation induced shortage.

MARKET CLEARANCE THROUGH PRICE ADJUSTMENT

It is widely believed that there is now a shortage of natural gas in the U.S. energy supply array. This shortage has two components. One component is the result of the FPC holding the field price of natural gas below the market-clearing price. Even at higher prices, a larger quantity of gas production would be demanded. The other component of the current shortage is the result of a larger quantity of gas being demanded at the artificially low price than would be demanded at the equilibrium market-clearing price. If the market price were raised to the equilibrium price, this second component would disappear. We are concerned with only the first component of the total shortage. We will call this the real shortage.[42]

In order to know the current magnitude of the real shortage, measured in millions of cubic feet per year, we need to know the elasticity of supply and the percentage deviation of the ceiling price below the market-clearing equilibrium price.[43] We do not have a good estimate of the elasticity of supply. Neither do we have a good estimate of the market-clearing equilibrium price. But we do have some feel for the likely range of the relative magnitude of the real shortage and the high and low boundaries of the supply elasticity. We use this information to estimate a set of alternative price increases necessary to generate sufficient new reserves to meet a set of assumed real shortages of current production. The magnitude of the required price increase is directly related to the assumed size of the real shortage and inversely related to the elasticity of supply. We express the real shortage as the percentage increase in output necessary to reach the market-clearing equilibrium output.

Since we are interested in the increase in reserves necessary to yield an increase in production sufficient to meet the real shortage, we need to know the equilibrium value of the reserves-production ratio. This condition is easier to state than to achieve.[44] There is, however, a fortunate circumstance

[41] When faced with confounding multicollinearity between key variables, such as that between oil and gas prices for the 1959–67 period, there are several things that can be done. One can get additional comparable data if it is available. One or more of the offending variables can be omitted. Constraints based on a priori information may be imposed. Or one may throw up his hands. Unfortunately, we are in the position of accepting this last alternative. Omitting the price of crude oil does not improve the results.

[42] The second component is just as real in the sense that if gas supplies were forthcoming at the artificially low ceiling price, there would be effective demand to absorb them. But the second component of the shortage is the result of our own policy decisions, not a consequence of the natural order of things. This is worthwhile remembering in discussions of end use controls or other regulations that we must impose in order to accommodate ourselves to the results of our own policies.

[43] Alternatively, if we knew the quantity demanded at the ceiling price, current supply, and the elasticities of supply and demand, we could determine the market-clearing equilibrium price.

[44] The magnitude of the effects of economic factors, such as prices, costs, tax rates, and the interest rate on the reserves-production ratio have been disputed by various observers. A recent contribution to this subject is the

that aids us here. For a long period of time, under conditions of considerable variation in price, interest rates, and the rate of growth of national income, the reserves-production ratio in crude oil was approximately constant at about 12:1.[45] At the end of 1968, reserves of nonassociated natural gas in the United States (not including Alaska) equaled 216,011,244 million cubic feet. Production of nonassociated gas equaled 14,739,944 million cubic feet. The reserves-production ratio was 14.7:1.[46] In order to calculate the increase in gas prices necessary to call forth the additional reserves required to support sufficient incremental production to satisfy various levels of real gas shortages, we assume that the free market equilibrium reserves-production ratio in gas is 12:1.

This assumption has several effects. It allows us to calculate the new gas discoveries necessary to support new production. With an equilibrium reserves-production ratio of 12:1, 12,000 cubic feet of new discoveries are necessary to support 1,000 cubic feet per year of new production. Also, if the current ratio is 14.7:1, then substantial progress toward eliminating the existing real shortage might be made by simply permitting operators to produce existing reserves at a higher rate. Under our assumption, it appears that about 3,700,000 million cubic feet of additional production could be achieved from existing nonassociated reserves.[47] This is an interesting calculation, but rather than rely on such additional production to provide the principal short-run alleviation of the current real shortage without making any fundamental policy changes aimed at a long-run solution to the problem, we feel that

---

CONSAD Research Corporation's report prepared for the U.S. Treasury and published by the Senate Finance Committee.

The simplest model of reserve-production behavior is one where firms demand reserves in order to supply current production and all reserves are purchased at time zero at constant unit cost c. The goal of firms is then to maximize

$$\sum_{t=0}^{t} p_t q_t (1+r)^{-t} - cR$$

subject to the constraint

$$f(q_1, q_2, \ldots, q_t) = R.$$

It can be shown that

$$\frac{R}{q_t}$$

is a function of prices and the interest rate. The problems with this formulation are that all reserves are not discovered at time zero, there are intertemporal substitutions possibilities between discovering reserves now and discovering reserves later, all reserves are not discovered at the same unit cost, public policy changes may affect price expectations so that the same observed price in one period may have different effects than in another period, and technological change has probably altered the functional form of the adding-up constraint.

[45]The reserves-production ratio in crude oil has recently been going down and now stands at 9.8:1. This is probably the joint result of public policy effects and high interest rates. It may also result (adjusting for public policy and interest rates) from exhaustion of the United States as a crude oil province, technological change in production techniques, and improvements in reservoir management. To the extent that the reduction in the reserves-production ratio in crude oil is a natural economic phenomenon and not a result of exhaustion of U.S. reserves or public policy, and to the extent that the behavior governing the stock of gas reserves held relative to the flow of current gas output is similar to that in oil, the necessary price adjustments estimated below are too large.

[46]American Gas Association et al., *Reserves of Crude Oil, Natural Gas Liquids and Natural Gas in the United States and Canada, as of December 31, 1968* (1969), tables I and II, pp. 120, 121. These figures exclude 62,864,813 million cubic feet of associated and dissolved reserves and 4,633,484 million cubic feet of production from these reserves.

[47]This idea will be regarded as bizarre and repugnant by those who regard a reserves-production ratio of, say, 20:1 as a moral commandment. The 3,700,000 million cubic feet figure is for purposes of illustration only. We do not know what the reserves-production ratio could be from a technical point of view, what it should be from some suboptimizing public policy point of view, or what it would be in a totally free market for natural gas reserves and production. It may be that we have been too conservative and that the true equilibrium reserves-production ratio is less than 12:1. Because of our relative ignorance about both the "true" equilibrium reserves-production ratio and the lag structure between current price changes and subsequent movements in the ratio, and because the ratio itself is affected by other variables as well as price, unconditional changes in the ratio may not be a very satisfactory indicator of the rate and amount of price change (or in some cases, perhaps, even the direction of price change) necessary to achieve an efficient allocation of resources over time.

any potential additional production available from this source should be used as a cushion to prevent price rises from being too abrupt during a transition period to a free market in natural gas.

The percentage increase in gas prices necessary to provide sufficient new reserves to permit enough additional current production to satisfy the existing real shortage ultimately depends on the relative size of the shortage and the elasticity of supply of new discoveries.[48] We feel that the elasticity of supply of natural gas is bounded by the values +0.1 and +1.0. Our estimate for the 1946-59 period was +0.69. This estimate, however, contained the transitory effect of the inventory of large gas prospects. This effect is no longer operating. Therefore, the current elasticity of supply is probably lower than +0.69. Table 8-3 is based on a range of elasticities. Because of the results of the 1946-59 estimations, and because of the argument just stated, we feel that the most weight should be given to the +0.5 elasticity of supply.

Table 8-3. Percent Increase in Price Necessary to Eliminate Assumed Real Shortages of Natural Gas

	Real shortages (% of current production)		
Elasticity of supply	10	25	50
0.1	100	250	500
0.5	20	50	100
1.0	10	25	50

Table 8-3 also contains alternative values for the relative real shortage.[49] These values—10 percent, 25 percent, and 50 percent—are percentages of current production. Thus, in 1968, if the real shortage had actually been 10 percent of 1968 production, the equilibrium output at the market-clearing price would have been (14,739,944 + 1,473,994) million cubic feet. We feel that the actual real shortage is now somewhere between 10 and 25 percent of current production.[50]

The wellhead price of gas for gas reserves delivered under new contracts in South Louisiana is about 20 cents per thousand cubic feet.[51] This is the upper bound for the Gulf Coast and Southwest area. It

[48]It also depends on the relations between gas and oil production. At higher prices of gas relative to oil, it would be less economically attractive to delay gas production in order to use gas supplies for pressure maintenance in oil production so that a larger fraction of oil reserves are recovered. To the extent that regulatory agencies would respond to such a change in incentives, the reserves-production ratio in associated gas would fall. We ignore this effect.

[49]There are two sources of demand for new reserves. One is the demand for incremental reserves to meet totally new or expanded uses. The other is replacement demand to offset declining production from depleted reservoirs in order to maintain output at current levels. For nonassociated gas, there are also two sources of supply. One is new discoveries of nonassociated reserves. The other is net positive extensions and revisions of already discovered reserves. For the period 1947-65, the ratio of new discoveries to production averaged 0.54. However, over this same period, total reserves increased because of the net positive effect of lagged extensions and revisions to previous years' discoveries. The price responsiveness of extensions and revisions has been discussed by J. Daniel Khazzoom in *An Econometric Model of U.S. National Gas Supply*, FPC, Office of Economics, Docket no. AR69-1, Area Rate Proceeding, Southern Louisiana Area (September 1970). The implicit assumption underlying table 8-3 is that extensions and revisions will continue to augment new discoveries sufficiently to meet the total demand (incremental plus replacement) for gas reserves.

[50]Paul MacAvoy, in "The Effectiveness of the Federal Power Commission," *Bell Journal of Economics and Management Science*, vol. 1, no. 2 (Autumn 1970), pp. 271-303, has estimated that the total excess demand for new reserves in 1967 at the going ceiling price was 11.7 tcf. This equaled about 64 percent of 1967 production. In his paper (chap. 7), MacAvoy estimates that the real shortage of natural gas may be as low as 25 percent of current supply. This estimate does not include the effects of any recent demand shifts caused by concern about pollution. We guess that much of the recent interest in natural gas as a clean fuel is price specific and that at higher prices for natural gas other alternatives such as emission control devices would become the economically efficient technology for reducing discharges. We admit that this is only a guess. In any event, the portion of excess demand that is relevant to a discussion of how far market price must rise in order to elicit sufficient supply response to clear the market is that portion of total excess demand that corresponds to effective demand at the market-clearing price. This is the portion of total excess demand at the artificially low price that we define as the real shortage.

[51]The "effective" price also depends on the conditions of sale other than price. Because conditions of sale are not standard, this price is not directly comparable to the prices ruling intrastate markets. See Milton Russell's paper (chap. 9).

is from this general area that it is reasonable to expect substantial new reserve discoveries that can be readily brought to market.

If the elasticity of supply of new discoveries is about +0.5 and if the relative real shortage, expressed as a percentage of current production, is 10 to 25 percent, then the increase in the average wellhead price of natural gas is 20 to 50 percent. In South Louisiana, this would mean a price of 24 to 30 cents per thousand cubic feet. In the rest of the Gulf Coast and Southwest area, the price would be substantially lower (depending on distance to market, quality, and size of field) but substantially higher than current levels.[52] Competitive market forces can be relied on to determine area prices that are no higher than the marginal costs of the last unit of production demanded. This will be the case whether the discovering and producing company is a large or small independent producer or an affiliate of a pipeline company.

SUMMARY AND CONCLUSIONS

Gas and oil are joint products in exploration, development, and production. Public policy that affects one must also inexorably affect the other. Under conditions where there is a ready market for both outputs, discoveries of each are affected symmetrically by changes in the price of the other. During the first 60 percent of the post-World War II era, before gas production and discoveries were completely adjusted to the development of the network of interstate pipelines, this symmetry was not the case. The existence of an inventory of large gas prospects, combined with the uncertainty associated with whether a project was gas or oil and the premium that pipelines were willing to pay for gas in large discoveries, caused the cross elasticity of gas prices on oil discoveries to be positive. But an inventory of small prospects dominated the effects of oil prices on average discovery sizes in oil exploration. Small gas discoveries associated with small oil discoveries were even further away from the historical mean of gas discoveries than the small oil discoveries were away from the historical mean of oil discoveries. This result swamped the price effect on drilling, and the cross elasticity of oil price on gas discoveries was negative. It is reasonable to believe that both cross elasticities are now positive, but so far it has not been possible to estimate their magnitudes.

Both the oil and gas markets are out of equilibrium. We produce too much crude oil in the United States and not enough gas. The result is that the social marginal opportunity costs of production of each output are not equal to the demand prices or social marginal valuation. The appropriate remedy is to raise the price of gas and lower the price of oil.[53] The problem in the case of oil is how severely domestic oil production and discoveries would contract if the U.S. price fell to a level closer to the world price.[54] The problem in the case of gas is how far the U.S. price of gas would have to rise to eliminate the shortage in the U.S. market.

Under alternative assumptions about the elasticity of supply and the relative size of the real shortage of natural gas, the percentage increase in average gas prices necessary to close the shortage was

[52] The equilibrium price may have a trend attached. One example of a trend is provided by a simple demand and supply system where quantity demanded, q_d, and quantity supplied, q_s, are defined as follows:

$$q_d = p^{-a}(1+g)^t$$
$$q_s = p^{\beta}\ (1+e)^t$$

where g represents growth in demand due to population and income growth, etc., and e represents the net shift in supply due to technological change and exhaustion of the United States as a petroleum province. The time trend of equilibrium price will be positive if g is greater than e. Of course, if the prices of number two fuel oil and other crude oil products were to fall substantially as a result of opening the United States to the world oil market, the demand for gas would shift to the left, the initial equilibrium price would be lower, and it might be some time before the time trend carried the equilibrium price for new contracts to levels exceeding those now observed. This effect would be limited or offset by the cross effect from lower oil discoveries on gas discoveries.

[53] An intensive policy review of the public regulations that keep oil prices high has recently been completed. See Cabinet Task Force, *The Oil Import Question*; and Burrows and Domencich, *Oil Import Quota*.

[54] The limit, of course, *is* the world price.

estimated. The central estimate of the supply elasticity of gas discoveries was +0.5 and the most likely range of the relative shortage of gas was 10 to 25 percent of current production. Under these circumstances, the market-clearing equilibrium price of gas is 20 to 50 percent higher than current levels.[55] Such a price increase may be considerably better than the next best alternative.[56]

If this is the case, and if public policy makers do not want to further complicate and inhibit the efficiency of resource allocation in energy markets, then we need to start at once to develop the appropriate administrative machinery. The United States needs a comprehensive energy policy that will accommodate substantial price flexibility in both the oil and gas markets.

[55]These estimates were made without considering cross effects in demand and supply from any possible changes in the real price of crude oil. In effect, the price of crude oil was assumed constant. This was for two reasons. First, the basic estimates in table 8-3 are very rough, and adding a layer of cross effects for apparent precision would actually add little of value. Second, and most important, it does not appear that any policy changes that would open the U.S. market are in the offing. Thus, the U.S. price of oil will probably remain constant, *or rise*.

[56]It has been estimated that alternatives to U.S. field-produced natural gas may be available in major consuming markets at prices representing a 35 to 50 percent increase over current charges. Such percentage increases at the consuming stage of the gas market correspond to 50 to 100 percent increases at the wellhead. Moreover, such alternatives may not be available in sufficient quantity to clear the market. Non-price rationing through formal or informal end-use controls of some kind would probably still be necessary. See R. Bruce Foster's paper (chap. 3).

9 PRODUCER REGULATION FOR THE 1970s

*Milton Russell**

The current area price regulation of the field price of natural gas by the Federal Power Commission (FPC) has been subjected to considerable criticism over the past few years. The purpose of this paper is to evaluate the prospective contribution to the public interest of a continuation of this FPC regulation of the late 1960s and of three alternative policies. The alternatives chosen for review are (1) a reform of the current price-fixing method, (2) a basic restructuring of price regulation away from a cost base, and (3) virtual price deregulation with suitable public interest safeguards. Numerous alternatives might have been chosen, but most, with some nuances of difference, could be fitted into some combination of those considered. The focus throughout is on the general relationships involved; this paper does not deal with specific application of regulatory policy through FPC procedure.

An evaluation of current and alternative policies requires specification of the goals of regulation against which the evaluation is laid. Such a specification, within the context of a review of the reasons given for gas producer regulation, is presented below. The goals accepted here are those consistent with the effective working of a competitive market: optimal allocation of resources and factors, both among alternative uses and over time; functional distribution of income; and minimum social cost in achieving whatever modifications of market behavior or results are desired. Current regulatory patterns are also described briefly below.

Regulation, whatever its goals, is superimposed on the underlying basic forces in the regulated industry and works both on and through them. Some consideration of the forces affecting supply and demand for natural gas in the field is therefore crucial in understanding current and prospective industry response to regulation. A model of the natural gas production industry is developed from this basis to show the effect of alternative regulatory policies on the jurisdictional and nonjurisdictional gas markets. This model indicates that past regulation has contributed to the current jurisdictional gas shortage and has led to inefficient allocation of resources. The paper concludes with a treatment of possible reforms of and alternatives to current regulation.

THE PUBLIC INTEREST IN NATURAL GAS FIELD PRICE REGULATION

Operationally, a sufficient reason for the regulation of the sale for resale of natural gas in interstate commerce is that it is required by law; this follows from the 1954 Supreme Court interpretation of the Natural Gas Act.[1] To consider policy, however, especially policy for the future, something more

*Professor of Economics, Southern Illinois University.

A number of persons contributed to this study through their perceptive comments, but special assistance was provided by Laurence A. Toenjes, formerly of Southern Illinois University and now of St. Mary's College, Moraga, California; Charles G. Stalon of Southern Illinois University; and the staff of the FPC. The form and substance of this study and its conclusions remain, of course, my responsibility.

[1] 52 Stat. 821 (1938).

than a legalism is required. The public interest in the natural gas industry must be defined so that a standard exists for evaluating the results of current and alternative regulatory methods.

In this study the public interest is assumed to be identified with efficiency in the allocation of resources both in production and in consumption. Policies are judged desirable insofar as they lead to maximization of potential economic welfare as measured by willingness to purchase the resulting total economy product valued in discounted present value terms.[2] This concept of the public interest ignores but does not preclude government action to redistribute income. It requires instead that income redistributive actions be identified explicitly so that alternative means of accomplishing the same end may be considered and their costs compared. Other possible elements of the public interest are similarly ignored in the analytical portion of this paper, not because they lack importance or relevance, but because they do not lend themselves to explicit or quantitative formulation. Implicitly, however, such considerations as environmental quality, regional development, and dispersal of economic power are very much related to this analysis, for in principle the strictly economic costs of achieving such other goals can be determined from it. Once such costs are estimated, then informed judgment can be applied to determine whether the goals sought are worth their cost and, in any event, whether preferable policies are not available to achieve the same ends.

Control of monopoly behavior, alteration of the intertemporal distribution of gas consumption, and redistribution of income are three commonly used arguments for the regulation of the producer segment of the natural gas industry. Each of these rationales is considered below, in terms of the public interest criterion stated above.

Regulation for Monopoly Control

The existence of significant monopoly power leads to misallocation of resources within and among industries, and hence the public interest (as defined) may require economic regulation of a monopolized industry. This conclusion is familiar and need not be demonstrated here.

The application of the "monopoly" rationale to producer regulation requires that market power of significant scope exist within the gas field markets. Surprisingly, this basic issue has not been readily resolved, because institutional and technological complexities in the industry make its analysis difficult. The conclusion reached in most studies is that while the industry does not fit the carefully drawn competitive model on either the demand or supply side, it is workably competitive. The pragmatic test of availability of alternative gas supplies and of alternative sales outlets is met for most pipelines and most gas producers during most periods. Nevertheless, random elements in both demand shifts and major reserve discoveries, long-term gas reserve contracts, immobility of pipelines, and long decision-investment-production lags do bring transitory periods of market power to nearly all producers and purchasers and perhaps virtually permanent market power to a scattered few. Such imperfections in the market most likely do not alter its basic configuration. Therefore, the most productive analysis of the industry would probably proceed, as this one does, on the assumption that generally competitive conditions hold, though with transitory and/or localized imperfections.[3] These imperfec-

[2] One limitation of this approach is important, though it may not be immediately evident. This goal involves comparative statics–not only does it fail to describe the path from some given point to another, but more importantly it does not take into account the possible alternatives foreclosed by choosing one path rather than another. Another limitation should also be emphasized. Throughout the analytical portion of this paper the problem of the "second best" is ignored. That is, in the absence of specific evidence to the contrary, it is assumed that allocation of resources more in keeping with the competitive norm results in an improvement in social welfare. Though the failure of actual conditions outside the industry to conform to the pattern of marginal cost equals marginal revenue is obvious, this approach still appears to be the best available starting place for analysis.

[3] Some industry observers have noted that, while the industry might meet the formal requirements for generally competitive behavior, the producing firms nevertheless may have incentives to alter their behavior in a way leading to noncompetitive results. Two possibilities are suggested: nonaggressive behavior toward substitute energy sources in order to protect investments in these industries and parallel behavior (perhaps unconscious) to restrict gas supply in order to foster the virtually unanimous producer goal of revision of producer regulation. Convincing evidence of the existence of such behavior has not been presented, but, admittedly, behavioral patterns of these sorts are virtu-

tions certainly may call for correction by specific government action but in themselves can scarcely provide justification for general control of the conditions for sale of *all* sales of natural gas in the field for resale in interstate commerce.

Regulation to Alter Use Rate

A difference between the discounted present value of a nonreproducible resource to society and to its private owners creates another possible rationale for regulation of the natural gas production industry. Such a difference would lead to a socially suboptimal allocation of the natural gas resource over time. The view that such misallocation results from the action of market forces has been expressed by some classes of consumers that have argued that consumption in some end uses was "wasteful" because the gas was underpriced in those uses.[4] The following arguments may be used to support this position: the assertion that the capital market is inadequate to present a viable waiting alternative to the private holder; that the social time preference is lower than that of the private sector; that some gas reserves are sold at marginal, rather than full, costs because they are jointly produced with oil; and, perhaps most tellingly, that the segmented ownership of a common pool of natural gas at best forces all owners to act on the highest discount rate among the individual owners and at worst leads to competitive exploitation. The important counter assertion is that the future value of natural gas reserves is overestimated because the market, for some reason, has underestimated the potential functional replacement or supplementation of domestic natural gas by alternative resources, such as atomic power, liquefied natural gas (LNG) imports, and coal conversion.

Should the socially optimal rate of resource exploitation be judged to diverge from the observed rate (neglecting decision and transaction costs), some governmental action to bring them into equality becomes justifiable. A judgment that the discounted present social value of the future use of the resource is higher than its current price would imply that the exploitation rate was too fast, and vice versa. Note, however, that price regulation is not a suitable instrument of use-rate control. A subequilibrium price that restricts the quantity supplied is, however, an important political adjunct to administrative rationing based on other goals entirely. A political case can more easily be made for restriction of gas use for some ends if there is not enough gas to satisfy all demands.

Regulation to Redistribute Income

The argument for field price regulation that has achieved perhaps the greatest public support is that income should be redistributed away from the producing sector and toward the consuming sector. In its crudest statement, this goal rests directly on the value judgment that it is better for consumers, even at the margin, to get something that is worth more to them than they have to pay for it than for the producing sector to get more for something than its cost of production. Since gas reserves themselves have no cost of production, only of discovery, a return of the original value of the gas reserve is thought to be unnecessary to retain productive factors within the industry. The conclusion is that the value of the "original endowment" can be transferred to consumers without loss in production. This view ignores the role of rent in the efficient allocation of an economy's resources. Because rent does perform allocative functions, income redistribution accomplished by ceiling price regulation has significant output effects and social costs.

The rental portion of the price paid for a natural gas reserve, except for the institutionally established minimum royalty rate (see below), is based on the differential value of one reserve as compared with the marginal reserve. An increase in price makes exploitation of previously submarginal reserves

---

ally impossible to document. The ambit through which such behavior could function, if it existed, is restricted both in magnitude and duration. All things considered, then, the undemonstrated possibility of a restricted amount of noncompetitive behavior, within a workably competitive industry structure, seems far too weak a reed on which to hang monopoly regulation of the entire industry.

[4]Often this issue appears in the guise of pipeline rate structure proceedings where it is alleged that rates discriminate against firm-load, residential, and commercial customers.

economic and hence increases the total rent paid to previously economically exploitable reserves. The differential *between* two reserves does not grow when price is increased, however, nor does it shrink when a lower price makes previously economical reserves noncommercial. With any elasticity in the supply function, then, a change in price changes both the total quantity of rent and the quantity of gas supplied, despite the fact that the full rent payment is not necessary to maintain the intramarginal reserves in production.

Positive rent, moreover, plays a role in the allocation of resources. First, the differential rent allocates the prospective natural gas reserves among prospective exploiters of those reserves by causing them to bid against one another for the right to exploit—presumably the firm that can make best use of the reserve can afford the highest rent to the reserve owner. Second, the payment received by resource owners at the margin of economic production (the minimum royalty, for example) indicates the market estimate of the discounted future value of the natural gas reserve. If price were reduced by the amount of this marginal rent, the use rate for gas would rise beyond the optimal level or else some other form of rationing would be required. Income redistribution, while to a limited extent attainable by means of ceiling price regulation, falls outside our constrained criteria for the public interest. As seen above, however, it can be evaluated in terms of its economic costs.

The Three Proximate Goals Summarized

The three proximate goals of ceiling price regulation can be evaluated when the public interest is identified as the maximization of economic welfare, as defined. The first of the goals, adjustment of the results of monopoly, is consistent with optimal allocation of resources. It is, in rough measure conceptually achievable through use of the ceiling price instrument. The condition of general monopoly power, however, has not been shown to exist. The second proximate goal, optimal intertemporal allocation of the natural gas resource, is likewise consistent with the public interest, but again it has not been demonstrated that the market results are faulty in this respect. More important, the chosen policy instrument would appear to have an obliquely positive effect at best and a distinctly negative influence on achieving this goal at worst. The income redistribution goal is outside the economic welfare criterion, though paradoxically the ceiling price instrument can achieve the redistribution goal. Success in this matter, unfortunately, is inconsistent and inimical to the satisfaction of the public interest in maximum economic welfare through production and distribution efficiency. Unintended redistribution within the consumer sector also results. In pursuit of each proximate goal the use of the ceiling price instrument is thus flawed, given the public interest as defined. By itself, however, recognition that use of the price control policy leads to inconsistent results does not lead directly to a rejection of ceiling price control. The larger policy decision revolves around the issue of *how much* gain and *how much* loss ceiling price control creates in this specific industry at some specific point in time. This topic is discussed later.

FPC FIELD PRICE REGULATION

The area price method of regulation of the field sales price, after a series of false starts and much litigation, was established by the late 1960.[5] Briefly stated, this regulatory method treats the entire natural gas industry within a given geographic area as the regulated unit and seeks to provide the industry, as it is thus circumscribed, with a price equal to the average cost of producing natural gas, including a fair return on industry investment. Individual companies are allowed to make contracts with any available buyers at any price at or below the regulated price for the area. No regulatory notice is taken of the extraordinary profits of the successful firm, or of the losses of the unsuccessful enterprise. Except for an inability to abandon a sales contract without permission (readily granted for depleted

[5] *Area Rate Proceedings*, AR 61-1 (Permian Basin), Opinion no. 468, 34 FPC 159; *Permian Basin Rate Cases*, 390 U.S. 747 (1968).

reservoirs), information filing, certification requirements, and some uncertainty as to the inviolability of the established ceiling (contract) price, the individual producing firm thus has an incentive structure similar to the opportunities and constraints of a firm facing a market-determined price. Consequently, the potential gas producer can consider his prospective costs and his anticipated physical yield from a given investment and, on the basis of the anticipated price, make his decision as to whether or not to invest in a given prospective gas exploration and discovery venture.

The FPC has developed a multi-price system within this general framework. So-called "old" gas is gas coming from reserves whose discovery, development, and production—in concept—were not influenced by federal regulation of field price. Associated gas (gas produced as a joint product with liquid petroleum) is treated as old gas because it shares the same incentive pattern. Old and associated gas are costed on a historical basis by areas. The goal under this procedure is a price that, for the group of producers as a whole, returns historically expended costs plus a reasonable return on capital. This is an "equity," rather than a "functional," concept of a ceiling price.

The ceiling price for new gas within an area has been determined by the FPC by estimating costs on a national basis and making some adjustments based on the area. The goal is a price equal to the current cost of producing gas. Current levels of rent payments are thus included in the "cost" of new gas.

The derivation of the ceiling prices described above is beset by difficulties, not the least of which is the analytical impossibility of allocating costs of a joint product. Natural gas and liquid petroleum are often produced from the same well; the allocation of costs between these joint products is strictly arbitrary. Even in the case of a field that produces only natural gas, some portion of the exploration cost must be assigned to liquid petroleum because of the non-zero probability that, ex ante, exploration was undertaken that could have led to petroleum production, and that, ex post, geological information elicited may be beneficial to future liquid petroleum exploration. The arbitrariness of the allocations required has been smothered in complex costing formulas that have been condoned legally because they are not obviously discriminatory or unreasonable, but no economic comfort should be taken from this fact.[6]

SUPPLY AND DEMAND FOR NATURAL GAS

Regulatory action works through and does not replace the market forces of supply and demand, and hence industry supply and demand conditions clearly have a significant bearing on industry regulation. "Supply" and "demand" in this paper are used in their formal sense: schedules of quantities that would be supplied or demanded, *ceteris paribus*, at various prices. The terms "quantity supplied" and "quantity demanded" will always refer to a desired specific quantity at some certain price. The unmodified word "demand" will refer to the pipelines' derived demand for a flow of gas production from the field. The word "supply" is defined in a similar fashion: the function relating price to the optimal annual production of natural gas from reserves that would exist at that price. This definition recognizes that it is the flow of gas that is important—not reserves per se—and that the rate of exploitation of reserves can vary due to changes in the price, interest rate, expectations, and so forth. Given these definitions, assumptions about gas demand and supply over the next decade can be made more explicit.

This paper has been drafted on the explicit assumption that there will be no significant cost breakthroughs in either natural gas production or in the competing segments of the energy industry.[7] Growth in gas demand will come from intensive, not extensive, expansion of markets, because most of the potential geographic markets are already being served. Most of the increase in demand in mar-

[6] Clark A. Hawkins, in *The Field Price Regulation of Natural Gas* (Florida State University Press, 1969), presents a cogent analysis of the cost allocation problem in natural gas producer regulation.

[7] Substantial treatment of prospective supply-demand conditions is found elsewhere in this volume. These brief remarks are designed solely as specific background for the policy discussion that follows.

kets already served will be due to higher per capita energy consumption, to higher population levels, and to possible net substitution of natural gas for other fossil fuels. The significance of this latter source of demand will depend on three conditions: the relative price of alternative fuels at the consuming points,[8] the level of concern for the environment, the cost of reducing air pollution from coal and heavy fuel oil and thermal pollution from nuclear power generation.

The supply of natural gas over the next few years—that is, the quantity that producers will have the reserves to produce when those reserves are being exploited in an optimal manner—will depend on a number of factors. The two major conflicing forces affecting supply, assuming mobility of capital and other resources, are changes in technology and the deterioration of the resource base. If cost-reducing improvements in technology keep pace with the deterioration of the resource base, assuming that interest rates and other factors remain the same, the same supply schedule would exist in the future that has existed in the recent past. This means that the same quantity supplied would come forth at the same price in the future as in the present.

To say that the supply curve may shift upward, shift downward, or remain the same in the future is to say nothing of course about the *shape* of the supply curve itself. Clearly the relationship between quantity supplied and price is positive. At any given time a higher price is necessary to make more gas available if the optimal relationship between quantity produced and reserves is not violated. For the analysis below, fortunately, more explicit identification of the supply function is unnecessary. The first premise of this paper, then, is the demand assumption that during the 1970s there will be a rather steady but not explosive increase in the demand for domestic natural gas. The second premise is that there will certainly be no decrease and possibly a small increase in the supply of natural gas from domestic sources. Given an upward sloping supply curve at any point in time, the *equilibrium* price of gas will *rise* if the demand curve (retaining the same shape) moves outward *further* than the supply curve.

THE FIELD MARKET MODEL: JURISDICTIONAL AND NONJURISDICTIONAL SALES

Analysis of the industry and of its regulation depends on understanding the way that the field market for natural gas operates.[9] This market, with its long-term contracts, joint output with petroleum, multitudinous permutable contract clauses, and significant levels of risk, is strikingly complex and baffling to the policy maker. Abstraction to a model that focuses on the relationships among demand, supply, regulated price, and the separable jurisdictional and nonjurisdictional demands allows the analyst to concentrate on some of the basic issues of relevance to the evaluation of regulatory alternatives.[10]

[8]The relative price of gas at the consuming point depends vitally on trends in the cost of transporting the gas. Four factors with possible contradictory influences will be at work in the gas transmission industry. First, a large portion of the early postwar interstate pipeline investment will be depreciated out during the 1970s, and the reduction in the charges for depreciation will be reflected in lower revenue requirements for the pipelines. Along with that development, however, there is some prospect that safety requirements will lead to replacement of large sections of the old gas transmission plant—at high cost. Moreover, much of the low interest rate debt of the pipelines will be rolled over at strikingly higher capital costs. This development will give a boost to the average cost of capital to the industry beyond that already occurring due to the higher cost of new issues. As a final trend factor, unit capital and operating expenditures will probably fall with increases in the transmission capacity of pipelines.

[9]The field market model presented in this section is formulated and presented in enlarged form in Milton Russell and Laurence A. Toenjes, *Natural Gas Producer Regulation and Taxation* (Institute of Public Utilities, Michigan State University, forthcoming).

[10]The Natural Gas Act states that the FPC will regulate the "sale in interstate commerce of natural gas for resale for ultimate public consumption for domestic, commercial, industrial, or any other use, and to natural-gas companies engaged in such transportation or sale. . . ." (52 Stat. 24 [1938]). As interpreted, the total language of the act has been held to deny to the FPC jurisdiction over the sales of gas for intrastate consumption and also to deny jurisdiction over direct industrial sales by the producer to the ultimate consumer. Nonjurisdictional sales have made up approximately half of total gas sales in recent years, as measured by delivered volumes.

The model below is based on the assumption that there is a single supply function. This assumption means that a significant number of producers, if not all, are indifferent to whether sales are jurisdictional or nonjurisdictional; thus, they sell their gas for the best contract terms available in either market. Price, in this context, is the value of the package of conditions found in the contract, evaluated in cents per thousand cubic feet (Mcf) not merely the stated Mcf price. It is also assumed in this model that transmission facilities exist that interconnect consuming and producing areas nationally. While the gas produced in each area might flow to particular regions because of cost advantages or historical relationships, at the margin, interregional transfers are possible, and hence a change in supply or demand in one location soon is felt throughout the system.

Total consumption demand is divided into jurisdictional and nonjurisdictional demand components in the comparative static field market model used here. In figure 9-1, *AM* and *BR* are individual demand curves representing, arbitrarily, nonjurisdictional and jurisdictional demand for natural gas. These functions have no empirical content except that they follow the premise of economic behavior and exhibit negative slope—more gas is demanded at a lower than at a higher price, all other things held equal. *BGU* is the horizontal summation of *AM* and *BR*, and hence the total demand for gas.

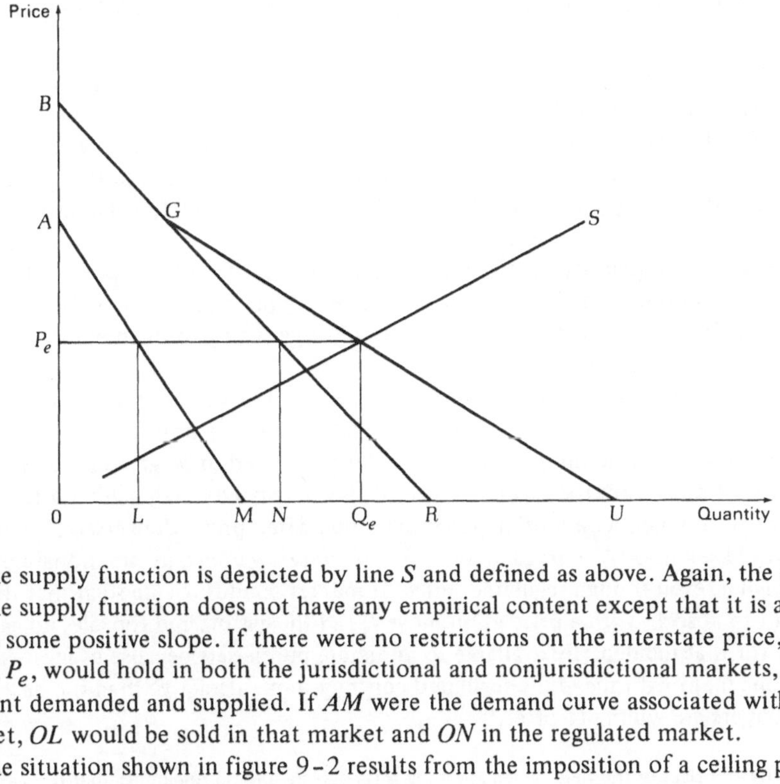

FIGURE 9-1.

The supply function is depicted by line *S* and defined as above. Again, the shape and form chosen for the supply function does not have any empirical content except that it is assumed that supply exhibits some positive slope. If there were no restrictions on the interstate price, the market-clearing price, P_e, would hold in both the jurisdictional and nonjurisdictional markets, with Q_e the total amount demanded and supplied. If *AM* were the demand curve associated with the nonjurisdictional market, *OL* would be sold in that market and *ON* in the regulated market.

The situation shown in figure 9-2 results from the imposition of a ceiling price for jurisdictional gas, P_r, below the equilibrium price, P_e. The resulting total quantity demanded, Q_d, is greater than quantity supplied, Q_s.[11] P_e continues to refer to the market-clearing price that would be established in the absence of an effective ceiling price, and thus it is determined simultaneously by the intersection of the supply function and the combined demand function (*BGU*).

Gas is sold from most of the producing states in both the jurisdictional and nonjurisdictional markets, and sellers dedicating new reserves are free to sell into either market as they choose.[12] If it is

[11] Other papers in this volume document the existence of excess demand in this industry during the late 1960s and through 1970.

[12] The one major exception to this freedom is production from federal offshore reserves, which by definition must be interstate and therefore jurisdictional.

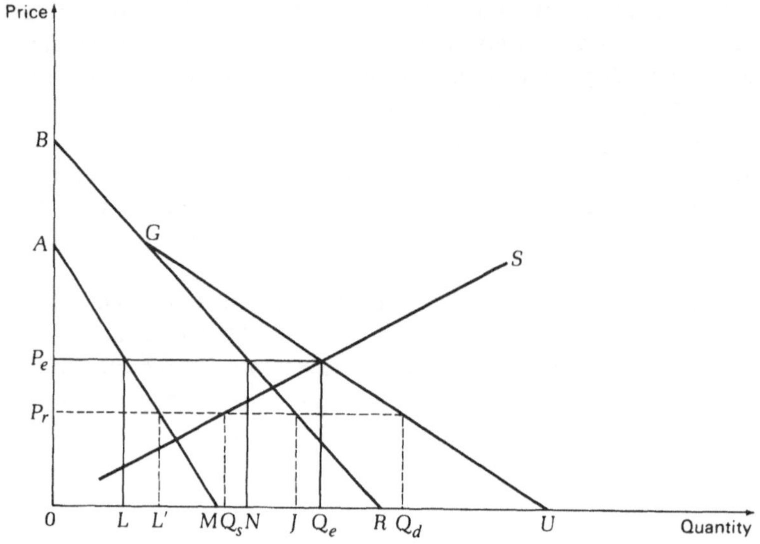

FIGURE 9-2.

posited that P_r is less than P_e and that gas is sold in both markets, virtually the same price, P_r, will exist for both jurisdictional and nonjurisdictional customers of a given state, subject to possible slight price differences associated with the ascertainable seller costs of selling in a regulated environment. Thus, in figure 9-2, with P_r below P_e, amount OQ_s will be supplied, with OL' going to the unregulated market and the remaining amount, $L'Q_s$, being available for purchase by jurisdictional consumers. Note particularly that whenever there is excess demand (due to P_r below P_e), the supply function together with P_r determines the total quantity sold; it is not determined by the intersection of the supply function and the demand function. The imposition of a ceiling price on a previously unregulated market, *expands* both absolutely and relatively the amount of gas going to the nonjurisdictional market and has the opposite effect on jurisdictional sales.

The mechanism for nonregulated price convergence toward the regulated price (restricting the analysis always to the situation where gas is sold in both markets) follows the usual pattern: if the unregulated price, including the value of nonmonetary contract terms, tended to be greater than P_r, producers would, over time, shift sales from the regulated into the higher-priced unregulated outlets. The resulting increase in quantity supplied would bring down the unregulated price. Conversely, no profit-maximizing producers would sell into the unregulated market if, *ceteris paribus*, jurisdictional purchasers were offering a higher, even though regulated, price. If market conditions are such that the nonjurisdictional market can absorb, with a price at or above P_r, all the gas offered for sale, no gas will be sold interstate except for that incidentally available to interstate purchasers because of legal restrictions or peculiar conditions that might make the regulated contract more attractive than the best terms nonjurisdictional buyers are willing to offer.[13]

[13]Observers of actual contract prices in the field, when applying this analysis, should remember that it is cast in the comparative static equilibrium framework. A change in supply, demand, or regulated price can leave the general level of nonjurisdictional price significantly above a regulated price for the period of adjustment. In practice, adjustment may take several years because of the long response lags in this industry. During this adjustment period one would expect to find most new gas sales being made to nonjurisdictional purchasers. If sufficient reserves cannot be dedicated to the nonjurisdictional market to bring the equilibrium price in that market down to the regulated price, then the resulting unregulated price will settle somewhere between the regulated and the unconstrained equilibrium price. *All* gas will thus go to the nonjurisdictional market. *Note*: Price is used as a proxy for the expected value of all elements of the gas sales contract; comparison of contracts on the basis of cents per Mcf alone, without evaluating delivery conditions, penalty clauses, timing, location, etc., can be misleading. Producer expectations as to future increases of the ceiling price, for example, may raise the expected real value of a jurisdictional contract written with escalation clauses tied to ceiling price.

The nonjurisdictional purchasers are thus able to purchase all the gas they desire at or slightly above the regulated ceiling price, if gas is sold in both markets, and at least at a price below the equilibrium price if it is not. This nonjurisdictional market does not suffer any long-term regulation-created shortage. The jursdictional market, on the other hand, is left with only the gas that is not desired by the nonjurisdictional buyers at the regulated price. In figure 9-2, under regulation, the jurisdictional buyers want to purchase *OJ*, a desire that is frustrated by lack of gas. Thus, the imposition of a ceiling price below the equilibrium price leads to a *reduction* in total gas consumption, a *shortage* in the jurisdictional market, and an increase in sales to nonjurisdictional consumers. The effect of this regulation-induced gas shortage on the total efficiency with which the nation's energy resources are allocated is considered below.

THE EFFECT OF FPC AREA PRICE REGULATION ON ECONOMIC WELFARE

Maximization of economic welfare was identified above as the general goal of public policy with reference to the natural gas industry. Other worthy goals were not rejected but simply ignored except as they are implicitly valued in terms of the economic welfare that must be given up in their satisfaction. Welfare maximization was taken to be implied by the maximization of the efficiency of the use of factors and of the distribution of products, both established by the competitive market test. While such a static and partial framework is subject to many varied and well-known defects, it is used here to provide a starting point for policy analysis.[14]

In line with the discussion on public interest and price regulation, allocational benefits can be expected from ceiling price regulation insofar as the deleterious effects of monopoly or monopsony in the field markets are reduced. The competitiveness of the field market for gas is examined elsewhere in this volume, and thus attention will not be given the subject here. It is sufficient to note that to the extent that the existence of monopoly and monopsony is transitory and/or localized, the benefits from its control by means of industrywide ceiling price regulation are similarly small.

The allocational and welfare effects of regulation other than those reflected in restriction of market power require considerably more comment. In the following discussion, consideration is given first to the effects of a sub-equilibrium ceiling price and then to the effects of regulation even if quantities supplied and demanded are equated.

Welfare Effects of a Sub-Equilibrium Price

The existence of regulation-created excess gas requirements has a serious effect on the efficiency with which the nation's resources are allocated. The discussion above regarding the field market model demonstrated that all the shortage would be focused on jurisdictional consumers. Those consumers who want an assured gas supply must, therefore, obtain nonjurisdictional status for their supply. Noneconomic shifts in plant location to producing regions is thus encouraged, resulting in less efficient use of resources and hence in a reduction in total economic welfare.

The market-clearing price is higher than the regulated price when the ceiling price is effective. Consumers who *do* get gas thus pay less for it than it is worth to some nonconsumers who, because of regulation, are precluded from bidding it away. The level of total satisfaction to the nation is reduced

[14] As one example of the defects of this framework, significant departures from the Paretian optimality conditions in this industry prevent the analyst from saying anything unambiguously about welfare effects of further movements toward or away from those conditions. The past effects of state production and conservation regulations on liquid petroleum production and the effect of continued import restrictions on crude petroleum are only two such matters that come to mind. Unfortunately, the discussion of the "theory of the second best" has failed to do much to clarify the policy muddle into which the policy maker has been thrust in these circumstances. While recognizing the implications of nonoptimality elsewhere, in this paper the welfare assumption is that movement toward the Paretian conditions will lead to greater welfare *except* when the movement away is deliberately chosen.

by the difference between the production forgone in higher value uses and the production achieved in lower value ones. This distortion and hence loss is maximized when the industry simply denies gas to new customers while providing old customers with all the gas demanded at the regulated price.

The final efficiency effect of a regulation-induced shortage of natural gas is the substitution of less desired energy sources or other alternatives for the gas not produced. The unsatisfied desire for energy from gas will be fulfilled by some other energy source, or alternative consumption items will be substituted for it. The inferior substitute will, of course, result in total welfare being reduced as compared with the production and consumption of appropriate quantities of gas. This is another way of viewing the misallocation of the gas reserve over time. Because of the restriction on quantity supplied, a suboptimal amount of gas is consumed in the present.

The policy issue for the future is whether a gas shortage, with its many misallocational effects, is inherent in area price regulation or whether it was instead a product of particular circumstances in the late 1960s and early 1970s. The short answer is that a gas shortage would appear to be a continuing tendency under area price regulation. Two factors combine to yield this conclusion. First, it is unlikely that productivity increases will be sufficient to offset both the deterioration of the resource base *and* the continued growth in demand. Consequently, the real cost of natural gas is likely to rise over the next decade, and regulatory lags are likely to leave the ceiling price, on the average, below the equilibrium price. Second, regulatory uncertainty in the industry tends to make producers underestimate their profit-maximizing level of output. This last point deserves further comment.

Investment in the exploration and development of natural gas production properties is relatively risky. The producing firm faces a broad array of possible outcomes in physical product terms from each investment that it makes. A supply function represents, then, the estimated value of the unit marginal cost at each price, with a large stochastic element. To this already risk-sensitized industry, regulation adds uncertainty, because the firm has little guide to the future regulated price. The producing firm, since 1960, has never been certain of the price it would ultimately receive for gas found, contractually sold, and even transferred. The firm has never been absolutely sure that the FPC would not at some future date declare previously established prices suspect, lower them, and subsequently require refunds.[15]

The quantity of investment that will come forward at a given regulated price depends on the expectations of future revenue from that investment. Given uncertainty, and with negative experience as a guide, the expected value of the regulated price will be lower than its nominal value, and the industry will produce less than the amount that would be consistent with its supply function and the regulated price.

Whether or not the industry's expectations regarding regulation are verified in practice makes no difference in the short run. In the longer term, it could be presumed that the failure of the negative expectations would bring excess profits into the industry; the existence of these profits would alter the behavioral function relating the observed and estimated ceiling price. Until that time, however, even if the ceiling price is exactly equal to the equilibrium price, it is likely that there will be excess demand created.[16]

It is not entirely germane at this point, but it might be noted that a shortage of gas leads to mixed results in equity terms. Clearly, those consumers who *do* get the gas at the regulated price benefit by

[15] There is some evidence that the industry considered the lowering of the interim area prices in just this light. Resources had been committed to exploration programs on the basis of an interim regulated price, and the industry suddenly found that the product of the exploration already undertaken would be allowed an even lower, rather than higher, price. Notwithstanding the legal niceties and certainly adequate forewarning, the industry *felt* betrayed and perhaps has reacted accordingly.

[16] On this point see Edmund S. Mills, *Price, Output and Inventory Policy* (John Wiley and Sons, 1962), pp. 98–99. Mills dealt with a slightly different problem, but his conclusions are consistent with the argument above. He stated: "There is also a moral for the theory of price control policy to be learned from this analysis, though it is not restricted to the example considered in this section. . . . The riskless model tells us that the monopolist will behave like a perfectly competitive firm by producing and selling the amount that makes marginal cost equal to price. In

the saving engendered. Those consumers who are denied gas, however, lose an amount equal to the difference in value to them of the substitute and the gas they would have had in the absence of the regulation-induced shortage. The producers of gas, as a whole, have their incomes reduced because of the decline in price and output, but for the mobile factors in the industry the loss is only the difference between the return in the gas industry and in the next best occupation. For the owners of non-mobile factors, however, rent is squeezed out at the margin and a net redistribution of capital toward the consuming sector is accomplished. No conclusion can be drawn as to the redistribution effects between income levels. The social desirability of these shifts is a matter of values, and no unambiguous conclusion can be drawn about the total social impact.

Area Price Regulation and Resource Misallocation

The area price method leads to a misallocation of resources even abstracting from the welfare losses associated with a gas shortage. This misallocation results from the failure to recognize the different values of different gas reserves within pricing areas, from the failure to reflect interarea value differentials appropriately, and from the use of the multi-price system.

Let us consider the first of these conditions—differences in the value of natural gas at the wellhead within a given pricing area. The value of a particular reserve of gas to a particular pipeline within a region is a function of its location vis-à-vis a pipeline (the farther away, the greater the gathering cost to the pipeline), its pressure, its size (since large deposits are less costly to process, contract, and to gather and transport), its rate of allowed variability of depletion, and the impurities and British thermal unit (Btu) content of the gas entrapped. The value of these characteristics can be different for different potential pipeline purchasers according to the conditions exhibited by those pipelines. In a free market, each of these reserves presumably could command a price equal to its value to the various pipelines, and in the bargaining process contracts would be fashioned that would match reserves and pipelines in an optimal fashion. In consequence each reserve would presumably be priced somewhat differently and not uniformly.

Consider now the effect of the imposition of a ceiling price equal to the average cost of production within the area. The effect of the ceiling price will be to restrict the level of output from high-value reserves. Both in a regulated and in an unregulated environment the producer of high-value gas would determine his level of output by equating his rising marginal cost to marginal revenue (price in either competition or regulation where the ceiling is below the equilibrium price); when the full value of a unit of his gas is not reflected in the regulated price, his output will be reduced. The demand for gas from the low-unit-value reserves will be shifted upward due to the shortfall in supply from high-unit-value reserves, and consequently more gas will be produced from the low-value area. Factors will thus be shifted to less productive uses, and the total want-satisfying power of the economy will be reduced.

Welfare losses from the regulatory failure to reflect appropriate interarea differences arise from essentially the same conditions as those discussed above. The value of a gas reserve to the ultimate consumer depends on the total cost to the consumer *at the consuming point* of a Btu of heat from that reserve. Location is a particularly important determinant of this cost because of transportation expense. The use of national cost patterns for new gas and historical costs for old and associated gas have created a pricing pattern that eliminates some of the quantity supplied from high-cost (but high-value) reserves, and forces the jurisdictional consumer into greater transportation and other costs.

While failure to recognize interarea value differentials has serious welfare implications in the existing producing regions, potentially even more serious implications arise because of the disincentive to invest in high-risk but potentially very valuable locations. Efficient allocation of resources demands that a trade-off be available to the producer between greater risk with a higher-value-based price, and lower risk with a lower-value-based price.

the risky situation, however, . . . the monopolist will produce only the amount he is certain to sell. This is sure to imply shortage on the average. . . ."

The final source of regulation-induced misallocation of resources is associated with the multi-price system.[17] Two allocational losses arise from the multi-price system. First, in this era of rising prices the average price of gas to the consumer is lowered below the marginal cost of its production; hence "too much" is consumed in the present. Second, the consumer whose pipeline receives proportionately more of its gas from an old gas area will consume too much, while the consumer from a predominantly new gas pipeline will consume relatively too little.

Optimal intertemporal allocation of any diminishing resource requires that proper allowance be made for its value in future use—and hence rent plays a proper allocating role. In this sense, the rent received by producers is a proxy for a portion of the social cost of consuming gas at any point in time. When that rental return is frustrated—as it is when old and associated gas prices are held below the current value of the resource, and even below that price for which the parties contracted—the optimal allocating rule of marginal social benefit equal to marginal social cost is violated, and too much is consumed.

Optimal allocation of gas among uses would lead to its contribution to want satisfaction being equal in each alternative at the margin. With differential pricing based on the date of dedication of a contract or on the type of well from which the gas is produced, there will be differences in prices in different consuming markets that have no economic justification. In these circumstances, the consumer with low-priced gas could extend his use of gas further at the margin than could the consumer with high-priced but identical gas. Since equal value would not be derived from each use, again optimal allocation conditions would be violated.

Regulation and Allocational Efficiency

Area price regulation clearly has made some contributions to economic welfare by improving the allocation of resources through some restriction of monopoly power. The discussion above demonstrates, however, that these regulatory gains have come at a considerable cost. The regulatory gas shortage has led to both social welfare loss and to unintended welfare redistribution. Even without the creation of a shortage, rigid ceiling prices have led, and undoubtedly would continue to lead, to distortions in the allocation of resources.

Since the current regulation has been shown not to be an unalloyed good that in every way is optimal, alternatives must be considered. While there are a multitude of possibilities, consideration of three seems to be indicated at this point. The first alternative would be to reform the area price procedure to avoid some of its most costly, least productive provisions. The second alternative would be to restructure regulation to eliminate its greatest failures in optimal resource allocation. The final alternative would be virtually to deregulate the industry altogether. Elements of each of these alternatives will be considered briefly in turn.

REFORM OF CURRENT AREA PRICE REGULATION

One policy decision open to the nation is to continue area price regulation in essentially its past form. This judgment would rest on the conclusion that the noneconomic benefits of area price regulation, when combined with its (very small) economic benefits, outweigh both its other costs and the losses it causes in economic efficiency. The noneconomic benefits sought need not be catalogued and evaluated here. They might result, perhaps, from such matters as approval of the resulting regulatory redistribution of income, a desire to substitute political for market allocation of the gas reserve by end use, a desire to maintain an instrument for planning control over this sector of the energy market,

[17] A description of the multi-price system and the justifications for it can be found above and in other papers in this volume. While attention is most often focused on the vintage of gas as creating price differentials under this system, a lower price for associated gas has perhaps more important misallocational effects in this context.

or merely from the feeling on the part of the general public that this industry should be regulated. However made, in policy terms the decision to retain the essential elements of area price regulation carries with it the obligation to maximize its benefits and to minimize the resulting welfare losses.

Several opportunities for improved regulatory performance exist within the basic structure of area price regulation. Two sources of resource misallocation appear amenable to relief: unnecessary producer uncertainty and inordinate regulatory lag and expense.

The major element in relieving unnecessary producer uncertainty would be to assure producers that there would be no future reductions in ceiling prices already legally established. It is very unlikely, given the prospective future energy situation, that cost-based regulation could ever justify a reduction in ceiling prices. The *cost* to consumer interests of a guarantee not to do what probably would not be done anyway is very small. The *value* of such a guarantee to producers may be appreciable, both as a symbol of good faith and as a relief from real anxiety. The result would be an increase in the supply of gas and more efficient allocation of resources.

Regulatory lags create a second form of uncertainty and lead to both misallocation and unnecessary regulatory expense. The prospective producer is uncertain of the future price of gas at the time he must invest in exploration and development for future production. Not only must he forecast market conditions (which will place an upper limit on his price), but he must also forecast the prospective behavior of the regulatory agency. This multiple uncertainty leads to reduced supply. A policy outlining criteria for future ceiling price determination would simplify and speed final determination of the ceiling price and reduce uncertainty. The argument against such a policy is that it is rigid in the face of changing conditions. The rigidity issue would be partially resolved if there were included an amendatory procedure requiring an appropriate interval for planning stability—say, four years—before the criteria could be changed, but allowing changes then.

An index-plus-trend approach might fit the criteria sought above and reduce the direct costs of regulation as well. Another alternative would be some scheduled rate of increase in the regulated price to account for cost increases due to the expected relatively greater increase in demand than in supply over time. This approach could perhaps also be effective in preventing the development of gross discrepancies between the quantities supplied and demanded, while retaining regulation premised on costs of production. Any such increase chosen would have to be less than the carrying cost of gas to prevent speculation with gas reserves in the ground. Periodic review would be required for assurance that the rate of increase remained appropriate to developing cost conditions.

In the context of price revisions under the area price formulation, there is one component of the cost of gas that might be considered for more frequent, if not periodic, revision. This sensitive factor, capital cost, is an important portion of the final price of gas. The cost of capital is particularly useful as an indicator for price adjustment for three reasons. First, external shifts in the cost of capital are observable in the relatively free and well-functioning capital market. Second, though difficult to observe, any shifts in relative capital cost to the gas production industry would indicate a market evaluation of prospective ceiling prices. A change in the differential between the basic cost of capital and the cost of capital for natural gas producers might indicate a need for a revision in the regulated price of natural gas. Finally, most of the costs of production of natural gas move along with the costs for liquid petroleum. The mobility of capital within the petroleum industry is clearly greater than it is within the economy as a whole. Therefore, should, the economy-wide cost of capital rise, it is likely that a prospective relative rise in petroleum prices would bring a shift of investment toward potential liquid petroleum reserves unless the capital cost allowance in the natural gas price were adjusted. The capital cost component of the new gas price is thus particularly sensitive and deserves special regulatory treatment because it is volatile, important, observable, and crucial in making producer investment decisions.

The essence of the changes suggested above is that area price regulation can be improved—quantity demanded and supplied more nearly equated and resources better allocated—without a drastic alteration in the current regulatory distribution of income and without a reduction in control of monopoly.

These changes would not be fundamental, however. The benefits and economic efficiency costs of cost-based area price regulation would remain essentially those discussed earlier.

RESTRUCTURING OF PRODUCER REGULATION

Restructuring of regulation would be necessary if a policy decision were made to emphasize economic welfare while retaining control over the price of jurisdictional gas. This policy would rest, of course, on a judgment that the additional economic well-being gained would be more valuable to society than the benefits given up in a shift away from cost-based regulation. The first goal of the regulation to achieve this aim is to maximize efficiency of operation of the industry. To that end, administrative fiat is replaced where possible by a market test; where fiat unavoidably remains, decisions reflect the results to be expected from quasi-market forces as a proxy for optimal allocation.[18] The second goal is to avoid or to minimize monopolistic exploitation of consumers, while recognizing that there undoubtedly will be supranormal returns to some factors.

The focus of regulation under this program would shift from costs in the field to value at the consumption point. A restructuring of the incentive system to producers would be in order to minimize the real social cost of gas in consumption. Such incentives would be designed, first, to bring forth the "right" amount of gas and, second, to get it at a minimum cost. Area price regulation as thus conceived would be divided into the two functional problems of determining interreservoir differentials and of determining how far out on the margin reserves should be exploited, that is, what the base price for gas should be.

Interarea Differentials

The most important element in determining the value of a given reservoir to consumers is its location.[19] The ultimate consumer must pay for the service of transporting natural gas from the point it enters the transmission line until it reaches the burner tip. The length of the journey is the major factor in establishing the cost of this transportation. Therefore, the value of a unit of gas is inversely related to distance from the ultimate consumer, by a factor determined by unit-distance transportation costs. Gas transmission facilities are also fixed in space and therefore cannot be moved in search of new reserves. A new deposit of natural gas near an existing pipeline that has excess throughput capacity would be worth more to the ultimate consumer than would that same field the same distance from the consumer but isolated from an existing transportation route. Therefore, the value of reserves depends somewhat on the historical accident of pipeline location.

The locational value of gas from a particular area to a particular consumer can be reasonably accurately ascertained. This does not in any way, however, similarly establish a value of potential gas production from that area to the abstraction called the "gas consuming public of the United States." Gas consumers as a whole are found throughout the United States, and for each subgroup of the gas-consuming public, location will likely dictate a different value for any given undedicated gas reserve. Therefore, while location conceptually is one of the determinants of the "value" of a given gas reservoir, unambiguous estimation of that value could break down because of the nonunique character of the relationships between gas supply and gas consumption areas.

Fortuitously, the geographical distribution of the major gas reserves and consumers makes a modification of this location value differential approach possible with only a limited sacrifice of clarity and equity. The major gas-producing regions within the major gas-exporting states are arranged in a roughly symmetrical, fan-shaped pattern, with the South Texas deposits forming the handle of the fan. Ignor-

[18] Efficiency is equated with the market allocation of resources only when the factor and product markets are workably competitive throughout the economy. Market imperfections in this and related industries obviously make the association of the market test and optimal allocation less than perfect but close enough to be useful.

[19] Other factors also affect the value of a reservoir but are not amenable to area price regulation. To this extent, resource misallocation could continue under even this restructured regulation.

ing the nonjurisdictional markets within the region, the largest consuming areas are mostly located outward from all producing portions of the region. A reasonably satisfactory locational differential would thus be one that just offset the transportation cost between any existing or potential producing area and the locationally marginal area actually tapped for jurisdictional sales.

The Base Price for Gas

The basic area price for gas, as noted above, must reflect the supply cost of sufficient gas to meet the quantity of gas that consumers are willing to purchase at the resulting price at the burner point. The base price is rather easy to define, but the practical application of this concept to the determination of an actual price is another matter. Not only are some of the economic variables required for this determination unknown, but some are unknowable. Ascertained current costs of production in the base supply area, as adjusted for cost trends and judgmental factors, might be an appropriate starting point.

Revisions of Area Prices

Revision of both the interarea differentials and of the base price may be necessary as conditions change. The differentials would likely change only because of large enough discoveries outside the gas-exporting regions to alter the whole character of the supply situation. Hence, only infrequent revisions would be required. Review of the base price for gas could proceed on the basis of observed contract rates.[20] Under price regulation designed to maximize economic welfare there would be no attempt to keep the ceiling price lower than the equilibrium price. At the equilibrium price in a free market there should be some distribution of contracts with prices both above and below "the" equilibrium price. If contract analysis showed that in the base price regions virtually all contracts were written at the ceiling price, it could be assumed that regulation was artificially truncating the expected price distribution and therefore that the ceiling price was too low. On the other hand, should relatively few prices be written at the ceiling price, there would be evidence that the ceiling price was not effectively cutting off such monopolistic gains as existed. The base price might then be revised downward. The necessary revisions in the base price would likely be small *once* the ceiling price was brought closer to the equilibrium price than it appeared to be during 1970.

To minimize their peculiar problems of uncertainty, producing firms under this regulation might well be guaranteed that investment decisions based on expectations of a given regulatory price would not be aborted by administrative fiat. The firm would thus be given some protection against the regulator, if not against shifting market forces. Reasons for this apparently one-sided protection were discussed above. An appropriate policy to fulfill this goal would provide that any downward revisions in ceiling prices would not take effect for some period, such as three years.

The Multi-Price System

The issue that has not yet been considered is the ceiling price for old and associated gas. It was pointed out above that effective allocation of resources in an economy depends on the consumers each paying for a commodity approximately its marginal social cost. To the extent it is required by regulation and not by free contract, the differential pricing of gas not only misallocates resources over time and among consumers, but it also transfers income with no obvious policy justification. While on the surface there may be an appealing equity argument that producers who have obtained gas reserves at a low cost should not be allowed to profit from the social appreciation of reserve value, there is a similar argument that particular groups of consumers should not obtain consumer surplus on the basis of the time at which they began to purchase gas. The same issue has even more force when applied to associated gas because here it is strictly fortuity that dictates which consumers will be benefited. Not

[20]Contract analysis might also be expected to turn up patterns of monopsonistic pricing, which, if found to be significant, might call for government response.

even contract time is a factor. If the income distribution issue is thus a standoff, then allocational efficiency becomes more appealing as the goal to be sought, and a one-price system is allocationally superior to the multi-price scheme.[21]

FIELD PRICE DEREGULATION

The problems with producer regulation, even if it were reformed or restructured in the manner outlined above, have led to renewed consideration of the deregulation of the field price of natural gas. One variant of such a policy can be evaluated in terms of its effect on economic welfare and on other policy goals. Deregulation of the gas field price that is sufficient to remedy most of the problems with current regulation would not require the dismantlement of all aspects of producer control. Since any deregulation proposal must be posited on grounds that workable competition exists in most markets at most times, some efforts to monitor the competitiveness of market behavior must be retained. Continuing elements of an appropriate regulatory process even under "deregulation" would include:

1. An information reporting system so that data continued to be made available on jurisdictional field sales.
2. A system of continuing surveillance to ascertain noncompetitive behavior. This system might well include automated contract analysis that, along with the values determined by the commission, would signal if a contract were out of line with expected market behavior. Investigations could then be launched to determine whether the observed variations from expected values are explainable by market conditions.
3. Supervision of contract terms where other than arm's-length bargaining might exist between the parties.

Even if these requirements were retained, the elimination of broad gauged area price regulation clearly would have important effects on the industry. These can be estimated broadly as the reversal of the distortions created by regulation, as a redistribution of economic welfare, and, to the extent anti-monopoly surveillance fails, as an increase in monopoly rents and concomitant reduction in output. Deregulation of the industry under conditions of gas shortage would bring short-run changes due to the existing disequilibrium; these change-over effects should be separated from the effect of deregulation of a market in equilibrium.

Short-run Effects

The immediate effect of deregulation in the presence of a gas shortage would be a substantial increase in the price level on undedicated reserves and an upward revision in the prices of old contracts for which such revisions are possible. A spot market for gas production would likely develop quickly to help smooth the transition for firms temporarily either short or long on gas. The quantity supplied would presumably increase somewhat because of the greater price incentive to expand production at both the extensive and the intensive margins. In response to the higher price, the quantity demanded by the market as a whole will fall, though contractual obligations and investment in gas-using capital equipment would again force a lag in the adjustment. The effects in particular gas-using and gas-producing regions and in submarkets will perhaps be different than for the industry as a whole but no less significant.

Most of the short-run response to deregulation thus would be dissipated in price changes. Time would be required before the underlying structure of both the producing and the consuming sides of the industry could respond to new incentives. Over time, of course, prices can be expected to settle down to a point above the level to which they were restricted but below the initial increases.

[21] If the nation decides that it wishes to reduce the relative income of the gas-producing portion of the public it can do so with less welfare loss by using other techniques. A regulatory process that operates on price at the margin is a particularly clumsy social instrument to accomplish income redistribution.

Long-run Effects

Industry adjustments to deregulation in the longer run, on the assumption that monopoly elements are small, transitory, or controlled, will lead to an improved allocation of resources and a redistribution of income. This means a higher total amount of economic welfare, shared somewhat differently by the members of the economy. The improved allocation of resources will result from the reduction of noneconomic factors affecting production and consumption decisions. No purpose would be served by attempting to enumerate here all the possible ways in which price control has led to lower real income, but some examples (in addition to those presented above) might be helpful.

Exploration and development investment decisions in a deregulated industry would be based on anticipated value to potential consumers of any reserves found, as well as on expected physical output and cost. One would expect, then, a considerable shift in exploration toward higher-risk, high-value areas, and a relative reduction in the exploration for gas less valuable at the prospective consuming point. On other levels, the reduction in regulatory uncertainty will shift the gas supply curve to the right because uncertainty can be considered a cost to the gas producer. In sum, at any given price *more* gas would be produced without regulation than with (ignoring monopoly), not only because resources will be allocated more efficiently, but also because more resources will be committed to the industry.

On the consuming side, the relocation of gas-using firms to gas-producing regions to avoid jurisdictional supplies would tend to be reversed because there would be less reason to fear a gas shortage in the nonproducing states. The development of substitutes for gas paradoxically might be slowed by the higher gas price under deregulation for the same reason. Some of the impetus for developing gas substitutes has come from fears of gas shortage, not from the natural gas price. Deregulation will allow the market to clear, and the equilibrium price may well be below the cost of gas from substitutes or from nondomestic sources. The incentive to develop alternative energy sources will consequently decline. On the other hand, higher gas prices would tend to force some bulk fuel users out of the gas market. Since the range between the price of environmentally inferior fuels and gas would be widened by the increase in gas prices, further incentives will be generated for development of pollution control devices and higher costs for their use justified.

The effect of deregulation also would be to shift income from some consuming groups to others, from consumers as a whole to producers as a whole, and from consuming states to producing states. This redistribution of income does not automatically imply a concurrent redistribution upward by income level, but the likelihood is that such redistribution would result. Some potential consumers will be made better off by deregulation because under regulation they were unable to get as much gas as they wanted. Others of course will be harmed, but again the total negative redistributional effect will be partly offset by the increase in total welfare due to improved energy industry efficiency. These redistributive effects will be further moderated by tax effects: in general, the income tax and any other non-lump-sum tax will share a portion of the change in income in each direction.

The long-run effect of deregulation is likely to be an industry that will be able to respond more flexibly to changes impinging from shifts in its environment. This flexibility implies more variations in observed industry results but also a lower expectation of really cataclysmic change. The initial effects of deregulation depend on the degree of distortions that have built up in the industry and on its ability to adjust smoothly. At present (1970) the distortions apparently are beginning to get serious. High capital intensity exists, along with long decision-investment-production lags, and hence the initial effects of deregulation are likely to be large and, to some portions of the community, painful. The ultimate effect of deregulation would be, however, an increase in economic welfare, as defined, for the community at large.

CONCLUSIONS

Public policy toward the natural gas industry, as reflected in field price regulation by the area price method from 1960 to the present, has led to misallocation of resources and to gas shortages in the jurisdictional consumer markets. This regulation has also, however, led to redistribution of income among various sectors of the economy and apparently has satisfied other policy goals. The gas shortage and misallocation were minimal until about 1968 because sufficient flexibility existed within the industry to adjust to the lower imposed price through reserve drawdowns and other factors. The costs of pursuing the existing regulatory method have become increasingly high over time, and the benefits received have become less demonstrable. Hence, it now appears there is no question but that consideration of some revision is in order.

Three alternative revisions of field price regulation have been examined here, using the criterion of optimal resource allocation. The first of these was some reform of the established area price method; the second was a basic restructuring of regulation to reflect the value of gas at the consumption point; the third was deregulation of price while retaining control over noncompetitive behavior.

Even with modifications suggested to reduce producer uncertainty and to minimize time lags, the cost-based area price scheme leads to very significant economic losses to the community. Most obviously, a gas shortage has arisen under essentially this same regulation, and there is no indication that regulation based on costs can, in the future, adjust to changing market conditions to prevent its development again. It is strikingly evident that any gas shortage from regulation will fall virtually *solely* on the jurisdictional consumers. In the more general case, average cost regulation interferes with efficient operation of the energy industry even though it may repair minor distortions in resource allocation due to such market power as might exist.

The optimal *regulatory* alternative, if we wish to provide an organization of production and distribution of the kind that will result in efficient allocation of resources, is to create conditions whereby the value of gas to consumers determines the price of gas both on average and in specific locations. This goal can be advanced by combining a base price with an area differential determined by value in consumption. Such a method utilizes a market-like incentive structure with publicly administered prices rather than market-determined prices. Monopoly exactions can be controlled easily where they exist, but the rigidity that prevents the exercise of monopoly power also prevents easy adjustment to changing conditions. If regulation it is to be, then consumer value-determined price, rather than one set on the basis of "cost," is certainly preferable.

Price deregulation, with effective reporting requirements to facilitate the exposure of monopoly or monopsony conditions in the field market, was the final alternative considered. If it is possible under such a policy to control such monopoly as exists, resources will be allocated efficiently, leading to maximum total economic welfare. The distribution of resulting income will, of course, lead to significant rent to producers in a period of rising gas demand. The alternative would be excessive consumer surplus to those potential gas purchasers fortunate enough to obtain gas supplies. On the basis that the public interest is identified with maximum economic welfare, and that redistribution or other goals are handled within another context, the public interest would call for deregulation of the field price of natural gas. "The" public interest is broader than economic welfare alone, however, and thus the policy maker must step outside this sharply limited context and consider other matters before a final decision is made.

Producer Regulation for the 1970s

What guidance can the above analysis offer on the ultimate choice among regulatory alternatives for the 1970s? A series of considerations seems important, presented below in broad strokes shorn of careful qualification.

The most basic conclusion is that gas field price regulation of whatever type is expensive to the economy, and, since it is expensive, its benefits should be obvious and substantial before regulation is undertaken.

A second conclusion important for policy is that the welfare redistribution that takes place under regulation is not what might be expected from superficial observation: there is redistribution from producing areas to consuming areas, of course, but also from some consuming areas and groups to others. Employment is shifted from consuming states to producing states as firms locate in the latter in part to obtain secure gas supplies. Exploration for gas closer to the consuming points and for gas that otherwise is more valuable to consumers is damped by some forms of regulation, while exploration for lower-value gas is relatively encouraged, making owners of those latter reservoirs better off and consumers and other reservoir owners worse off. Some of the biggest losers from regulation are the owners of the rights to produce natural gas—the royalty owners—and among the largest royalty owners are the states of Texas and Louisiana and the U.S. government. The distributional effects of these losses are unknown. As to income level, some income is surely redistributed from high-income oil royalty owners to low-income consumers, but it is also redistributed from owners of gas company stocks (including charitable foundations and pension funds) to high-income gas consumers. Moreover, welfare is transferred from low-income people who are unable to get gas because of the gas shortage to high-income consumers whose consumption of gas is higher because the regulatory system allows them to receive and to pay less for gas than the gas is worth to others. In summary, income redistribution is achieved through regulation of the producer end of the natural gas industry, but the efficiency cost of that redistribution is high, the amount of the redistribution achieved is unknown, and even the direction of the redistribution cannot unambiguously be ascertained. (Though at a guess, the direction of change is probably from producing to consuming areas and from higher incomes to lower.)

Governmental policy toward related and associated industries must also be considered in formulating a natural gas policy. The efficiency with which the natural gas resource is allocated depends importantly on the rate structure of the natural gas pipelines and distributors and on the certification policy followed in the construction of facilities in these branches of the industry. Again, natural gas producer policy must be articulated with petroleum policy generally; import controls on liquid petroleum, state and federal natural resource taxation, leasing policy in the federal domain and in Alaska, and merchant marine support through the Jones Act all affect optimal gas regulation policy. Finally, natural gas is very much only one part of the *energy* industry. Governmental policy toward coal and nuclear power, as well as toward liquid petroleum, must be considered when decisions affecting natural gas are in prospect.

The final conclusion follows from the first three: in truly complex policy matters, especially where redistributional effects are thought to be large, an important element must be the public reaction to the *image* of alternative policies. Natural gas regulation has created a significant social cost by placing an essentially economic matter into the sphere of public conflict. Divisive forces have been fostered in the nation that might as easily have been avoided. But, now that the regulation exists, it satisfies elements (mainly consumers) who, whatever the facts, think themselves better off because of it. Removal of regulation now, after the nation has more or less accustomed itself to the existence of regulation and therefore has paid in social discontent for its implementation, would perhaps lead to as much social division as the institution and enforcement of regulation in the first place.

In summary, natural gas price regulation of any of the above types lowers the efficiency with which resources are allocated when the alternative is an unregulated workably competitive industry. The reforms suggested here would make regulation less costly than the regulation in existence in 1970, and regulation restructured along the market value pattern would be loss costly still. Income redistribution is a doubtful goal of natural gas regulation policy; it provides little basis for choice among the regulatory alternatives, including deregulation. Any policy decision must take related industry policy into account. Political factors, defined in terms of what people perceive to be in their own and the national interest, might tend to favor maintenance of a regulatory structure for its own sake, whether or not it were particularly effective in fulfilling some accepted public goal.

It is from these, and a web of other considerations, that a public policy toward regulating the price of natural gas in the field must be drawn.

10 THE FUTURE OF NATURAL GAS REGULATION WITH AN INADEQUATE SUPPLY OF DOMESTICALLY PRODUCED NATURAL GAS

*James R. Nelson**

The meaning and consequences of the word "deficiency" must depend in part on trends of both the supply of and the demand for natural gas. A given relative deficiency will have a different absolute value depending on whether demand is rising or supply is shrinking. In an industry with increasing-cost margins, such as those that attend the production of natural gas, the price adjustment required to overcome a deficiency should increase with growth in absolute domestic shortage.

ENVIRONMENTAL ISSUES AND DEMAND FOR NATURAL GAS

The last year or two have produced near panic in some quarters due to changes in both the supply of, and the demand for, natural gas. Even in combination, new discoveries and extensions have apparently failed to yield contributions approaching those of most of the period since World War II. And on the demand side, the appearance of intense concern about pollution has opened the way to possibly unprecedented growth in natural gas markets. Except for hydroelectricity generated without reservoir capacity at waterfalls that no tourist wants to visit, natural gas appears to have the fewest environmental drawbacks of any source of fuel or energy. Meanwhile, the environmentally less desirable competitors of natural gas are in supply difficulties for this and other reasons, including attacks on pumped storage plants, new health questions raised with respect to nuclear energy, the aftereffects of new coal mine safety legislation, and concern for the environmental impact of open pit mining. To top it all, Texas oil production is closer to capacity than it has been for any extended period since the 1940s, new supply difficulties seem to arise for Middle East oil faster than old problems are resolved, and low sulfur oil is almost as sought after for fuel purposes as natural gas. It would almost appear that the only way to reconcile the literal greening of America with the environmental capabilities of available forms of energy would be to let grass grow in the streets.

This chaos at the heart of a swirl of confusion extends beyond the immediate capabilities of the shrewdest of regulators or the freest of markets. An approach to the environment that introduces rigid absolute standards, insists on action now, demands national uniformity, and worries not at all about open-end cost is an approach to the environment that is simply incompatible with rational energy policy, or with rational policy for any one source or type of energy. Even if these rigidities are alleviated, all approaches to the economics of environment must involve controversy in view of the newness of the field, the inadequacy of scientific research to unearth even some of the most important questions about quality of environment and how it is influenced, and the sophistication of the problems that have already been unearthed. Natural gas policy must be aimed at an unpredictably

*Professor of Economics, Amherst College.

moving target. It cannot be aimed sensibly unless some kind of environmental guidelines can be worked out to reduce the extent of the unpredictability.

Any discussion of environmental issues leads to more general questions of geography. The answer to some pollution problems may finally be to eliminate or at least to reduce them. The answer to others may be to spread them as widely and hence as thinly as possible. For still others, the answer may be to concentrate the nuisances all in one place.

But geography is important to the natural gas industry for reasons quite remote from environmental factors. If a free national market for natural gas existed, the need to import natural gas on a major scale would already have been preceded by a tilt in *relative* natural gas prices, as all absolute prices rose, with prices in producing areas going up the most. The advent of imports would presumably tend to stabilize prices in importing areas even if they rose in the rest of the country—just as the advent of gas produced from coal would presumably tend to stabilize prices near the site of synthetic gas production. The present real world is a world of price regulation, but on interstate prices only. Intrastate sales of gas are not controlled by the Federal Power Commission (FPC) and may not be controlled at all. So the theoretical (and actual) tilt in relative price away from producing areas should be accompanied by a reverse tilt in relative availability. If interstate pipelines are forced to ration users, many of the users will tend to shift toward unrationed intrastate supplies.

It is here that environmental policy and natural gas policy might join hands to proceed toward an objective only indirectly related to either: the freer import of end products in whose production natural gas is especially important. If this is impractical, then they might still join hands to sponsor a *common* policy—free market or regulated—to eliminate the locational anomalies bound to flow, in a period of increasing scarcity of domestic supplies, from an industry now two-thirds controlled and one-third free.

INTERRELATIONSHIPS OF NATURAL GAS AND PETROLEUM

To proceed beyond the foregoing generalizations, the first requirement is a careful look at the remarkably complex interrelationships of natural gas and petroleum. On the supply side, the two may be joint products and hence completely complementary yet economically differentiable; they may be competitive, or they may be independent. On the demand side, the cross elasticities are uneven. The internal combustion engine has always been the prime outlet for American petroleum output, with scarcely even a mention of natural gas until very recently. Conversely, natural gas is as good for practically every stationary heating purpose as petroleum, and better for many. As an influence on both sides of the market, public policy has been exceptionally inconsistent with petroleum and natural gas. Since the early 1930s the primary objectives of domestic petroleum policy have been first to raise the price of crude oil and, second, to hold it up by limitations on domestic and imported supply. But since 1954 the natural gas industry has been on notice that its interstate sales were subject to price control (in presumption and in fact, to *maximum* price control) by the FPC.

As if all these complexities were not enough,

> The major gas producers, who account for well over two-thirds of the current national production, are integrated oil companies with interests in oil refineries and in the marketing of oil products in this country and all over the world.[1]

But the relationship is confusing even here. For the natural gas industry, unlike the oil industry, is mainly characterized by vertical disintegration. Natural gas pipelines own only about a tenth of the reserves on which they draw, and they typically sell to local distribution companies for final resale, except for sales to small numbers of direct wholesale consumers.

[1] Testimony of John N. Nassikas, FPC Chairman, in *Natural Gas Supply Study*, Hearings before the Subcommittee on Minerals, Materials, and Fuels of the Senate Committee on Interior and Insular Affairs, 91 Cong. 1 sess. (1970), p. 126.

Therefore the first stage of this analysis must be to sort out what can be said about natural gas in isolation from oil from what cannot be said and then to determine the extent, if any, to which natural gas regulatory problems can be meaningfully considered apart from oil regulatory problems. The first thing that can be said is that the key to the natural gas industry in the United States is *heat*, whereas the key to the petroleum industry is *transportation*. Thus it is misleading to indulge in this kind of comparison:

> . . . inequities that may have resulted in the past with respect to gases of varying qualities and contents, are insignificant in comparison to the gross discrepancies which exist today between the price of 1,000 Btu of gas at the wellhead as compared to 1,000 Btu of oil at the wellhead.
>
> On a Btu basis (heat content), natural gas production in the United States now exceeds crude oil production. While this rapid rise in natural gas production and consumption relative to crude oil production has resulted in a significant increase in the producer's revenue, this increase has not been of the order of magnitude necessary to stimulate exploration for gas. This stems from the fact that in 1964, for example, one million Btu's of natural gas had a value of only 29.7% of the value of one million Btu's of crude oil at the wellhead. It is apparent, therefore, that this substantial upward trend in the production of Btu's in the form of gas rather than oil has resulted in substantial increases in energy consumption without commensurate increase in the producer's revenue. Gas is vastly underpriced in relationship to all other energy sources.[2]

It is an unfortunate fact that an obvious way to attack a price structure is to attempt to penetrate behind it in order to discover some "real" standard of value. In the time of Adam Smith, "corn" (in the sense of "small grains") may have provided a good approximation. Today, British thermal units (Btu) do not. An alternative measuring rod might be miles of transportation provided by a standard passenger car per cubic foot of tank space. It would of course be unfair to make a direct comparison between petroleum and natural gas at atmospheric pressure and doubly unfair to rest a comparison on crude oil rather than gasoline. But the ratios are nonetheless indicative. Even if automobiles could safely be operated with natural gas at highest pipeline pressure of 840 pounds per square inch, the Btu's contained in a barrel of oil that has a volume of about 6.5 cubic feet would occupy a volume of about 100 cubic feet. (At atmospheric pressure, the comparison would be 6.5:6000, or almost 1 to 1,000.) If the market demands heat, then the Btu is the proper standard of value. If the market demands a convenient source of motive power, then price per Btu may be largely irrelevant.

The very thing that makes petroleum so especially useful for nonheating purposes–illumination, transportation–is also the quality that most distinguishes it, economically, from natural gas: it is much cheaper to move. This difference can be established directly by a mere comparison of costs per mile per Btu, or costs per mile per any other attribute. But for present purposes, the most interesting comparisons are indirect. For example, the largest single industrial use of natural gas is as a fuel for thermal generation of electricity. This use alone accounts for about one-sixth of all consumption of natural gas in the United States. In 1967 the West South Central states (the center of U.S. production of both oil and natural gas) used over 1.3 trillion cubic feet of natural gas in the production of electricity but only 2,117 tons of coal and 362,680 barrels of oil. Use of natural gas in the West South Central states was 40 percent of the comparable use of natural gas in the entire United States, while coal use was less than 0.01 percent of the national total and oil use 0.2 percent of the total.[3] New England, farthest from all sources of fossil fuels, consumed 25 percent of all the oil used in the United States in generating electricity, just over 2 percent of the coal, and 0.3 percent of the natural gas.[4] The transportation discrepancy was even greater than these figures indicate, because the New England fuel consumption was concentrated on residual oil obtained from Venezuela or even more remote origins.

[2] *Natural Gas Supply Study*, written statement of R. M. Leibrock, past president, Permian Basin Petroleum Association, p. 258.

[3] Data derived from FPC, *1969 Annual Report* (1970), table 5, p. 11.

[4] Ibid.

So the Btu is an inappropriate measure of the competition of natural gas and oil because it leaves out the significant question: *Where? For what purpose?* Even with the development of bigger-diameter pipelines operating at higher pressures, natural gas is far more of a localizing factor than petroleum. A really energy-intensive industry should move to the source of natural gas. It has no such incentive to move to the source of petroleum.

Having discarded the conventional oil-natural gas comparison via price per Btu as geographically and functionally inadequate, we must reintroduce two further comparisons that tell a good deal about the past and possible future of the demand for natural gas.

The first of these comparisons relates to the competition between natural gas and coal. In a study of the natural gas industry written in 1949, when the revolutionary expansion of long-distance natural gas pipelines was close to its peak, it is pointed out that natural gas could be transported for 1.35 cents per million Btu per 100 miles; the comparable figure for coal in 1946, before a long series of postwar rail rate increases, was 2.35 cents.[5] Using the same conversion factors, the average price of natural gas at its point of origin would have been less in 1946 than the average price of bituminous coal.[6] This resulted in the penetration of natural gas pipelines into areas much more remote from natural gas than from coal supplies (quite apart from the advantages of natural gas in terms of cleanliness and convenience) and also caused the exceptional upward pressure on natural gas prices as the major long-distance pipelines were trying to ensure their sources of supply. Since coal was the preeminent source of heat energy in the United States before World War II, the immediate postwar period was obviously one of intense disequilibrium in the competitive positions of coal and natural gas and in the demand for natural gas itself.

The second of these comparisons relates to the possibly revolutionary effects of liquefied natural gas (LNG) on the convenience of natural gas as a source of fuel for transport purposes:

> The biggest spur to the use of natural gas in transportation, industry sources say, has been the development of techniques to liquefy gas by making it supercold. When chilled to minus 260 degrees Fahrenheit, natural gas becomes a clear, odorless liquid resembling water. It also is greatly compressed, taking up only one cubic foot of space for each 600 cubic feet it occupies as a vapor.[7]

The LNG revolution is already well established on the oceans. As the preceding quotation indicates, it may possibly become established on the highways. What is not clear, as yet, is the possibility of low-cost overland movement of LNG. If LNG pipelines should prove to be permanently uneconomic, gasoline would still preserve an advantage *to* many points of use even though it forfeited the advantage *at* the point of use. Moreover, the process of liquefying and deliquefying natural gas imposes costs that are now of the order of 20–25 cents per thousand cubic feet (Mcf)[8] —which in themselves would at least double the cost of new natural gas discoveries at the wellhead. Therefore, if LNG is to be preferred at points of use, this preference would presumably have little to do with the relative cheapness of LNG as a motor fuel and a great deal to do with its much better performance with respect to pollution.

But regardless of the practicability of the LNG-fueled internal combustion engine in the contest for an economical low-pollution passenger vehicle, the very prospect of a totally new type of outlet for natural gas raises two new issues. The first is that of a quantum jump in demand, which was discussed in the first section; in this case, however, the quantum jump might be thought of as, in part, a

[5] Joseph A. Kornfeld, *Natural Gas Economics* (Dallas, Texas: Transportation Press, 1949), pp. 191–92.

[6] Robert E. Mead, president, Independent Petroleum Association of America, puts the wellhead value per Btu of natural gas in 1946 at only 37.4 percent of the minehead value of bituminous coal (*Natural Gas Supply Study*, p. 232).

[7] "Cleaner Auto Fuel," *Wall Street Journal*, February 4, 1970, p. 1.

[8] In FPC, Bureau of Natural Gas, *A Staff Report on National Gas Supply and Demand* (September 1969), table 15, p. 63, liquefaction costs for shipment rates of from 300 million cubic feet to 500 million cubic feet per day range from 19.2 cents to 22.4 cents per Mcf, and "receipt and regasification" costs range from 3.6 cents to 4.8 cents per Mcf.

response to a new technology and not simply as a causal force helping to make the new resource economic. The second issue, and the most immediately relevant to this paper, is that of a possible change in the regulatory climate that might be produced by effective LNG competition for the motor fuel market.[9]

The advent of such competition would strip away most of the arguments based on the history of the gas industry, or by analogy with the regulated gas distribution industry, for price regulation of producers of natural gas. Integrated petroleum companies are much more important as owners of natural gas reserves than natural gas pipeline companies or distributors. In the absence of general national price controls, the national government has never seriously considered *maximum* price regulation for gasoline or other oil products. Important elements in the development of present natural gas regulation are products of fear that an absence of arms-length bargaining could produce monopoly producer profits, or that the cost-plus elements in public utility regulatory procedures provide inadequate consumer protection. If a liquid product could be carried through to the final consumer, even at the cost of elaborate special handling, the argument for price control of natural gas at the wellhead would no longer be able to work back from the existence of a public utility monopoly at retail. This would not, of course, inevitably provide a conclusive case against regulation. The answer might even appear to lie in regulating the retail price of gasoline and other petroleum products. But at least the major difference in the demand structure for gasoline and for natural gas would have been reduced, and this might also reduce many differences in the technique of supply. Either or both would throw into even sharper relief the differences in the regulatory philosophy applied to the two industries.

On the supply side, the outstanding feature of the gas-oil record is the great disparity in the historical price behavior of the two products. The average value of crude oil at the well hit its low point between World War I and World War II at 65 cents per barrel in 1931; natural gas, on the contrary, averaged 7 cents per Mcf at the well in 1931 and did not reach its prewar low point until it descended to 4.5 cents in 1940. The record since 1938, by ten-year intervals, is given in table 10-1.

Table 10-1. Average Value at Well for Crude Oil and Natural Gas, 1938, 1948, 1958, and 1968

Year	Crude oil (*$/barrel*)	Natural gas (*¢/Mcf*)
1938	1.13	4.9
1948	2.60	6.5
1958	3.01	11.1
1968	2.93	15.2

SOURCES: For 1938, U.S. Department of Commerce, *Historical Statistics of the United States: Colonial Times to 1957* (1960), p. 360. For remaining years, *Natural Gas Supply Study*, Hearings before the Subcommittee on Minerals, Materials, and Fuels of the Senate Committee on Interior and Insular Affairs, 91 Cong. 1 sess. (1970), p. 232.

Although the price of crude oil has been increased since 1968, its price history since the mid-1930s can be summarized with remarkable brevity. First, average values were close to $1.10 per barrel at the wellhead in the late 1930s. Then there was a very rapid price run-up beginning with $1.22 in 1945 and continuing with $1.41, $1.93, and the $2.60 already noted for 1948. This much higher price was held through the early and mid-1950s in spite of greatly expanded drilling activity and much higher capac-

[9] In a speech delivered in Honolulu on February 19, 1969, Commissioner Carl E. Bagge of the FPC said:

... the most likely use of the LNG commodity in the immediate future will be to fuel buses and other commercial and passenger vehicle fleets. ...

When used for these new purposes [including others named in the speech] there is a sharp distinction between LNG and natural gas. LNG in this perspective is not the life-blood of a gas utility. It is not the backbone of a service requiring public oversight. LNG ... is a distinct physical commodity. And it has distinct economic attributes.

I believe that the LNG commodity is a potentially competitive fuel, and that it will indeed be a competitive fuel. Its markets are those characterized by aggressive competition. Its marketers will be those prepared for competition. Government's response is critical—it must not hinder this competition. And to effect an appropriate government response, I submit that the transportation and sale of the LNG commodity should not be subject to Federal Power Commission regulation (*Natural Gas Supply Study*, p. 104).

ity by rigid output control—especially in Texas. The output control still exists, but not as rigidly; drilling activity has dropped off, and crude oil prices have shown very modest gains in the last twenty-odd years. State regulation operated as a ratchet to prevent upward price swings caused by demand pressure from being reversed—at the cost of great excess capacity.

The natural gas record would have differed from this even without a totally different regulatory approach. For one thing, "average value at well" means less, historically, for natural gas than for oil. Appalachia, with high wellhead prices, had some importance in the 1930s and is relatively insignificant today. And prior to the construction of long-distance pipelines, much natural gas in the Southwest was practically valueless. A further difference arises from the fact that interstate pipelines must satisfy the FPC that they have obtained assurance of long-term deliveries of gas before they are even allowed to begin construction. In spite of the prevalence of numerous types of escalator clauses, the mere existence of such long-term contracts probably gives natural gas prices a certain amount of relative short-run stability. They have increased proportionately more than crude oil prices since 1938, and they are still increasing. But long-run contracts may insert some time lag into this increase.

There is a temptation to argue that the significantly lower rate of growth in consumption of liquid petroleum, as compared with natural gas (petroleum annual growth rate in the United States, 1946-69 was 4.4 percent; natural gas annual growth rate, 7.3 percent)[10] is attributable to this different price history. In view of the concentration of gas as a source of heat and the continuous relative decline of coal from 1946 through 1968, such an argument would have to be presented with caution. It is nevertheless true that the relative output of gas associated with oil has declined, leaving a heavier burden to be carried by nonassociated gas. This, in part at least, is a reflection of the lower rate of growth of oil production and consumption.

Until recently, state regulation of oil probably had more to do with these various patterns than federal regulation of natural gas. But the exact contribution of state regulation is by no means self-evident. Indeed, some of the effects of state regulation tend to be offsetting and therefore to defy any casting of net balance. These effects will be discussed in the following paragraphs.

One effect was an increase in supply and reserves of natural gas that could be attributed to the stimulus to oil drilling caused by state regulatory policies that managed to hold the higher price line after 1947-48. Throughout the postwar period, until 1952, wellhead revenues from oil were over ten times wellhead revenues from gas.[11] Even if so-called directional drilling (for oil *or* gas) could be guaranteed always to hit what it aims for, the income effect of higher revenues from oil exploration might have overcome the substitution effect of trying for oil as soon as the oil price led the way. In fact, directional drilling partakes of the uncertainties of all exploratory drilling. As table 10-2 demonstrates, directional drilling for natural gas has been more successful than directional drilling for oil; nevertheless more natural gas wells are the product of drilling for oil than of drilling for natural gas—partly because the scale of the oil industry, including its exploration and drilling effort, is so much larger; partly for the precise reason that directional drilling for oil is less successful than directional

Table 10-2. Estimated Number of Productive U.S. Wells Originally Authorized as Oil or Gas Wells, 1966 and 1967

	Authorized as oil wells			Authorized as gas wells		
Year	Total	Oil producers	Gas producers	Total	Oil producers	Gas producers
1966	1,387	1,110	277	507	86	421
1967	1,372	974	398	146	12	134

SOURCE: J. Daniel Khazzoom, "Gas Production Directionality," *Public Utilities Fortnightly*, vol. 84, no. 13 (December 18, 1969), table 2, p. 24.

[10]Statement of Robert E. Mead, *Natural Gas Supply Study*, p. 232.

[11]Statement of Commissioner Lawrence J. O'Connor, Jr., of the FPC, *Natural Gas Supply Study*, app. c, p. 169.

drilling for natural gas. Although this table presents data for a period that should reflect more success in directionality than would have been possible twenty-five years ago, oil well authorizations produced 675 gas wells in 1966-67 versus 555 gas wells that were the result of original authorizations to find gas.

This first result of high oil prices in the late 1940s must have been reduced, if not eliminated, by the negative effects of prorationing–and of a bare ability to hold the price line–in the years thereafter. M. A. Adelman presents material that may bear on this question. His data show, as one would expect, that production of nonassociated gas rose relative to associated gas over the years between 1946 and 1961: for the United States as a whole, from 61.5 percent gas-well gas to 68.4 percent gas-well gas; for the states east of the Mountain States, from 67.1 percent gas-well gas to 73.8 percent gas-well gas. But the ratio of nonassociated to total gas reserves actually declined east of the Rockies over this period, from 76.4 percent in 1946 to 73.5 percent in 1961, and was practically stagnant for the entire United States at 72.5 percent and 72.2 percent.[12] Nonassociated gas that moves from the well by interstate pipeline is covered by long-term contract, but oil production, which largely determines the volume of associated gas production, is not. It would appear reasonable to conclude that prorationing of oil over the period covered by the Adelman study was responsible for stretching out the payout period for associated gas to the length of the period for nonassociated gas.

This conclusion may be affected by the fact that Texas has had both the most severe prorationing of oil production since World War II and the highest percentage of intrastate sales that are outside FPC jurisdiction. Over the quinquennium 1963-67, 55.4 percent of all natural gas production from the Texas Gulf Coast was consumed within Texas, as against an intrastate average for the country as a whole of 33.9 percent and intrastate percentages for other major producing areas that ranged from 20.2 percent to 42.1 percent.[13]

It is impossible to shake this kaleidoscope into any obviously meaningful pattern. But the following summary does no violence to the reported facts: The first result of effective reimposition of state output controls after World War II was to maintain a price level for crude oil that was very high by prewar standards and high even by the standards of most other price increases between the late 1930s and the late 1940s. This produced a rush of drilling that attained a momentum of its own and carried most of the way through the 1950s. Meanwhile, the cutbacks in output that price maintenance required were biting into the incentive originally offered by the price increases. Moreover, for a complex of reasons probably related to the cutbacks in output, prices of domestic crude oil went into the doldrums after their initial surge and their subsequent strong regulatory support. The main effects of these events on natural gas production were probably indirect, through their influence on drilling activity rather than direct via their influence on immediate production of associated gas. Even for drilling activity, much of the positive incentive was dissipated–prior to the growth of offshore drilling–by inadequate or nonexistent state standards for efficient drilling (for example, by granting allowables on the basis of formulae including such parameters as the number of wells).

Finally, the New Era of state oil regulation is beginning to stress incentives for secondary recovery of petroleum. These are unlikely to do anything to increase natural gas production, and, by way of reinjection in some fields, they may actually reduce natural gas sales. With repressuring, the oil-gas relationship comes full circle–from casinghead or other associated gas as a complement of oil production, to gas, from whatever source, as a productive factor expended or invested to permit further oil production.

The era of federal regulation might as well be dated from the imposition of compulsory oil import quotas in 1958 as from the Phillips case in 1954. For the aspects of state regulation that were reviewed above, in a sense, all come home to roost with a combination of: (1) rigid quantitative limits appli-

[12] M. A. Adelman, *The Supply and Price of Natural Gas*, Supplement to *The Journal of Industrial Economics*, Oxford, England (1962), tables IV-A and IV-B, pp. 52-53.

[13] FPC, Bureau of Natural Gas, *Staff Report on National Gas*, table 21, p. 89.

cable to most of the United States governing the importation of crude oil and all its by-products except residual fuel oil; (2) increasing restiveness on the part of Canadian authorities, who have expressed disinterest in the U.S. approach of more gas and less oil from Canada (a perfect result, incidentally, of the schizoid character of U.S. regulation of oil and natural gas) and would apparently prefer to approach oil and gas matters via a tie-in sale; (3) the appearance on the scene of new methods of shipping natural gas, which for the first time make imported LNG a possible supplement to, or competitor of, domestic production. President Nixon has decided in favor of the retention of oil import quotas. But it still seems appropriate to explore various assumptions with respect to the interrelationships of oil, gas, and their regulation in foreign trade. Here are three of the emerging problems:

1. Middle Eastern and North African oil fields are exceptionally productive. Western European and Japanese markets for imported oil have been highly, and increasingly, receptive. Practically all countries have experienced significant price inflation since World War II. Most of the major new oil fields in less-developed countries have been brought into production by foreign oil companies–in an era when foreign oil companies are, at best, no more popular than foreign anything else. All of these points, and more, could be added easily to a dual total of high export prices and high royalties for major producing countries. But the period between the end of World War II and the Suez crisis of 1956 probably carried prices, and eventually royalties, beyond any level they could otherwise have reached. The potentially outward-moving supply curve of the U.S. oil industry in this decade was truncated by state regulation to fit boom price levels. During the critical period of postwar oil development abroad, this regulation served as the ridgepole to hold up the entire structure of prices everywhere. The ridgepole was finally detached with the imposition of compulsory oil import quotas by the United States in 1958. But even this drastic move toward nonsupport of world prices had its counter influence: the U.S. market, the largest in the world and fed by some of the highest-cost oil production, offered a potential demand elasticity for imports that might otherwise have induced keen business competition for markets and some governmental competition for revenues via moderation in unit royalties. The world price structure was detached from that of the United States at the expense of removing the best available incentive for reduction in world export prices.

This contains a definite moral for natural gas policy. Imports of natural gas into the United States are now free of duty and free of quotas as well. But they are subject to license by the FPC. On its past record, the commission is interested in imposing *maximum* prices for natural gas. At the behest of the U.S. Supreme Court, it is willy-nilly practically compelled to do so. On the basis of recent public statements by several of its members, the FPC is tending now to favor maximum freedom of competition (as contrasted with any kind of direct or indirect price control) in the production and sale of natural gas. Whether one consults the past record or the recent statements, the conclusion must be that neither position is compatible, in principle, with either the domestic or foreign oil policies of the United States. Landed prices of liquid natural gas in U.S. ports are still generally well above delivered costs by domestic pipeline. If the battle for freedom of importation is to be fought, the time to fight it is *before* there is an immediate threat that importation will hold down or even force down domestic prices. The alternative is the oil cycle: first, domestic prices rise; then they temporarily carry world prices at least part way with them; and then the subsequent "normal" decline in world prices is arrested, perhaps indefinitely, by the intervention of producing countries to absorb economic surpluses in royalties.

2. Once regulation is forced to move to the frontiers to protect the price structure of the domestic industry, or seeks to move to the frontiers to continue its previous protection of consumers, the standards of price control become snarled in the perils of international diplomacy. The Civil Aeronautics Board and the Maritime Commission have already had experience with the problems of regulation without sanctions. The FPC would probably also learn some of the new problems created by overseas involvement. One of the more important ones is this: either the commodity being supplied from abroad is characterized by a near-horizontal supply curve and near-perfect competition or it is not.

If the first of these two sets of conditions is met, regulation is relatively easy—because detailed regulation is unnecessary. If either of the second set of conditions should prevail (that is, monopoly or a rising supply curve), any regulatory measure would interfere with monopoly profits or with economic rents. Whatever the analytical differences to an economist, they will often look the same to a foreign politician. The result is more likely to be loud cries of "imperialism" than successful regulation.

3. Movements beyond the water's edge may create unprecedented regulatory complications even if they do not involve foreign countries. An interesting example has already arisen in connection with overseas shipments of LNG from Alaska. Contracts have already been signed for shipments to Japan. Hawaii has not yet made such arrangements. Hawaiian spokesmen blame the Jones Act of 1920, which reserves all shipping routes between American ports for U.S.-built and U.S.-manned vessels. A spokesman for the Honolulu Gas Company estimates that Jones Act requirements add 60 percent to the investment required for LNG service from Alaska to Hawaii and 39 percent to total transportation cost per Mcf.[14]

The first of these three problems is one that in part, has been created by the petroleum industry and by state regulation of output in the United States. The second has no specific analogue in the oil industry, but could in general be approximated in the combination of politics and economics with which international oil companies and their governments have always had to deal. The third could become even more of an oil than a gas problem if direct tanker shipment of oil from North Alaska should prove to be the potentially lowest-cost method of bringing petroleum to the United States. In every case, the problems are such that they are better subjects for international diplomacy or Congressional action than for domestic regulation.

REGULATION OF SOURCES OF NATURAL GAS SUPPLY

This section will be specifically devoted to the natural gas industry, and to the type of regulation—or the absence of regulation—best adapted to sources of natural gas supply other than domestic gas or oil wells. Since both economic and administrative aspects would differ from alternative to alternative, each of the main possibilities will be considered separately before all are summarized together. These will be taken in ascending order of probable remoteness from immediate practical importance—and also, as it happens, in ascending order of possible ultimate importance. The first candidate for consideration is overland imports of natural gas from Canada and Mexico. The second is LNG shipments, especially from noncontiguous areas such as Venezuela and North Africa. The third is synthetic pipeline gas ("manufactured gas" reborn—in a new form).

Natural Gas Imports from Canada and Mexico

The possibility of substantial imports of natural gas from Mexico can be dismissed quite rapidly on the basis of present knowledge. Net imports to the United States from Mexico (imports less exports) amounted to just over 0.25 percent of U.S. consumption in 1968. Although sales of natural gas within Mexico more than trebled between 1960 and 1967, reserves increased by only 17 percent; therefore the ratio of reserves to production declined from about 30 to about 20.[15] Finally:

> The natural gas resource base of Mexico, although substantial, is estimated to be only one tenth to one sixth of the natural gas resource base of Canada.[16]

As this quotation indicates, Canada is quite another matter, and for three reasons: its own prospects as a natural gas producer, its location athwart any possible overland route from the natural gas of

[14]Statement of Paul C. Joy, on behalf of James F. Gary, President of the Honolulu Gas Co., *Natural Gas Supply Study*, p. 98.

[15]FPC, Bureau of Natural Gas, *Staff Report on National Gas*, table 14, p. 57; table 37, p. 100.

[16]Ibid., p. 58.

northern Alaska to the lower forty-eight states, and its possibilities as a producer and exporter of petroleum.

Since the Alaskan position is marginal to the discussion in the rest of this section, it will be outlined and dismissed first. Table 10-3 shows the present status of Alaskan natural gas reserves compared with those of the contiguous forty-eight states.

Table 10-3. Potential Gas Reserves for Alaska and Contiguous Forty-Eight States, by Category

(trillion cubic feet)

	Probable	Possible	Speculative	Total
Alaska	22	18	387	427
Contiguous 48 states	238	317	245	800
Total	260	335	632	1227

SOURCE: Adapted from FPC, *1969 Annual Report* (1970), table 20, p. 46.

Thus speculation on Alaska involves a much larger total than all probable natural gas reserves for the entire United States. Conversely, less speculative reserves in Alaska are not a substantial fraction of the national total. The problem of mobilizing Alaska's natural gas reserves may therefore become at least as important as the problem of transporting Alaskan oil—or prove to be of relatively minor importance if speculation about Alaskan reserves proves to be unfounded.

Canadian possibilities are even greater than the speculative estimate for Alaska. In early 1969, the Canadian Petroleum Association estimated natural gas reserves in the Western Canada Sedimentary Basin at 270 trillion cubic feet (tcf) (or slightly more than all U.S. "probable" reserves), reserves for the remainder of Canada at 455 tcf, and total reserves at 725 tcf[17]—or almost equal to the estimate for all degrees of probability for the 48 contiguous states of the United States. The official estimate of production, consumption, and export availability of Canadian natural gas is given in table 10-4 on two different assumptions.

Even on the basis of Case A, which rests on oil and gas development that is already well advanced, the export potential by the latter 1970s is expected to approach 50 percent of production. On the more speculative assumptions of Case B, about two-thirds of Canadian production would be available for export during the 1980s. These two sets of statistics may be compared with total U.S. marketed production of natural gas in 1968, which was 19.3 tcf.[18] Thus, the Canadian export contribution,

Table 10-4. Amount of Canadian Natural Gas Produced and Available for Export

(billion cubic feet)

	Actual	Estimated			
	1966	1975	1980	1985	1990
Case A[a]					
Net production	1,089	2,412	2,980	3,533	4,190
Available for export	432	1,165	1,359	1,449	1,626
Case B[b]					
Net production	1,089	2,525	3,958	5,886	8,132
Available for export	432	1,275	2,300	3,700	5,400

[a]Canadian gas production limited to Western Canadian Sedimentary Basin reserves.
[b]Case A, plus production from "frontier area" reserves.

SOURCE: National Energy Board, *Energy Supply and Demand in Canada and Export Demand for Canadian Energy* (Ottawa: 1969), pp. 63-64.

[17]National Energy Board, *Energy Supply and Demand in Canada and Export Demand for Canadian Energy* (Ottawa: 1969).

[18]FPC, *1969 Annual Report*, p. 43.

which has recently been of the order of 2 percent of U.S. consumption, may be expected to expand very considerably while still providing only a moderate fraction of total U.S. requirements.

The most interesting feature of the Canadian position is the difference in orientation between the National Energy Board in Canada and the FPC in the United States. This difference has already produced one case that was passed back and forth across the border before it could be decided, and it should produce more such cases. But discussion of this case can be postponed in favor of a few preliminary remarks about the policies of the Canadian National Energy Board.

The first noteworthy feature is the different approach taken by the National Energy Board on the question of price control on internal sales of natural gas. The board began operations in 1959, and the act creating it contains a whole section entitled "Traffic, Tolls and Tariffs," which begins with the statement: "The Board may make orders with respect to all matters relating to traffic, tolls, or tariffs."[19] Yet the board reported, as recently as 1969:

No formal proceedings with respect to the regulation of rates, tolls, and tariffs have been held. Discussions continued, however, on rate matters between the Board and companies under its jurisdiction and it appears likely that rate hearings will be necessary in the near future. A firm of chartered accountants, experienced in regulatory work, has been retained to conduct a rate study of a gas pipe line company under the Board's jurisdiction in co-operation with the staff of the Board. A similar study had previously been initiated relating to one of the oil pipe line companies.[20]

On the other hand, the board has been kept busy by Part VI of its Act—"Exports and Imports." This part begins with the statement:

Except as provided in the regulations, no person shall export any gas or power or import any gas except under the authority of and in accordance with a license issued under this Part.[21]

Two paragraphs further along, there is the provision:

Upon an application for a license . . . , the Board shall satisfy itself that

(a) the quantity of gas or power to be exported does not exceed the surplus remaining after due allowance has been made for the reasonably foreseeable requirements for use in Canada having regard to the trends in the discovery of gas in Canada; and

(b) the price to be charged by an applicant for gas or power exported by him is just and reasonable in relation to the public interest.[22]

The international meaning of all this emerged in a case that passed successively from the board to the FPC and then back to the board. In Canada, the applicant for an export permit was the Westcoast Transmission Company. To simplify discussion of the case, the essentials of the various natural gas supply contracts between Westcoast Transmission Company and the El Paso Natural Gas Company, the U.S. demander, are summarized in table 10-5.

Since the tariffs about which the various decisions revolved all contained both a contract demand charge and a commodity charge, table 10-5, which shows a lower base annual load factor for the contract of February 28, 1966, than for later contracts, somewhat exaggerates the excess of the prices finally envisaged under this contract as compared with its successors. But the National Energy Board seems generally correct in its observation that the Second Amendatory Agreement price was "approximately midway between the prices previously put forward by the companies but rejected by one or other of the respective national agencies."[23]

The sequence of events was as follows:

First, on February 28, 1966, Westcoast and El Paso rolled their 1954 contract for 300,000 Mcf per day into a new contract for 500,000 Mcf per day. The new contract provided for a price increase of

[19] National Energy Board Act, 1959, c. 46 as amended, pt. IV, par. 50.

[20] National Energy Board, *Annual Report for the Year Ended 31 December 1968* (Ottawa: 1969), p. 21.

[21] National Energy Board Act, par. 81.

[22] Ibid., par. 83.

[23] National Energy Board, "In the Matter of the Application under the National Energy Board Act of Westcoast Transmission Company Limited: Reasons for Decision" (Ottawa: February 1968, mimeo.), p. 9.

Table 10-5. Natural Gas Supply Contract Provisions, between Westcoast Transmission Company and El Paso Natural Gas Company

	Original contract, 1954	New contract, February 28, 1966	First Amendatory Agreement	Second Amendatory Agreement
Volume (*Mcf/day*)	300,000	500,000	–	–
Incremental volume (*Mcf/day*)	–	200,000	200,000	200,000
Annual load factor (*percent*)	90	85	95	95
Base price at stipulated load factor (*¢/Mcf*)	22.0	27.0	–	–
Incremental base price (*¢/Mcf*)	–	30.33	29.5	30.5
Annual price escalation (*¢/Mcf*)	–	0.225	–	–
Incremental price escalation (*¢/Mcf*)	–	0.375	–	(New price of 31.5¢ after 5 years)
Term	Superseded	25 years	20 years	20 years

NOTE: Costs are given in U.S. cents.

SOURCE: National Energy Board, "In the Matter of the Application under the National Energy Board Act of Westcoast Transmission Company Limited: Reasons for Decision" (Ottawa: February 1968, mimeo.)

5 cents per Mcf on the average, or of 8.33 cents per Mcf on the incremental 200,000 Mcf, plus annual price escalation of 0.225 cents per annum on the average or 0.375 cents per annum on the increment.

Second, this new contract was approved by the National Energy Board in March 1967.[24]

Third, this new approved contract was taken before the FPC and disapproved in Order No. 526 of August 10, 1967.[25]

Fourth, the alternative contract terms suggested by the FPC were embodied in a First Amendatory Agreement, which in turn was disapproved by the National Energy Board.[26]

Fifth, the final Canadian disposal of the matter came with a board decision in February 1968 approving the terms given in table 10-5 under "Second Amendatory Agreement."

As such things go, the total elapsed time of two years is less than might have been expected. So the Westcoast rate case is not presented as an outstanding example of regulatory lag. But it does illustrate the new dimension of natural gas regulation introduced by reliance on imports from Canada on such a scale that the American importer may be expected to establish the Canadian price and not simply to follow in the wake of Canadian buyers.

The National Energy Board's clearest statement of export price policy was contained in its refusal to accept the First Amendatory Agreement in December 1967. The board then announced this position:

> Consideration of the application as it relates to export price rests on three main points:
> (1) the export price must recover its appropriate share of the costs incurred;
> (2) the export price should, under normal circumstances, not be less than the price to Canadians for similar deliveries in the same area;
> (3) the export price of gas should not result in prices in the United States market area materially less than the least cost alternative for energy from indigenous sources.[27]

The board found that the First Amendatory Agreement passed the first two of these tests but failed the third. As a result of its own study of delivered costs of natural gas from the San Juan Basin in New Mexico and from British Columbia to the Portland, Oregon, area, the board concluded that

[24] National Energy Board, "Report to the Governor in Council in the Matter of the Application under the National Energy Board Act of Westcoast Transmission Company Limited" (Ottawa: March 1967, mimeo.).

[25] 38 FPC 311.

[26] National Energy Board, "In the Matter of the Application under the National Energy Board Act of Westcoast Transmission Company Limited: Reasons for Decision" (Ottawa: December 1967, mimeo.).

[27] Ibid., p. 7-1.

"the proposed export price is low in relation to the value of the gas in the market area." This was not only the basis for the board's disapproval of the agreement; it was also the basis for an extensive essay. This began with the allegation that Opinion No. 526 of the FPC had used the "in line" approach to stipulate a price by Westcoast to El Paso that was in line with *Canadian* (not U.S.) prices. The board tartly observed at this point:

Such an application to international trade of theory developed for regulation of trade in gas within the United States is unacceptable. . . .

One of the board's objections was to the FPC's disallowance of an escalator clause in the absence of any arrangement whereby a Canadian company could obtain a higher price from a U.S. regulatory body at some later date. But, more fundamentally,

. . . the in-line theory is wholly incompatible with the principle which is basic from the Canadian viewpoint that exports of gas from Canada should not be approved unless the price to be received bears a reasonable relationship to the value of the gas in the market to be served, as well as being not less than the price to Canadian customers. . . . The in-line principle makes the Canadian floor price the United States ceiling price.[28]

Or, as the general position was pungently summed up—with the addition of an interesting corollary—by the president of the Independent Natural Gas Association of America:

The chances are that Canada will expect a good price for its products, either in top value dollar per Mcf or concession by this country on other matters either related or unrelated. . . .[29]

Given that Canada expects to find a market for half, or more, of its natural gas production in the United States; given that many responsible American sources have been telling all who will listen—including Canadians—that the U.S. natural gas supply is already beginning to fall short of demand and is doomed to fall even further behind; given U.S. attempts to restrict the importation of Canadian crude oil in spite of a considerable after-duty price discrepancy; and given both the theory and the practice of international trade—what else would one suppose Canadian export price regulation on natural gas to be for but to ensure arms-length bargaining and to determine that full opportunity cost, or value of service, has been obtained from the U.S. buyer?

Is there anything in this area that U.S. regulation can, or should, accomplish?

If all Canadian production of natural gas were in the form of associated gas, then the National Energy Board and the FPC would constitute themselves a bilateral regulatory monopoly and lock horns whenever possible. If all Canadian production were already available to the market, with exploration and development costs all incurred and nothing to be added but customers, then the FPC could exert substantial leverage on behalf of U.S. consumers, the U.S. balance of payments, or whatever other nationalistic objective it chose. With Canadian natural gas development probably still in its infancy, the range of U.S. options is much less. The final result of the intergovernmental confrontation over the various Westcoast-El Paso contracts was to bring the National Energy Board halfway back toward the FPC's position on prices, to achieve what might be called a bureaucratized value of service rate. Doubtless the FPC could continue to win partial victories of this kind. But, to the extent that it managed to push Canadian prices below U.S. prices, which presumably reflect no more than marginal costs, the result would be a loss—lower-cost Canadian production forgone and American consumers forced to follow American producers further up an increasing cost curve.

But the most important aspect of the Canadian position relates to the very title of the Canadian government organization involved in the Westcoast case. The organization is the National *Energy* Board. Part VI of the National Energy Board Act, "Exports and Imports," deals only with gas and power; but paragraph 87 of the act states: "(1) The Governor in Council may by proclamation extend the application of this Part to oil." From the Canadian standpoint, it would seem to be the most elementary logic to insist on a package deal when faced with a maximum-price-controlled U.S. natural

[28] Ibid., pp. 8-12, 8-14, 8-15.

[29] Testimony of Walter E. Rogers, *Natural Gas Supply Study*, p. 56.

gas market apparently starved for imports and a minimum-price-controlled U.S. oil market doing everything possible to keep out imports. This would make good sense in terms of Canadian politics. It would make excellent sense in terms of international economics. And it would clearly expose a fact that is scarcely a secret now: institutionally, analytically, and politically the United States has almost nothing resembling a national energy policy and even less resembling an international energy policy. Even if natural gas and oil have had rather separate U.S. markets in the past, this may change with LNG. Moreover, from the standpoint of a supplying country, the two logically belong together.

In this sense, U.S. relations with Canada on natural gas supplies and prices may be significant even beyond their direct quantitative importance because of the precedents they set for dealings with other countries. In a number of ways, Canadian precedents would be useful precedents. Canadians are likely to have a more sober and realistic view of "value of service" and of the economic rents this may entail than some other possible sources of supply. Geographically, Canada is in a position to offer acceptable gas and oil prices in several areas of the United States while still reaping excellent internal returns. The typical nervousness of U.S. officials about defense considerations can be disregarded in this case. But it would seem only reasonable to assume that Canada would make concessions below full value of service for natural gas only if it received concessions with respect to increasing sales in the premium-priced U.S. crude oil market.

Importation of Liquefied Natural Gas

Liquefied natural gas has all the glamour that Giant Power had forty years ago. But unless the process of liquefaction contains more potential for technical breakthroughs than has yet been reported, the denouement may be quite different in the two cases. Transmission of electricity lends itself to economies of scale practically without limit. The production and reconversion of LNG presumably does not. LNG's biggest transportation advantage is its relative ease of tanker movement compared with natural gas, but this becomes a very considerable disadvantage when compared with crude oil—not only because of the expensive terminal operations required for LNG, but also because of the far more expensive tankers required for the line haul.

This pessimistic appraisal of the cost prospects of LNG is, of course, open to the rejoinder that many a so-called decreasing cost curve is really a learning curve, and everyone working with LNG still has a lot to learn. True. But still there do not seem to be the static possibilities for decreasing costs associated with larger-diameter pipelines. There may be the dynamic possibilities for decreased costs that have already been realized in such industries as gas and oil well drilling or the fabrication of large-diameter high-pressure pipe, but this is improbable.

Whether or not this forecast proves to be correct, it is useful to summarize an estimate of the recent cost situation of LNG. This is done in table 10-6, which shows some evidence of economies of scale at the terminals. The New York import is assumed to be 300,000 Mcf per day, whereas the

Table 10-6. Estimated Cost of Delivering Liquefied Natural Gas to New York and Los Angeles

(cents per Mcf)

	New York, from:				Los Angeles, from:		
					Cook Inlet, Alaska		
	Maracaibo, Venezuela	Skikda, Algeria	Brega, Libya	Port Harcourt, Nigeria	U.S.-built and flag	Foreign vessels	Maracaibo, Venezuela
Liquefaction	21.8	22.0	22.2	22.4	19.2	19.2	19.4
Marine transportation	14.0	24.2	29.0	30.6	22.9	14.8	29.6
Receipt and regasification	4.8	4.8	4.8	4.8	3.7	3.7	3.6
Total as gas	40.6	51.0	56.0	57.8	45.8	37.7	52.6

NOTE: Estimated cost is the total cost, excluding gas costs and including profit.

SOURCE: Abstracted from FPC, Bureau of Natural Gas, *A Staff Report on National Gas Supply and Demand* (September 1969), table 15, p. 63. This table is attributed in the source to James T. Jensen of Arthur B. Little, Inc., "The Prospects for Tanker Shipments of Liquefied Natural Gas into the Continental United States" (paper presented to the American Institute of Mechanical Engineers, Los Angeles, February 22, 1967).

Los Angeles assumption is 500,000 Mcf per day—which is the only obvious reason for the lower terminal figures for shipments from Maracaibo. But if this and other influences are allowed for by the drastic assumption of a 50 percent across-the-board cut in all costs, the result would require a delivered price equal to a standard wellhead price in the United States *even with no value whatever attributed to the natural gas at the foreign wellhead.*

Therefore the first thing to be said about LNG is that it offers no competitive threat to domestic natural gas sold to large industrial users located close to the wellhead. Many of these users buy gas that has not been shipped across state lines and therefore falls outside the jurisdiction of the FPC; hence the significance of this conclusion for regulatory policy is not clear. If major industrial users of natural gas want to hold their costs down, they will probably have to follow the pioneering trail of aluminum companies and find supplies isolated from the main currents of world demand and commerce. They will get no help from LNG.

The effect of LNG on other users, including the vast mass of residential, commercial, and industrial users in the northeastern states and in other areas remote from major gas supplies, is likely to be minor in terms of price reductions but major in terms of holding back price increases that might otherwise occur. Existing sources of information are less helpful than they might be on the cost and price structure equations involved. For example, the source for table 10-6 reports that the "weighted average wholesale pipeline price of gas," as of the time the underlying data were compiled, was about 42 cents per Mcf to the New York City area, and about 37 cents per Mcf to the Los Angeles area.[30] Since this "weighted average" is compounded of a demand charge and a commodity charge, an analysis of the future of LNG would have to start with final consumption data and continue by comparing interruptible consumers and appropriate categories of firm load in terms of the marginal costs of supplying them with extra natural gas by pipeline and by regasified LNG.

In the absence of such data, it may be useful to do two things: first, to compare certain published estimates of LNG costs relative to pipeline costs and, second, to attempt some generalizations about LNG cost structures as compared with pipeline cost and price structures.

LNG Costs and Pipeline Costs. Pertinent statements on these costs were made during the hearings before the Subcommittee on Minerals, Materials, and Fuels of the Senate Committee on Interior and Insular Affairs in 1969.[31] According to Hollis M. Dole, assistant secretary of the interior for mineral resources:

> The delivered cost of LNG to either the eastern or western seaboard will be about 40 to 55 cents per Mcf, depending on the distance transported, exclusive of purchased gas cost which may be from 10 to 15 cents Mcf. Thus, the total delivered cost may range from 55 to 70 cents per Mcf.
>
> To import 1 trillion cubic feet per year of LNG, which is about the U.S. annual increase in demand, would require a total investment of about $1.5 billion and would involve a fleet of 30 tankers. LNG may help mitigate peak demands, but a significant contribution from it is several years away.

John O'Leary, director of the Bureau of Mines, expressed the following opinion:

> I would assume that currently we could bring liquefied natural gas into the eastern seaboard from the 55- to 65-cent range.
>
> The trade-off price there in New York is now about 42 cents, and in Baltimore it would be about 45, and in Washington about 45, Philadelphia 40 to 45, Boston perhaps 55.

The following exchange took place between Senator Frank Moss of Utah, Chairman of the subcommittee, and John Nassikas, FPC chairman:

> *Senator Moss.* . . . you made reference to the fact that El Paso Natural Gas was proposing to bring in LNG from Algeria to the East Coast to begin in 1973.
>
> Do you know what the proposed price of LNG per thousand would be delivered to the East Coast?

[30] FPC, Bureau of Natural Gas, *Staff Report on National Gas*, table 15, p. 63.

[31] *Natural Gas Supply Study*. The quoted testimony that follows appeared on pp. 15, 31, 152, 172–73, and 255.

Mr. Nassikas. Fifty to 55 cents. This is, as I recall it, on the eastern seaboard at New York City.
Senator Moss. Which would be somewhat higher than the pipeline costs at the present time.
Mr. Nassikas. Yes.

Later in the hearings, Senator Henry Bellmon of Oklahoma questioned Commissioner Lawrence J. O'Connor, Jr., of the FPC.

Senator Bellmon. What is the [El Paso Natural Gas] company going to sell the LNG to customers for? What will the price be?
Mr. O'Connor. It is a 50-cent price onboard, and a 55-cent price regasified. It is comparable to the city gate price of the pipeline.
Senator Bellmon. What is that price?
Mr. O'Connor. That varies from 35 cents in this area to as high as 60 in Boston.
I think this same company made an estimate of what it would cost them to build a new pipeline to serve this same load in New York, which is a firm load, and assuming that they would pay an additional 3 cents per thousand cubic feet in the field to get a billion cubic feet a day, they came up with a 50-cent price with a new line to provide this service.

After reviewing the testimony presented during the hearings, Howard Boyd, president of the El Paso Natural Gas Company informed Senator Moss that "careful studies show that—if available—domestic gas volumes approximating those our project proposes for delivery to East Coast markets would reach those markets at about the same cost as would gas derived from the Algerian product."

Although some of these quotations suffer from inadequate specification of "where" or "when," they all seem to point in the same direction: LNG is now competitive at the city gate for urban gas distribution systems only at extreme pipeline range (for example, Boston). If costs are measured on a replacement-plus-prospective basis and not as embedded historical averages, LNG may be competitive with pipeline natural gas in several of the major metropolitan areas along the northeastern coast.

Cost and Price Structures: Pipelines and LNG. All the emphasis, so far, has been on the possible contributions of LNG in cheapening transportation costs—*by* LNG, including avoidance of the negative effects of pollution, and *of* LNG, including the possibility of opening up wholly new overseas sources of supply. As a description rather than a forecast, this emphasis is misplaced. Most of the LNG installations in the United States are for liquefaction and storage of gas delivered by pipeline or at the wellhead—not by LNG tanker.[32]

Storage is an integral feature of the public utility natural gas industry. At the end of 1968, underground natural gas storage facilities in the United States had a capacity of 4.8 tcf, equivalent to about 25 percent of 1968 production.[33] The technical reason for this is that natural gas pipelines are a highly capital-intensive industry. A high proportion of costs are constant whether the load factor of the pipeline is high or low. The economic reason for storage is the fluctuation of demand for natural gas from season to season. In many supply areas, the marginal cost of storage is less than the marginal cost of expanding the pipeline and also less than the price obtainable from the extra peak sales made possible by the extra capacity. Even the FPC's use of the word "tilt" to describe the transfer of some pipeline capacity costs onto the commodity charge may express an economic as well as a political reality, for, if natural gas pipeline rates slavishly copied the two-part tariffs on electricity, distributors or customers might find it to their economic advantage to over-store natural gas.

If the 600:1 ratio of gas to liquid that prevails in the natural gas industry has inaugurated a new tanker fleet, it has also set the stage for a new structure of natural gas rates at wholesale and retail. Underground storage of LNG has not yet been fully successful. But aboveground storage is clearly in demand, expensive aboveground tanks and all. The reason must be that the superiority of LNG stor-

[32]In FPC, Bureau of Natural Gas, *Staff Report on National Gas*, twenty-four LNG plants are listed as being operational, under construction, or planned in the United States (table 39, pp. 101–03). The largest, at Port Nihiski, Alaska, is a facility for the outbound shipment of LNG to Japan. Others are in such definitely landlocked communities as Oak Creek, Wisconsin; Buhl, Minnesota; Hastings, Nebraska; and Indianapolis, Indiana.

[33]FPC, *1969 Annual Report*, p. 43.

age over underground natural gas storage is more than enough to overcome its extra cost per unit of volume. The weaknesses of imported LNG as a competitor for baseloads immediately disappear if the alternative is LNG liquefied from domestic gas. This doubtless explains the fact that the first formal application to the FPC for a license to import LNG under long-term contract has not come from El Paso Natural Gas, the Philadelphia Gas Works, or other proponents of LNG as a base load supply. It has come, instead, from the Distrigas Corporation of Boston, which proposes to serve the Boston Gas Company and other Massachusetts gas utilities for peak-shaving purposes. (Even if the purpose of the supply had not been announced, the prices charged for it would be sufficient to restrict its use. The annual seasonal price variation covers a range of from 68 percent per million Btu between April and October to a peak of 85 cents in January and February.)[34]

In terms of cost *structure*, then, the contribution of LNG in metropolitan areas all along the northeast coast should be to reduce effective marginal capacity costs and to raise effective marginal commodity costs. The paradoxical result might be to validate the FPC's "tilt" toward loading the commodity charge. Thus, perhaps not quite accidentally, does today's technical progress come to the aid of yesterday's politics.

In this discussion, the economic contribution of LNG has been viewed rather narrowly. The unexplored horizon of LNG, as Commissioner Carl Bagge has pointed out on many occasions, lies in the direction of new uses (for example, transportation) and the avoidance of social disutilities (for example, air pollution). Regardless of individual reactions to the reported decline of natural gas reserves in both 1968 and 1969, there can be no question about the impact of decisions such as the following.

> In a precedent-setting decision, the Illinois Commerce Commission ruled that if the [Commonwealth Edison] company failed to comply with a detailed schedule of pollution control requirements, the rate increase [of 4.5 per cent] would be cut in half. . . . The schedule for compliance by Commonwealth Edison requires the company to convert its Ridgeland station from oil to gas burning facilities by August 31. . . .[35]

If both the extent and the type of these new uses transcend all previous experience, then U.S. regulation of producer prices of natural gas is indeed in a new environment.

Predictions for LNG. Now that the time has arrived to cast the balance on the future impact of LNG, a sane first approximation would be to recall the history of forecasts of the nuclear power industry. The soothsayers' divining rods have almost invariably pointed in opposite directions from one observation to the next, and practically every swing in predictions has been excessive. It took LNG a long time to recover from a catastrophe in Cleveland during World War II, and it may take it a while to recover from what will doubtless prove to be optimistic elements in recent forecasts of costs. But at some future date LNG will be important. Therefore, let us boldly predict, for the Year X.

A complete description of the LNG supply curve to the American buyer would require knowledge of the future of oil output for leading offshore producers, of the coefficient of associated gas to be attributed to this output, of the relationship of nonassociated to associated gas reserves in various important producing areas, of the net demands at various origin price levels of consumers in other countries, and of the development of the technologies of liquefaction and tanker shipment. The one sure conclusion to emerge from this is that the American public had better prepare itself for another shock: having moved from the lowest-cost to perhaps the highest-cost source of oil, we are repeating the same journey in natural gas. In terms of natural resources, we are becoming an old country.

In the absence of the Utopian supply of data outlined in the previous paragraphs, a few comments may still be made. One is that Western Europe is shifting from its previous role as almost purely a consumer of oil and natural gas. Things have now reached the stage for a natural gas pipeline to be constructed from The Netherlands to Italy. Meanwhile, new sources of oil and gas continue to appear

[34] "Gas-short New England Due More Foreign LNG," *Oil and Gas Journal*, v. 68, no. 8 (February 23, 1970), p. 80.

[35] "Utility Rate Rise Tied to Pollution," *New York Times*, July 11, 1970, p. 26.

throughout the world. So it does not follow that a major new U.S. demand for foreign natural gas would send wellhead prices up as fast as they rose in the United States after World War II.

But, with a product such as natural gas, transport costs become overwhelmingly important if the distance between origin and destination is as wide as the Caribbean or the Atlantic. With these costs probably beyond the control of the United States and almost certainly beyond the control of countries of origin, the demand for the product will appear to have a much lower elasticity as seen by the seller than as seen by the buyer. The seller can see only a remote retail market, separated from him by layers of costs that are practically fixed in absolute terms. Therefore, to say that foreign supplies delivered in the United States should have relatively high price elasticity is not to say that this is how suppliers will see the market. Conversely, of course, a price increase that could mean a bonanza for a supplying country might scarcely be noticed, percentage-wise, by ultimate buyers.

The best "regulation" of the LNG market that the United States could provide would simply be to continue the present policy of permitting duty-free imports. From the standpoint of pollution control, natural gas comes close to being the optimum product. From the standpoint of market dynamics, it is also under present circumstances a very difficult product to reduce to a tame and predictable market performance. In a highly dynamic field, costs are likely to decrease rapidly and unexpectedly. The United States is now a high-cost energy country. It cannot afford to cut itself off from the future development of the energy market.

In a stricter sense of the word "regulation," the growth of LNG could provide a new world to redress the balance of the old. Prices of oil imports into the United States are far above most marginal costs of production-plus-transportation. A regulator of *energy* imports–not a civil servant charged with keeping out foreign oil, or beating down the price of foreign natural gas–could even achieve a kind of second-best triumph by distributing oil import quotas by the package. Rights to import oil might go to those who bid the most LNG per barrel of oil for the privilege. Even without this kind of scheme, some way should exist to relate oil import policy to natural gas imports. The two products are obviously partial substitutes in consumption. They are often also rigid or associated complements in production. A foreign minister of power or energy could be excused for confusion or cynicism if confronted with two sets of U.S. officials determined, respectively, to keep out one of his products by raising its price or by other methods and to hold down the price of another product to improve its chances of admission to the U.S. market.

It is probably true that LNG from foreign sources will add still more competitors to the producer-selling market for natural gas and thereby weaken the case for the establishment of regulatory maximum prices yet further. But the net addition of competitors should not be exaggerated. The business of foreign production and shipment is a business for a few giants, not a horde of dwarfs. And the governments of exporting countries are likely to be, at most, in an oligopolistic position.

The net effect of these and other points may be to improve the economic case for one kind of regulation (over maximum import prices) and to weaken the economic case for another (over domestic wellhead prices). Control of import prices ranges from the difficult to the impossible. Even when easiest, it may not be worth the social friction it causes. With specific reference to natural gas, low-cost foreign producing countries may feel they are more entitled to the economic rents accruing from output of the product than are relatively prosperous consumers in the United States. But, in spite of these objections, there remains a fragment of an undemolished argument for some kind of maximum price regulation on imports to prevent a "rent and royalty" escalation in foreign countries that might maximize short-term gains but cause the long-run decay of the industry.

Pipelines are most vulnerable to LNG competition just where they receive maximum gross revenue per thousand cubic feet moved–in consuming centers farthest from the source. These areas of maximum vulnerability include some major markets, scattered from Boston to Los Angeles. Any major shifts to LNG in these markets would create a danger of upward average cost and price spiralling on the part of the pipelines, starting with higher unit average accounting costs as business shifted to LNG, and ending–in the absence of care and common sense–in higher prices, still lower consumptions, and so on.

If LNG ever seriously penetrates the transportation market, the previous advice to adopt an energy policy would have to be broadened with the further advice that a general energy policy would not be enough. Oil and natural gas might have to be treated "precisely alike" to the utmost extent that this phrase is meaningful.

Synthetic Pipeline Gas

The gas industry in the United States began as a processor of coal, and it may end the same way. If only half the known coal reserves of the United States were assumed to be recoverable through conversion to natural gas, the result would still be a coal reserve equal in natural gas content to forty times the present proven U.S. gas reserves.[36] In view of this abundance of riches, certainly known to exist before much had been discovered about other reserves of fossil fuels, it is disconcerting to be told that "the economic feasibility of producing pipeline quality gas from coal or lignite has yet to be demonstrated on a large scale in the U.S."[37]

But this problem may prove to be, at least in part, a product of lower natural gas prices that cannot continue at past levels into the future. In any event, there is presumably *some* level of natural gas prices at which coal could make a massive competitive entry. On the basis of preliminary research in a pilot plant, this level has been placed as low as 39.7-44.2 cents per million Btu, or about the same per thousand cubic feet of gas.[38] The tests on which these costs are based were conducted at Rapid City, South Dakota, using lignite. Most U.S. lignite beds are not only far from major natural gas producing areas but also far inland. Even bituminous coal is not found close to any coast. So it is possible that demand for LNG and demand for gas from coal might develop simultaneously and combine to form a fairly uniform roof over possible delivered prices for natural gas at points distant from the domestic wellhead.

The major lignite deposits of the United States unfortunately are also far from major population centers and generally are not well situated relative to present pipelines. So the marketing problem would be a great deal more serious for lignite-derived gas than for LNG or present natural gas. But important bituminous coal beds lie quite near to existing pipelines and not far from major consuming areas for natural gas.

The first difficulty to arise in discussing the relevance of gas from coal to federal regulatory policy with respect to natural gas is that a pilot plant, research and development type of activity yields no direct information on the presence or absence of economies of scale. For lignite, there is a possibility that a whole new network of pipelines would have to be built—with attendant dangers of monopsony. If scale economies derived from pipeline diameters were to be matched by scale economies in production of synthetic natural gas, then there might be problems of bilateral monopoly. If the question is pursued further into an argument by analogy, this last hypothesis comes close to the situation that has prevailed since the early days of the industry in the transportation and refining of crude oil. Although refineries are more likely to own inbound crude lines than outbound products lines, both types of ownership are common, without producing regulatory supervision of refinery margins or even very effective regulatory supervision of pipeline charges and profits.

Let us suppose that scale economies in the production of synthetic natural gas are present but not dominant. We need not suppose what is an obvious fact: namely, that U.S. reserves of bituminous coal and lignite are very widely scattered. Raw materials would have different costs at the outset, and their productivities would also differ depending on their physical and chemical characteristics, the process used to produce synthetic gas, and the efficiency of that process. Some economic rents would doubtless accrue to certain conversion plants in the process of reorienting the market for coal and lignite to

[36] FPC, *Staff Report on National Gas*, p. 70.

[37] Ibid., p. 65.

[38] Consolidation Coal Company, *Pipeline Gas from Lignite Gasification*, R & D Report No. 16, Interim Report No. 4 (Library, Pennsylvania: May 9, 1969), p. 3.

an entirely new type of demand. But some negative economic rents, in the form of unforeseen losses, would also appear at the same time for the same reasons. Both these pluses and these minuses would be mere by-products of a central situation—extreme fluidity during the early stages in the general market outlook for the product and in the specific apparent advantages of supplying the product by one or another of various methods of production. Regulation is by no means at its best when it is turned loose on a young and speculative industry. The regulatory record tends to be unimpressive, and the net impact on the development of the industry is not likely to be toward stimulating it.

Therefore the very absence of hard economic knowledge with respect to pipeline gas may in itself provide the main clue toward an answer to the question that now concerns us. If there is no special case derived from comparative statics for extending regulation to synthetic natural gas prices f.o.b. plant, there is a special negative case that interlocks with the probable dynamics of such an industry in both directions. A highly dynamic industry is a poor subject for regulation; and a highly regulated industry is a poor candidate for highly dynamic performance.

MONOPOLY VERSUS ECONOMIC RENT

So far the topic of economic rent has been approached from a number of different angles: the importance of quasi-rent in total returns once an oil or gas well is completed and producing; the element of rent in the return that a "born speculator" may receive from a risky industry; wide differences in the quantity, quality, and location of oil and gas resources as well as in the cost of producing them, and so forth. Less has been said about monopoly and monopsony in the production, sale, and purchase of natural gas. The excuse for this is that there probably is a good deal less to be said.[39]

Let us suppose, for the moment, that there is a major problem of unearned increment accruing to as yet unspecified persons or groups connected with natural gas production. This supposition would be compatible with the idea either of an increasing cost curve, looking forward to expanded production over any moderate period of time, or of increased costs (continuous upward shifts in supply curves) as time goes by. The supposition would also be compatible with the FPC's area rate decisions in both the Permian Basin and Southern Louisiana cases. The idea of freezing old prices while allowing higher prices on new production certainly reflects awareness of increased costs as an ingredient in generating economic rents, and may also reflect a desire to contend with increasing costs if output is expanding, or is to be expanded, through time. A two-price system for natural gas, on the analogy of a two-price system for copper during World War II, may appear to be an acceptable compromise between the economist's rules for allocation of resources and the politician's rules—or the political economist's rules—for income distribution.[40]

As economists in either the Ricardo or the Henry George tradition would point out, the trouble with this is that prices can do clumsily what taxation can do cleanly. Moreover, the allocational efficiencies of a two-price system at the producer level become lost in the shuffle, or leak out of the pipeline, before they reach the final consumer. The marginal cost argument for resource allocation is a two-edged argument. It indicates not only what the original seller should receive but also what the

[39]Three standard economic works on the natural gas industry—Edward J. Neuner, *The Natural Gas Industry* (University of Oklahoma Press, 1960); Paul W. MacAvoy, *Price Formation in Natural Gas Fields* (Yale University Press, 1962); and Clark A. Hawkins, *The Field Price Regulation of Natural Gas* (Florida State University Press, 1969)—all convey the feeling that the monopoly problem was in the past *de minimis* compared with the monopsony problem created by single pipeline buyers in the early days of long-distance pipelines, that monopoly is still no more of a problem in natural gas production than in many unregulated sectors of the economy (such as crude oil production and petroleum refining, activities carried on by many of the same firms that also produce natural gas), and that the economic issue that may require regulation—which all three authors seem to doubt—is the possibility of large-scale economic rents accruing to producers.

[40]Edmund W. Kitch, in "The Permian Basin Area Rate Cases and the Regulatory Determination of Price," *University of Pennsylvania Law Review*, v. 116 (December, 1967), identifies the author of the two-price idea that was eventually incorporated into the Permian Basin decision as Professor Alfred Kahn of Cornell University. He also cites Professor Kahn's testimony as supporting the linear descent of the Permian Basin pricing structure from copper price control during World War II (p. 212 and fn. 78).

final buyer should pay. One edge of the argument is only relatively more effective than one blade of Alfred Marshall's scissors.

As a practical matter, price regulation of natural gas was apparently the only way to get at a group that was believed to be exploiting the public via both crude oil prices and tax concessions for a number of years before, during, and after World War II. Except for one ill-advised campaign contribution, this price regulation would doubtless have been eliminated a decade and a half ago. But the practical environment has changed. Oil and natural gas are no longer viewed as the special claimants for tax exemption that they once were. Therefore, the tax route may be able to take over some of the burden now carried by regulated prices.

There is, of course, one final drawback to this approach. It assumes that economic rents are maximized for a given price level, whereas all that may be maximized in fact is the total cost of excess wells drilled. This problem has been acute for oil in the past. It is becoming less acute as exploration and development move offshore and into northern Alaska.

Finally, of course, the mere shift of new drilling to large blocks of publicly owned land should automatically help to take care of the economic rent problem. In principle, and abstracting from the necessarily speculative nature of predrilling guesses on economic reserves, competitive bidding for drilling rights on public lands should transfer all economic rents that matter—that is, economic rents ex ante, or surpluses beyond all required incentives—to public treasuries. If monopoly is not an issue in this case, then the recapture of economic rents is surely not an issue. Attempts to recapture them *in advance* may, of course, have special drawbacks compared with royalty payments spread over time.

SYNTHETIC, OR CHRONOLOGICAL, MARGINALISM

As pointed out in the previous section, maximum price control is a clumsy method of appraoch to problems of economic rent. But if all else fails, the least clumsy variant of this method is through the use of *chronological marginality* in applying maximum price control to natural resources.

The possibility of differential pricing to reflect increasing costs through time is particularly great if an increasing-cost industry typically sells under long-term contract or if the industry passes its product to final consumers through a decreasing-cost intermediary. Due to the importance of the natural gas pipeline as the indispensable middleman in the industry of supplying natural gas for interstate consumption, both of these conditions—long-term contracts, and a decreasing-cost intermediary—are present in natural gas. It is true that the FPC simply requires that the pipeline assure its long-term supply; it does not require that all specific contracts be of equal length or of equal price rigidity. But the technical supply requirement probably exerts some independent influence toward fixing the price, through time, of the quantity supplied. And the large-scale existence of decreasing costs between producer and consumer provides at least some room for maneuver if it is desired to raise wellhead prices on some output (for example, on newly committed output) without necessarily having to find some scapegoat to pay this higher marginal price or to risk the ire of consumers as a class by spreading it more or less evenly over all of them.

Under these conditions, it is possible to talk about price differentiation on the supply side as more or less the mirror image of price differentiation on the demand side. It is an elementary proposition of economics that a monopolist who could sort out his units of sale carefully enough might charge a different price for every unit sold and pocket practically the entire value of his product to consumers. A monopsonist may do the same thing to producers, on the supply side. A Robin Hood monopolist might charge all that the traffic will bear on the demand side and distribute the proceeds in the form of higher pay to his workers. A Robin Hood commission might force all the price reduction that suppliers could stand, on the supply side, and distribute the savings in some kind of average cost pricing. In both cases, the marginal conditions for an optimum would be infringed—by payments to factors above value of the marginal product in the first case, and by a price to consumers below the marginal cost of the product in the second case. But the firm would be operating simply as an intermediary in the first case, as the commission would be in the second.

There would be income effects, of course. Demanders who had to pay a limit price for each unit would be poorer than if they paid a single price equal to the price at the margin; hence they would normally consume less. Suppliers who received the highest price only at the margin would also be poorer than if they had received this highest price for all units; hence, in the face of even the slightest imperfection in capital markets, they would eventually supply less. But these income effects might be trivial when compared with the effects (economic and political) of the redistribution of income by the use of differential, split, or mixed-marginal and average, price systems.

If policy is to emerge from the classroom it may have to find ways of reconciling the allocational need to provide higher price incentives to producers at the margin of new discovery or development with a possible distributional need to deny these incentive prices to all production. There is also a further distributional need to intermingle high and low prices on their way through to final consumers so as to satisfy popular standards of equity and to discourage uneconomic use of the product, for example, by charging the highest prices to customers with the most elastic demands. All this is too complicated to recommend as an attempted long-term solution. This kind of thinking, after all, is what has produced the serpentine tangle of our present railroad rates. But chronological supply-side price differentiation, with or without measures to average or distort supply-side prices when they are finally passed along to ultimate demanders, may be the only effective way to escape from a present unsatisfactory static disequilibrium to a hoped-for future satisfactory static or dynamic equilibrium. Counsels of perfection often have only one drawback: you can't get there from here.

The principles discussed in the previous paragraph are not completely remote from past FPC practice with respect to the establishment of wellhead prices of domestically produced natural gas. As additions or amendments to this practice, it might be suggested even at the outset that more careful attention might be paid to marginal costs with respect to current demand levels and trends, to the marginal costs of potential new supplies as well as of existing sources, and to the marginal opportunity costs of suppliers who have alternative markets beyond the reach of the FPC. There would still remain the delicate problem of using differential pricing as a transitional device toward probable higher future prices and not as a self-perpetuating system with increasingly tyrannical requirements of its own.

And, to proceed one final step down the road marked "political economy"—if the kind of marginal pricing discussed in this section should dictate higher *average* delivered prices to foreign suppliers (as it probably would), what better guarantee could one have of its being transitory? Or even, perhaps, ephemeral?

PUBLIC UTILITY REGULATION AND COMPARATIVE ADVANTAGE

As economists, public utility commissioners tend to be landlubbers. They have a keen appreciation of the significance of internal frontiers, but know little about the outside world.

As public officials, these same commissioners are also faced with the need for a new type of approach at the frontier. At home, a commissioner orders. Abroad, if he is to achieve anything at all, he must negotiate.

In the special case of natural gas, even the internationally minded and diplomatic commissioner is faced with a basic dilemma: how to reconcile an attempt to hold down the price of imported natural gas, to a degree even beyond that automatically assured by duty free entry, when national petroleum policy is designed to increase the scarcity value or in other ways to raise the price of imported petroleum.

If the two commodities are associated in production, this dilemma from the demand side appears to be compounded by a further dilemma from the supply side: How can production of one product be stimulated while holding back another—where the two products are produced in fixed, or even linked, proportions?

National policy, to date, appears to rest on the assumption that two wrongs make a right—if marginal import cost of foreign oil is held far above marginal production cost plus transportation, then hold marginal import cost of imported natural gas down *below* marginal domestic cost. This recipe seems to involve the same kind of confusion as that which causes some elementary students to believe that two currencies can depreciate with respect to each other at the same time; the geographical locus is shifted in the middle of the problem. But the confusion is simply a by-product of a similar and more pervasive confusion with respect to domestic energy policy.

The obvious way out of this problem would be to permit batch imports—so much petroleum, so much natural gas or LNG; or so much petroleum, so much bid in the form of a maximum amount of natural gas to be supplied. This proposal has the typical theoretical disadvantages of a barter system versus a monetary system. But the danger in the energy sphere is that anyone who waits for the equivalent of a monetary system—in the form of abolition of quotas, or of equalized tariffs, or of no tariffs or quotas at all—is likely to wait a long time. We are only now discovering how infectious import quotas can be. Unless some general energy policy can be adopted, the crude oil example will loom larger before the natural gas industry as domestic wellhead prices rise and techniques for maritime movement of natural gas improve.

MICRO-ECONOMICS AND MACRO-POLITICS

If every resource of the U.S. government were to be devoted to the effort, royalties to foreign governments on natural gas imported into the United States could probably be reduced almost to zero. The principal income effect would probably also lie against many countries that, on balance, are not particularly popular in the United States at the moment.

So, given this hypothetical desire to roll back prices onto the foreigner, federal regulation would be an essential instrument for achieving it. After all, the U.S. government may not be able to recapture economic rents earned from the sale of foreign resources (although it could doubtless try harder than it ever has). Therefore its only method of creating an income effect in its favor would be to improve its own terms of trade by forcing back the prices of imports.

The very way in which this option is phrased should serve to indicate this writer's reaction to it. Even if the objective were praiseworthy—which is at least arguable—the implementation would involve the use of a steamhammer to crush a nut.

There is, however, a defensive case for permitting regulation of prices and quantities of imports. This is simply that the governments of foreign suppliers have almost the same incentive to try to cartelize natural gas exports as to try to cartelize crude oil exports. The latter activity has not been completely successful in spite of what has amounted to effective collusion in such forms as U.S. import quotas, so the former may not rank high among our international economic preoccupations. But the defense mechanisms should be kept in being on the assumption that they might at some time be needed.

BYGONES ARE FOREVER BYGONES

W. Stanley Jevons, from whom this phrase was stolen, was keenly interested in natural resource problems and in price discrimination as well as in marginal analysis. He was also interested in applying the law of mortmain to economic problems. All these interests are relevant to this concluding section.

Whatever the merits and demerits of federal regulation of U.S. natural gas prices during the 1960s, the world of the 1970s requires a new approach. Foreign supplies are bound to grow if the natural gas industry is to remain healthy, new technologies are likely, and the political and economic situation of the allied oil industry is changing. There is no point in recapitulating the details presented in previous sections. As a final summary, a maxim might be coined: as a guide to future regulatory standards, the newest form of natural gas technology is to let in as much fresh air as possible.